Lecture Notes in Mathematics 1553

T0254707

Subseries: Fondazione C. I. M. E., Firenze
Adviser: Roberto Conti

J.-L. Colliot-Thélène K. Kato P. Vojta

Arithmetic Algebraic Geometry

Lectures given at the 2nd Session of the Centro Internazionale Matematico Estivo (C.I.M.E.) held in Trento, Italy, June 24-July 2, 1991

Editor: E. Ballico

Springer-Verlag
Berlin Heidelberg New York
London Paris Tokyo
Hong Kong Barcelona
Budapest

Authors

Jean-Louis Colliot-Thélène
Université Paris-Sud
Mathématique, Bât. 425
F-91005 Orsay Cedex, France

Kazuya Kato
Department of Mathematics
Tokyo Institute of Technology
Oh-Okayama, Meguro-ku
Tokyo, Japan

Paul Vojta
Department of Mathematics
University of California
Berkeley, CA 94720, USA

Editor
Edoardo Ballico
Dipartimento di Matematico
Università di Trento
38050 Povo, Trento, Italy

Mathematics Subject Classification (1991): 14C15, 14C25, 14C35, 19E15, 14G40, 11J99, 11M99, 11G99, 11R23, 14M20

ISBN 3-540-57110-8 Springer-Verlag Berlin Heidelberg New York
ISBN 0-387-57110-8 Springer-Verlag New York Berlin Heidelberg

Printing and binding: Druckhaus Beltz, Hemsbach/Bergstr.
46/3140-543210 - Printed on acid-free paper

PREFACE

The CIME Session on "Arithmetic Algebraic Geometry" was held at Villa Madruzzo (Trento, Italy) from June 24 to July 2, 1991.

There were the following lecture series:

- Jean Louis Colliot-Thélène: Cycles algébriques de torsion et K-théorie algébrique
- Kazuya Kato: Lectures on the approach to Iwasawa theory for Hasse-Weil L-function via B_{dR}
- Christophe Soulé: Arakelov Geometry
- Paul Vojta: Application of arithmetic geometry to Diophantine approximations

Furthermore, the participants gave several seminars, namely:

- Dan Abramovich: Subvarieties of Abelian Varieties and Jacobians
- Luca Barbieri Viale: Birational Invariants Via Cohomology Theories
- Torsten Ekedahl: a) An Infinite Version of Chinese Remainder Theorem
  b) On the Density of Extensions of Generic Ramification Type
- Frans Oort: a) CM Liftings of Abelian Varieties
  b) A Conjecture by Coleman Jacobians over C having Complex Multiplication
  c) Newton Polygons and Abelian Varieties
- Angelo Vistoli: Bivariant Intersection Theory and Alexander Duality
- Christoph Wirsching: Quillen's Metric for G(2,4)

This volume contains enlarged versions of three of the lecture series. For an exposition of Soulé's lectures, see the very recent book by Ch. Soulé, D. Abramovich, J. F. Burnol and J. K. Kramer: "Arakelov Geometry", Cambridge Studies in Advanced Mathematics 33, Cambridge University Press, 1992.

The four main speakers took particular pains to start at a level understandable by motivated (but not specialist) graduate students and then they arrived at the frontier of research (and beyond). They provided further support for their audience in various ways, before, during and after this CIME Session. Just an example: J.L. Colliot-Thélène drew up, as early as 1990, a long list of references (both elementary and very advanced) complete with detailed comments on their content: the list was widely circulated. The authors of the three lecture series published here made the same effort for their written versions.

I owe special thanks to several people (including all the lecturers and participants) for their precious help (both on mathematical and practical matters) in connection with the organization of the conference and the production of this volume.

Edoardo Ballico

Table of Contents

Cycles algébriques de torsion et K–théorie algébrique
Cours au C.I.M.E., juin 1991

Jean-Louis Colliot-Thélène
C.N.R.S., Mathématique
Université de Paris–Sud
91405 ORSAY Cedex

L'objet de ce cours est de décrire certaines applications qu'a eues la K–théorie algébrique à l'étude des cycles de torsion sur les variétés algébriques, et plus particulièrement à l'obtention de théorèmes de finitude, sous diverses hypothèses arithmétiques sur la nature du corps de base. Ce domaine de recherches fut ouvert par Spencer Bloch en 1974, connut de nouveaux développements à la suite de la percée de Merkur'ev et Suslin en 1982, et a fait l'objet d'un rapport de W. Raskind en 1989 ([R1]).

Une série de travaux récents montre que le sujet est loin d'être épuisé, et il m'a donc semblé bon de faire à nouveau le point.

La première partie de mes exposés au C.I.M.E. fut consacrée aux résultats classiques, qu'on trouvera dans les paragraphes 1 à 5. Après un rappel des résultats de finitude connus sur le groupe de Picard (§ 1), on définit au § 2 les groupes de Chow et on décrit au § 3 le programme de Spencer Bloch pour contrôler la torsion dans ces groupes de Chow. Au § 4, le lecteur trouvera une démonstration simple du théorème de Roitman sur les zéro–cycles de torsion lorsque le corps de base est algébriquement clos, sans réduction au cas des surfaces. Passant au cas d'un corps de base fini, on esquisse au § 5 la démonstration simplifiée (Raskind et l'auteur) du théorème de Kato et Saito sur les extensions non ramifiées de variétés projectives et lisses sur un corps fini et le groupe de Chow en dimension zéro. On explique également le théorème de finitude (Sansuc, Soulé et l'auteur) pour la torsion en codimension deux.

Les paragraphes 6 à 9, qui développent mes deux derniers exposés au C.I.M.E., sont consacrés à des résultats récents (1989–1991) de Raskind et l'auteur [CR3], de Salberger [Sb2], et de S. Saito [S4], résultats qui portent sur la finitude de la torsion du groupe de Chow en codimension deux pour une variété projective et lisse X, définie sur un corps k arithmétique (local ou global), et satisfaisant l'hypothèse que le second groupe de cohomologie cohérente $H^2(X, \mathcal{O}_X)$ s'annule. Les premiers résultats de finitude pour cette torsion avaient été obtenus sous l'hypothèse additionnelle $H^1(X, \mathcal{O}_X) = 0$. C'est Salberger qui montra comment éliminer cette dernière hypothèse lorsque k est un corps de nombres.

L'approche de Raskind et de l'auteur, au-dessus d'un corps de nombres, est décrite au § 6. J'ai inclus dans ce paragraphe une brève esquisse de la méthode galoisienne, qui avait déjà permis dans le passé d'obtenir des résultats de finitude pour certaines variétés (Bloch [B4], l'auteur [C1], Gros [G], Ōkōchi, et plus récemment Coombes [Cb], sur un corps global ; Raskind et l'auteur [CR1], sur un corps local). L'approche plus récente de S. Saito [S4] fait l'objet du § 7. Je me suis astreint à bien dégager les énoncés valables au-dessus d'un corps quelconque. Cette démarche permet d'obtenir certains résultats de finitude au-dessus d'un corps de type fini sur le corps des rationnels. Au § 8, je décris les résultats que cette même approche permet d'obtenir au-dessus d'un corps local. Enfin, au § 9, je donne quelques indications sur l'approche de Salberger, renvoyant à sa récente prépublication [Sb2] pour plus de détails.

J'ai arrêté là ce rapport sur les cycles de torsion. Parmi les thèmes que je ne traite pas, mais qui méritent l'intérêt du lecteur, je citerai :

– Les conjectures et résultats très précis sur le groupe de Chow des zéro–cycles sur une surface rationnelle définie sur un corps de nombres (travaux de Sansuc et l'auteur [CS], et de Salberger [Sb1]), qu'on voudrait bien voir généraliser, au moins conjecturalement, à de plus vastes classes de variétés. Dans l'immédiat, on aimerait traiter le cas de surfaces satisfaisant

$H^2(X, \mathcal{O}_X) = 0$ et $H^1(X, \mathcal{O}_X) = 0$, cas où les approches précises de Coombes [Cb] (que je ne traite pas ici) et de Saito [S4] pourraient se révéler utiles.

– Les travaux en cours de Bloch et de ses élèves sur la torsion du groupe de Chow de certaines surfaces X au-dessus d'un corps de nombres avec $H^2(X, \mathcal{O}_X) \neq 0$ (sur un corps local, voir [R2]).

– Les exemples de torsion dans le groupe de Griffiths dus à C. Schoen [Sc].

Plan

Remerciements. L'influence des travaux de Spencer Bloch sur tout ce sujet est manifeste. Pour les discussions que j'ai eues avec eux dans le passé lointain ou proche, je salue S. Bloch, S. Saito, P. Salberger, J.–J. Sansuc, C. Soulé et tout particulièrement W. Raskind.

Je remercie la Fondazione Centro Internationale Matematico Estivo de m'avoir donné l'occasion de m'éclaircir un peu plus les idées et de fouler les rues de Trento et de Verone.

Une version préliminaire de ce cours a fait l'objet d'un exposé en Septembre 1989 au Centre Culturel Européen de Delphes, que je souhaite aussi remercier pour son invitation.

Notations.

Etant donné un groupe abélien A et un entier $n > 0$, on note ${}_nA$ le sous-groupe des éléments de A annulés par n, on note A/n le quotient A/nA. On note A_{tors} le sous-groupe de torsion de A, et on pour l premier, on note A_{l-tors} le sous-groupe de torsion l-primaire, i.e. le sous-groupe formé des éléments annulés par une puissance de l.

Etant donné un anneau unitaire R, on note R^* le sous-groupe multiplicatif formé des éléments inversibles de R.

Etant donné un entier n inversible sur un schéma X, on note μ_n le faisceau étale sur X défini par le schéma en groupes des racines n-ièmes de l'unité. Pour j entier positif, on note $\mu_n^{\otimes j}$ le produit tensoriel j fois de μ_n avec lui-même, on note $\mu_n^{\otimes 0} = \mathbb{Z}/n$. Pour j négatif, on note $\mu_n^{\otimes j}$ le faisceau étale $Hom_X(\mu_n^{\otimes -j}, \mathbb{Z}/n)$. Enfin, étant donnés un nombre premier l inversible sur X et $j \in \mathbb{Z}$, on note $\mathbb{Q}_l/\mathbb{Z}_l(j)$ la limite inductive des faisceaux $\mu_{l^n}^{\otimes j}$ pour n tendant vers l'infini.

Par $H^i(X, \mathcal{F})$ (sans indice) on notera le i-ième groupe de cohomologie sur le schéma X, pour la topologie de Zariski sur X, à valeurs dans le faisceau $\mathcal{F}$. On fera cependant parfois une exception , en notant $H^i(k, M)$ le i-ème groupe de cohomologie galoisienne du groupe de Galois absolu $G = Gal(\overline{k}/k)$ à valeurs dans un G-module continu discret M, ou, si l'on préfère, le i-ème groupe de cohomologie étale à valeurs dans un faisceau étale M sur le spectre Spec(k) du corps k.

§ 1. Groupe de Picard d'une variété

Dans ce paragraphe, nous rappelons quelques résultats très classiques sur l'équivalence linéaire des diviseurs, que nous confronterons dans les paragraphes suivants avec la situation bien plus complexe des cycles de codimension plus grande que 1.

Soit k un corps, X une variété algébrique lisse et irréductible sur le corps k. On désignera par $k(X)$ son corps des fonctions rationnelles, et par $\mathrm{Div}(X)$ le groupe des diviseurs de X, c'est-à-dire le groupe abélien libre de générateurs les points de codimension 1 de X :

$$\mathrm{Div}(X) = \bigoplus_{P \in X^{(1)}} \mathbf{Z}.$$

Chaque anneau local de X en un tel point P est un anneau de valuation discrète de corps des fractions $k(X)$, ce qui permet de définir un homomorphisme

$$v_P : k(X)^* \longrightarrow \mathbf{Z}.$$

On définit alors alors le groupe de Picard $\mathrm{Pic}(X)$ de la variété X comme le conoyau de l'application *diviseur*

$$\mathrm{div} = \bigoplus_{P \in X^{(1)}} v_P : k(X)^* \longrightarrow \mathrm{Div}(X),$$

c'est-à-dire qu'on a une suite exacte :

$$k(X)^* \longrightarrow \mathrm{Div}(X) \longrightarrow \mathrm{Pic}(X) \longrightarrow 0.$$

Le groupe de Picard admet plusieurs représentations. De la description ci-dessus, et de l'identification, sur une variété lisse (donc localement factorielle), des diviseurs de Weil (combinaisons linéaires à coefficients entiers de sous-variétés de codimension 1) aux diviseurs de Cartier (définis localement pour la topologie de Zariski comme diviseurs d'une fonction) on déduit facilement l'identification :

$$\mathrm{Pic}(X) \simeq H^1_{Zar}(X, \mathcal{O}_X^*) = H^1_{Zar}(X, \mathbf{G}_m),$$

où la cohomologie est la cohomologie de Zariski.

Par ailleurs, une version du théorème 90 de Hilbert due à Grothendieck montre que sur tout schéma, il y a une identification :

$$H^1_{Zar}(X, \mathbf{G}_m) \simeq H^1_{et}(X, \mathbf{G}_m) \simeq H^1_{fppf}(X, \mathbf{G}_m),$$

avec le groupe de cohomologie étale (ou encore avec le groupe de cohomologie $fppf$) à valeurs dans le faisceau défini par le groupe multiplicatif $\mathbf{G}_m$.

Lorsque X est une courbe projective et lisse C, les points de codimension 1 sur C s'identifient aux points de dimension 0, et l'on dispose d'une application degré :

$$\deg = \bigoplus_{P \in C^{(1)}} \mathbf{Z}P \longrightarrow \mathbf{Z}$$

définie par linéarité à partir de l'application qui associe au point fermé P de C le degré $[k(P) : k]$ de P relativement à k, et une formule classique dit que cette application est triviale sur les diviseurs de fonctions, i.e. induit une application degré de $\mathrm{Pic}(C) \longrightarrow \mathbf{Z}$.

Les énoncés suivants rassemblent les propriétés les plus importantes du groupe de Picard.

PROPOSITION 1.1. — *Supposons la variété lisse X absolument irréductible et complète (par exemple projective). Soit $k \subset F$ une inclusion de corps, et soit $X_F = X \times_k F$. Alors l'application naturelle* $\mathrm{Pic}(X) \longrightarrow \mathrm{Pic}(X_F)$ *est injective.*

Lorsque F/k est une extension galoisienne (qu'on peut supposer finie), la démonstration repose sur le théorème 90 de Hilbert : $H^1(\mathrm{Gal}(F/k), F^*) = 0$. Soient $\overline{k}$ une clôture algébrique de k, puis $G = \mathrm{Gal}(\overline{k}/k)$ et $\overline{X} = X \times_k \overline{k}$. On peut en fait montrer qu'on a une suite exacte fonctorielle :

$$0 \longrightarrow \mathrm{Pic}(X) \longrightarrow \mathrm{Pic}(\overline{X})^G \longrightarrow Br(k),$$

où $Br(k) = H^2(G, \overline{k}^*)$. L'image de $\mathrm{Pic}(\overline{X})^G \longrightarrow Br(k)$ est annulée par tout entier $[E : k]$ pour E/k extension finie sur laquelle X acquiert un point rationnel. En particulier, la flèche $\mathrm{Pic}(X) \longrightarrow \mathrm{Pic}(\overline{X})^G$ est un isomorphisme dès que X a un point k-rationnel. □

THÉORÈME 1.2. — *Supposons la variété lisse X absolument irréductible et complète (par exemple projective). Soient $\overline{k}$ une clôture algébrique de k, puis $G = \mathrm{Gal}(\overline{k}/k)$ et $\overline{X} = X \times_k \overline{k}$. Il existe une variété abélienne J/k (la variété de Picard $\underline{\mathrm{Pic}}^0_{X/k}$ de X) et un groupe abélien de type fini, le groupe de Néron-Severi $NS(\overline{X})$ de $\overline{X}$, tels que l'on ait une suite exacte*

$$0 \longrightarrow J(\overline{k}) \longrightarrow \mathrm{Pic}(\overline{X}) \longrightarrow NS(\overline{X}) \longrightarrow 0. \tag{1.1}$$

Cette suite exacte est G-équivariante. Lorsque X est une courbe, $NS(\overline{X}) = \mathbb{Z}$ et l'application $\mathrm{Pic}(\overline{X}) \longrightarrow NS(\overline{X}) = \mathbb{Z}$ est induite par l'application degré. □

PROPOSITION 1.3. — *Soit X une variété irréductible et lisse sur un corps k. Pour tout entier n inversible dans k, on dispose d'une application injective :*

$$\mathrm{Pic}(X)/n\,\mathrm{Pic}(X) \hookrightarrow H^2_{et}(X, \mu_n). \tag{1.2}$$

Pour obtenir cette injection, il suffit de prendre la suite exacte de cohomologie étale associée à la suite de Kummer

$$\begin{array}{ccccccccc} 1 & \longrightarrow & \mu_n & \longrightarrow & \mathbb{G}_m & \longrightarrow & \mathbb{G}_m & \longrightarrow & 1 \\ & & & & x & \longmapsto & x^n\,. & & \end{array}$$

On notera qu'un bout de la suite exacte en question s'écrit :

$$k[X]^*/k[X]^{*n} \longrightarrow H^1_{et}(X, \mu_n) \longrightarrow {}_n\,\mathrm{Pic}(X) \longrightarrow 0. \tag{1.3}$$

□

PROPOSITION 1.4. — *Soit X une k-variété lisse irréductible et U un ouvert non vide de X. On dispose d'une suite exacte :*

$$k[X]^* \longrightarrow k[U]^* \longrightarrow \mathrm{Div}_{X\setminus U}(X) \longrightarrow \mathrm{Pic}(X) \longrightarrow \mathrm{Pic}(U) \longrightarrow 0. \tag{1.4}$$

Ici $k[X]^*$, resp. $k[U]^*$, désigne le groupe des fonctions inversibles sur X, resp. sur U, et $\mathrm{Div}_{X\setminus U}(X)$ le groupe des diviseurs de X à support en dehors de U, qui est un groupe libre de type fini.

La démonstration de cette proposition de localisation est élémentaire. □

Ces énoncés sont à la base des divers théorèmes de finitude pour le groupe de Picard.

PROPOSITION 1.5. — *Pour toute variété irréductible propre et lisse X sur un corps k et tout entier $n > 0$, le groupe des points de n-torsion, ${}_n\operatorname{Pic}(X)$ est fini. Ceci vaut encore pour tout ouvert U d'une telle variété.*

(D'après Hironaka, en caractéristique zéro, toute k-variété lisse est un ouvert d'une k-variété propre et lisse).

Démonstration : En utilisant la suite de localisation (1.4) et le fait que $\operatorname{Div}_{X\setminus U}(X)$ est un groupe de type fini, on voit que l'énoncé pour U résulte de l'énoncé pour X. D'après la proposition 1.1, ${}_n\operatorname{Pic}(X)$ s'injecte dans ${}_n\operatorname{Pic}(\overline{X})$. La finitude de ce dernier groupe résulte alors du théorème de structure 1.2, du fait que le groupe de Néron–Severi est de type fini, et du fait que sur un corps algébriquement clos, les points de n-torsion d'une variété abélienne forment un groupe fini. □

Remarque 1.5.1. : Lorsque n est inversible dans k, en utilisant la suite (1.3), on obtient pour toute k-variété X et n inversible dans k, une surjection

$$H^1_{et}(X,\mu_n)/H^1_{et}(\operatorname{Spec}(k),\mu_n) \twoheadrightarrow {}_n\operatorname{Pic}(X).$$

Pour X absolument intègre, le groupe de gauche s'identifie à un sous-groupe de $H^1_{et}(\overline{X},\mu_n)$. Or, de façon tout à fait générale, les groupes $H^i_{et}(\overline{X},\mu_n)$ sont finis (en caractéristique, 0, SGA 4 XIX Springer LNM **305**; en général, Deligne, Théorèmes de finitude, in [SGA4 1/2]).

PROPOSITION 1.6. — *Si k est un corps fini, et X une variété propre, lisse et géométriquement intègre sur k, le groupe $\operatorname{Pic}(X)$ est un groupe de type fini. En particulier son sous-groupe de torsion $\operatorname{Pic}(X)_{tors} \subset \operatorname{Pic}(X)$ est fini.*

Démonstration : Cela résulte immédiatement de la proposition 1.1, qui donne une injection $\operatorname{Pic}(X) \subset \operatorname{Pic}(\overline{X})^G$ (où $G = \operatorname{Gal}(\overline{k}/k)$ (cette inclusion est en fait ici un isomorphisme), du théorème 1.2 (suite exacte G-équivariante (1.1) et engendrement fini de $NS(\overline{X})$) et de la finitude de $J(\overline{k})^G = J(k)$ qui est le groupe des points k-rationnels d'une k-variété algébrique sur le corps fini k. □

PROPOSITION 1.7. — *Soient k un corps p-adique (extension finie d'un corps $\mathbf{Q}_p$) et X une variété propre, lisse et géométriquement intègre sur k. Alors le groupe $\operatorname{Pic}(X)_{tors}$ est fini.*

Démonstration : Utilisant les énoncés 1.1 et 1.2 comme ci-dessus, on est ramené à voir que le groupe $J(k)_{tors}$ est fini. Mais comme J est une k-variété abélienne et k un corps p-adique, le groupe des points k-rationnels $J(k)$ est un groupe analytique p-adique commutatif compact. Il contient donc un sous-groupe ouvert U d'indice fini isomorphe à un produit R^q, où R est le groupe (additif) des entiers de k et q est la dimension de J. La torsion de $J(k)$ s'injecte donc dans le quotient fini $J(k)/U$. □

Remarque 1.7.1. : De cette proposition, on peut déduire la finitude de $\operatorname{Pic}(X)_{tors}$ lorsque X est une variété propre, lisse et géométriquement intègre sur un corps k de type fini sur le corps $\mathbf{Q}$ des rationnels (i.e. engendré, comme corps, par un nombre fini déléments). Choisissons en effet un nombre premier p. Comme le corps p-adique $\mathbf{Q}_p$ est de degré de transcendance infini sur $\mathbf{Q}$, on peut trouver une extension finie L de $\mathbf{Q}_p$ et des plongements de corps $\mathbf{Q} \subset k \subset L$. D'après la proposition 1.1, on a une inclusion $\operatorname{Pic}(X) \subset \operatorname{Pic}(X_L)$, et donc aussi $\operatorname{Pic}(X)_{tors} \subset \operatorname{Pic}(X_L)_{tors}$. La finitude de $\operatorname{Pic}(X)_{tors}$ résulte alors de celle de $\operatorname{Pic}(X_L)_{tors}$ (proposition ci-dessus).

Proposition 1.8. — *Soient k un corps p-adique (extension finie d'un corps $\mathbf{Q}_p$) et X une variété propre, lisse et géométriquement intègre sur k. Alors pour tout entier $n > 0$, le quotient* $\mathrm{Pic}(X)/n$ *est fini.*

Démonstration : Du théorème 1.2 on déduit que le groupe $\mathrm{Pic}(\overline{X})^G/n$ s'insère dans une suite exacte :

$$J(k)/n \longrightarrow \mathrm{Pic}(\overline{X})^G/n \longrightarrow T/n$$

avec T un groupe abélien de type fini. La structure de $J(k)$ rappelée dans la démonstration précédente implique immédiatement la finitude de $J(k)/n$ et donc celle de $\mathrm{Pic}(\overline{X})^G/n$. On sait que le groupe de Brauer $Br(k)$ d'un corps local est isomorphe à $\mathbf{Q}/\mathbf{Z}$. La finitude de $\mathrm{Pic}(X)/n$ résulte alors de celle de $\mathrm{Pic}(\overline{X})^G/n$ et de la suite exacte :

$$0 \longrightarrow \mathrm{Pic}(X) \longrightarrow \mathrm{Pic}(\overline{X})^G \longrightarrow {}_mBr(k),$$

où $m > 0$ est le degré d'une extension de k sur laquelle X acquiert un point rationnel. ☐

Remarque 1.8.1. : On peut donner une autre démonstration, plus générale. Soit k comme ci-dessus et X une k-variété quelconque. D'après (1.2) on a une injection

$$\mathrm{Pic}(X)/n\,\mathrm{Pic}(X) \hookrightarrow H^2_{et}(X, \mu_n).$$

Mais la finitude des groupes de cohomologie $H^i_{et}(\overline{X}, \mu_n)$ pour tout i rappelée plus haut, la suite spectrale de Hoschschild–Serre

$$H^p(\mathrm{Gal}(\overline{k}/k), H^q_{et}(\overline{X}, \mu_n)) \Longrightarrow H^*_{et}(X, \mu_n)$$

et la finitude des groupes de cohomologie galoisienne de $\mathrm{Gal}(\overline{k}/k)$ à valeurs dans des modules finis (Serre, Cohomologie Galoisienne, Springer LNM **5**) assurent la finitude de tous les groupes $H^m_{et}(X, \mu_n)$ pour X une variété sur un corps p-adique k.

Théorème 1.9. — *Soit k un corps de type fini sur le corps premier, et soit X une k-variété intègre propre et lisse, puis U un ouvert de X. Alors les groupes* $\mathrm{Pic}(X)$ *et* $\mathrm{Pic}(U)$ *sont des groupes de type fini. En particulier leurs sous-groupes de torsion sont des groupes finis, et pour tout entier $n > 0$, le quotient* $\mathrm{Pic}(X)/n\,\mathrm{Pic}(X)$ *est fini.*

"Démonstration" : La suite de localisation (1.4) permet de se ramener au cas de X, et d'après la proposition 1.1, il suffit de savoir que pour une k-variété abélienne J, le groupe $J(k)$ est de type fini. Lorsque k est un corps de nombres, c'est là précisément l'énoncé du théorème de Mordell–Weil. Le cas plus général d'un corps k de type fini sur le corps premier est traité par exemple par Lang dans son livre *Diophantine Geometry.* ☐

Remarque 1.9.1. : La démonstration complète du théorème ci-dessus représente l'un des grands succès de la géométrie diophantienne des années 30–50, et l'esquisse ci-dessus est loin d'en donner une juste représentation. En fait, il y a un théorème de Mordell–Weil *faible*, qui dit que pour un entier $n > 0$ et k de type fini sur le corps premier, le quotient $J(k)/n$ est un groupe fini. Ensuite on développe la théorie de la hauteur sur les variétés abéliennes pour déduire du théorème de Mordell–Weil faible le théorème *fort* que $J(k)$ est de type fini. Ceci du moins est le plan lorsque k est un corps de nombres. D'autres arguments sont nécessaires pour traiter le cas d'un corps de type fini sur le corps premier, et aussi pour établir le théorème de Néron–Severi affirmant que le groupe de Néron–Severi est de type fini. Au vu de l'analogie bien connue entre les arguments de Mordell–Weil et ceux de Néron–Severi, on peut se demander, déjà sur un corps de nombres, si une bonne théorie des hauteurs permettrait d'établir que $\mathrm{Pic}(X)$ est de type fini, directement à partir de la finitude de $\mathrm{Pic}(X)/n$, sans dévissage du groupe $\mathrm{Pic}(X)$.

Remarque 1.9.2. : Le principe de la démonstration du théorème de Mordell-Weil faible est le suivant, au moins sur un corps de nombres. On utilise l'injection

$$J(k)/nJ(k) \hookrightarrow H^1_{et}(k, {}_nJ)$$

et l'on montre en s'appuyant sur la propreté de J/k, que l'image de cette injection consiste de classes non ramifiées en dehors des places de mauvaise réduction de J et des places divisant n. On montre par ailleurs que ces classes non ramifiées forment un groupe fini.

Une autre façon de voir les choses, au moins pour n inversible dans le corps de type fini k, est d'insérer la flèche (1.2) dans un diagramme commutatif, où les flèches horizontales sont des injections,

$$\begin{array}{ccc} \mathrm{Pic}(\mathbf{X})/n\,\mathrm{Pic}(\mathbf{X}) & \longrightarrow & H^2_{et}(\mathbf{X}, \mu_n) \\ \downarrow & & \downarrow \\ \mathrm{Pic}(X)/n\,\mathrm{Pic}(X) & \longrightarrow & H^2_{et}(X, \mu_n), \end{array}$$

où $\mathbf{X}$ est un modèle régulier de la k-variété lisse X, modèle qui est de type fini au-dessus soit d'un corps fini si $car(k) > 0$, soit, si $car(k) = 0$, d'un ouvert non vide de $\mathrm{Spec}(\mathbf{Z})$ où n est inversible. La régularité de $\mathbf{X}$ assure la surjectivité de la flèche verticale de gauche, et la simple hypothèse que $\mathbf{X}$ est de type fini au-dessus de S assure la finitude des groupes $H^i_{et}(\mathbf{X}, \mu_n)$ (voir [M2], II, 7.1, qui s'appuie d'une part sur le théorème de finitude de Deligne in [SGA41/2], d'autre part sur le calcul de la cohomologie étale des anneaux d'entiers de corps de nombres). Il convient de constater que la démonstration ci-dessus vaut sans hypothèse de propreté pour X/k.

L'avantage de cette démonstration est, comme l'a noté S. Saito ([S4]), qu'elle s'étend en partie dans l'étude des cycles de codimension supérieure (§ 7 ci-après, théorème 7.5).

§ 2. Groupes de Chow

(Référence : Fulton [F] Chapitre I)

La définition donnée au paragraphe précédent du groupe de Picard admet une généralisation naturelle aux cycles de (co)dimension quelconque.

Suivant Fulton, voici les définitions et propriétés de base. La théorie ***homologique***, avec ses définitions valables pour des schémas éventuellement singuliers, est plus naturelle que l'ancienne théorie de l'équivalence rationnelle.

Soit k un corps et X une k–variété algébrique, i.e. un schéma de type fini et séparé sur k. Rappelons la bijection naturelle entre les points (au sens des schémas) du schéma X et les sous-schémas fermés intègres de X, associant à un point P son adhérence schématique $V(P)$, dont le point générique n'est autre que P. La dimension de P est alors par définition celle de $V(P)$. C'est aussi le degré de transcendance sur k du corps résiduel $k(P)$ de l'anneau local de X en P. Il sera commode de remplacer l'ancien langage des sous–variétés fermées par celui des points schématiques.

On définit le groupe des cycles de dimension i sur X comme le groupe libre $Z_i(X)$ sur les points de X de dimension i.

Tout k–morphisme propre $f : X \longrightarrow Y$ de tels schémas induit une application $f_* : Z_i(X) \longrightarrow Z_i(Y)$. Cette application est définie par linéarité à partir de l'application qui à un point P de X de dimension i associe le cycle 0 si le point $f(P)$ est de dimension plus petite que i, et associe le cycle $[k(P) : k(f(P))]P \in Z_i(Y)$ si $f(P)$ a même dimension que P, l'indice $[k(P) : k(f(P))]$ étant alors le degré de l'extension finie $k(P)/k(f(P))$ de corps résiduels.

Etant donnés une k–variété intègre Y de dimension d et $f \in k(Y)^*$ un élément de son corps des fonctions rationnelles, on peut définir le diviseur $\mathrm{div}(f)$ de f comme un élément de $Z_{d-1}(Y)$. Lorsque Y est normale, ses anneaux locaux réguliers aux points P de codimension 1 sont des anneaux de valuation discrète, définissant une valuation $v_P : k(Y)^* \longrightarrow \mathbf{Z}$. On définit alors, suivant Weil, $\mathrm{div}(f) = \sum_{P \in Y^{(1)}} v_p(f)P$. Pour Y intègre quelconque, il est encore possible de définir $\mathrm{div}(f) \in Z_{d-1}(Y)$, soit en ayant recours à des longueurs d'anneaux artiniens, soit en utilisant la normalisée $r : Y_n \longrightarrow Y$ de Y, et en définissant $\mathrm{div}_Y(f) = r_* \mathrm{div}_{Y_n}(f)$.

Une formule très utile dit que pour $p : X \longrightarrow Y$ un k–morphisme propre surjectif de variétés intègres (irréductibles et réduites) de même dimension, donc tel que l'extension de corps $k(X)/k(Y)$ est finie, et $f \in k(X)^*$, on a $p_*(\mathrm{div}(f)) = \mathrm{div}(N_{k(X)/k(Y)}(f))$, où N désigne la norme.

Etant donnés une k–variété algébrique X et un entier $i \geq 0$, on définit alors le groupe de Chow $CH_i(X)$ de dimension i comme le quotient de $Z_i(X)$ par le sous–groupe engendré par les $p_*(\mathrm{div}(f))$, pour tous les $p : Z \longrightarrow X$ k–morphismes propres $p : Z \longrightarrow X$ d'une k–variété intègre Z de dimension $(i+1)$ et pour toutes les fonctions rationnelles f non nulles sur un tel Z. Dans cette définition, on peut se limiter aux k–morphismes birationnels sur leur image. On peut de plus se limiter soit aux Z normales, soit aux sous-variétés fermées, mais non nécessairement normales, $Z \subset X$.

Lorsque la k–variété X est équidimensionnelle de dimension d, on considère aussi les groupes de Chow $CH^i(X) = CH_{d-i}(X)$.

Propriétés de base des groupes de Chow

1) *Fonctorialité covariante par k-morphismes propres.* Tout k–morphisme propre de k–variétés intègres $f : X \longrightarrow Y$ induit un homomorphisme $f_* : CH_i(X) \longrightarrow CH_i(Y)$. Ceci se voit en utilisant la formule ci–dessus.

2) ***Fonctorialité contravariante par morphismes plats.*** Si $f : X \longrightarrow Y$ est un k–morphisme plat de dimension relative n, on définit de façon naturelle des morphismes $f^* : CH_i(Y) \longrightarrow CH_{i+n}(X)$, soit, pour X et Y équidimensionnels, $f^* : CH^j(Y) \longrightarrow CH^j(X)$.

3) Si $f : X \longrightarrow Y$ est un morphisme fini et plat de degré d, le composé

$$CH_i(Y) \xrightarrow{f^*} CH_i(X) \xrightarrow{f_*} CH_i(Y)$$

est la multiplication par d.

4) *Suite de localisation.* Si $i : Y \subset X$ est l'inclusion d'une sous-variété fermée, et $j : U \longrightarrow X$ est l'inclusion de l'ouvert complémentaire de X et $i \geq 0$ un entier, on a la suite exacte :

$$CH_i(Y) \xrightarrow{i_*} CH_i(X) \xrightarrow{j^*} CH_i(U) \longrightarrow 0.$$

5) *0-cycles.* Soit X une k-variété propre. Le morphisme structural $X \longrightarrow \mathrm{Spec}(k)$ induit un homomorphisme $CH_0(X) \longrightarrow CH_0(\mathrm{Spec}(k)) = \mathbf{Z}$, appelé l'application degré, qui associe à (la classe d') un 0-cycle $\sum n_P P$ l'entier $\sum n_P P[k(P) : k]$. On note $A_0(X)$ le noyau de cet homomorphisme.

Le groupe $A_0(X)$ est un invariant k-birationnel des k-variétés intègres propres et lisses, comme on le voit ([F], 16.1.11) en utilisant des correspondances et le lemme de déplacement (valable pour les 0-cycles sur une variété lisse quelconque).

Sur un corps k algébriquement clos, pour toute k-variété projective et irréductible, le groupe $A_0(X)$ est un groupe divisible (ceci se voit par réduction au cas des courbes lisses projectives).

Pour X projective lisse et géométriquement intègre sur un corps k, de variété d'Albanese $\underline{\mathrm{Alb}}_X$ (c'est la duale de la variété de Picard $\underline{\mathrm{Pic}}^0_{X/k}$), on dispose d'une application canonique

$$\mathrm{alb} : A_0(X) \longrightarrow \underline{\mathrm{Alb}}_X(k)$$

dans le groupe des points k-rationnels de X. Cette application a un conoyau de torsion, et est surjective lorsque k est algébriquement clos, comme on voit en se restreignant à l'application $A_0(C) \longrightarrow \underline{\mathrm{Alb}}_X(k)$ induite sur une k-courbe projective et lisse $C \subset X$ convenable.

Pour terminer ce paragraphe, citons deux difficultés fondamentales rencontrées dans l'étude des groupes de Chow en codimension plus grande que 1. Soit X une variété projective et lisse, géométriquement intègre sur un corps k.

Première difficulté. Si F est un corps contenant k, et j un entier, $j \geq 2$, l'application naturelle $CH^j(X) \longrightarrow CH^j(X_F)$ n'est pas nécessairement injective, à la différence de ce qui se passe pour $j = 1$ (§ 1, Prop. 1.1). Le noyau de cette application est de torsion, mais il peut être non nul lorsque le corps k n'est pas algébriquement clos. Par ailleurs, lorsque F/k est une extension galoisienne de groupe G, même en supposant que X possède un point k-rationnel, l'application $CH^j(X) \longrightarrow CH^j(X_F)^G$ n'est pas nécessairement surjective (on peut simplement dire que le conoyau de cette application est de torsion).

On ne peut donc utiliser ici les arguments développés au § 1.

Deuxième difficulté. Comme il fut établi pour la première fois par Mumford, une démonstration toute différente étant donnée par S. Bloch [B2], p. 1.19 (voir aussi [BS]), même dégagé de sa partie *discrète*, i.e. de son image dans par exemple la cohomologie entière de X lorsque $k = \mathbf{C}$, pour $j \geq 2$, le groupe $CH^j(X)$ est en général loin d'être représentable par une variété algébrique. Ainsi pour k algébriquement clos non dénombrable et X une surface l'application naturelle alb : $A_0(X) \longrightarrow \underline{\mathrm{Alb}}_X(k)$ mentionnée plus haut peut avoir un énorme noyau (qu'on ne peut couvrir par les cycles supportés sur une courbe de X). On se reportera à la thèse de Jannsen (*Mixed Motives and Algebraic K-Theory*, Springer LNM 1400, §§ 9 et 10) pour une discussion complémentaire.

§ 3. K–théorie, cohomologie étale, et torsion dans les groupes de Chow (le programme de Bloch).

3.1. Corps et anneaux de valuation discrète

Etant donné un corps F, on définit le groupe K_2F comme le quotient de $F^* \otimes F^*$ par le sous-groupe engendré par les éléments $a \otimes b$ avec $a + b = 1$. Lorsque F est le corps des fractions d'un anneau de valuation discrète, de valuation $v : F^* \longrightarrow \mathbf{Z}$, de corps résiduel κ, on dispose du symbole modéré

$$\begin{array}{rccl} T: & K_2F & \longrightarrow & \kappa^* = K_1\kappa, \\ & \{a,b\} & \longmapsto & (-1)^{v(a)v(b)} cl(a^{v(b)}/b^{v(a)}), \end{array}$$

où $cl(c)$ désigne la classe dans k^* d'une unité $c \in A^*$. C'est l'analogue de la flèche $K_1F = F^* \longrightarrow K_0\kappa = \mathbf{Z}$ définie par la valuation v.

On supposera le lecteur familier avec les aspects élémentaires de la cohomologie galoisienne. Etant donné un corps k, de clôture séparable k_s et $G = \mathrm{Gal}(k_s/k)$, pour un G–module continu discret on notera $H^i_{et}(k, M)$ et parfois simplement $H^i(k, M)$ le groupe de cohomologie galoisienne $H^i(G, M)$. Pour n inversible dans k, on note μ_n le groupe des racines n–ièmes de l'unité dans k_s^*, et pour j entier > 0, on note $\mu_n^{\otimes j}$ le G–module $\mu_n \otimes \cdots \otimes \mu_n$ (j fois). On convient que $\mu_n^{\otimes 0} = \mathbf{Z}/n$ avec G–action triviale.

La théorie de Kummer, c'est-à-dire la suite de cohomologie galoisienne associée à la suite exacte

$$\begin{array}{ccccccccc} 1 & \longrightarrow & \mu_n & \longrightarrow & k_s^* & \longrightarrow & k_s^* & \longrightarrow & 1, \\ & & & & x & \longmapsto & x^n & & \end{array}$$

et le théorème 90 de Hilbert $H^1(k, k_s^*) = 0$ donnent l'isomorphisme

$$k^*/k^{*n} \simeq H^1(k, \mu_n).$$

On en déduit une application par cup-produit :

$$(k^* \otimes_{\mathbf{Z}} k^*)/n \longrightarrow k^*/k^{*n} \otimes_{\mathbf{Z}} k^*/k^{*n} \simeq H^1(k, \mu_n) \otimes_{\mathbf{Z}} H^1(k, \mu_n) \longrightarrow H^2(k, \mu_n^{\otimes 2}).$$

Un calcul purement algébrique, à base de normes (voir [S]) montre que cet homomorphisme annule les éléments de la forme $x \otimes y$, avec $x + y = 1$. Ainsi elle définit un homomorphisme

$$K_2k/nK_2k \longrightarrow H^2(k, \mu_n^{\otimes 2}).$$

Le théorème de Merkur'ev–Suslin ([MS], [S]) assure que cet homomorphisme est en fait un isomorphisme. Si k est le corps des fractions d'un anneau de valuation discrète comme plus haut, de corps résiduel κ, on dispose d'un diagramme commutatif

$$\begin{array}{ccc} K_2k & \longrightarrow & H^2(k, \mu_n^{\otimes 2}) \\ \downarrow & & \downarrow \\ \kappa^* & \longrightarrow & H^1(\kappa, \mu_n), \end{array}$$

où la flèche verticale de gauche est le symbole modéré et la flèche verticale de droite le résidu en cohomologie galoisienne, analogue en degré supérieur de la flèche évidente $H^1(k, \mu_n) \longrightarrow H^0(\kappa, \mathbf{Z}/n)$ i.e. $k^*/k^{*n} \longrightarrow \mathbf{Z}/n$ induite par la valuation.

3.2. La méthode de Bloch

Soit X une k-variété algébrique intègre, et $n > 0$ un entier inversible dans k. On considère le diagramme commutatif suivant de complexes, où X^i désigne l'ensemble des points de X de codimension i :

$$\begin{array}{lccccc}
\text{(degrés)} & 2 & & 1 & & 0 \\
\mathcal{C} & \bigoplus_{x\in X^{i-2}} K_2k(x) & \to & \bigoplus_{x\in X^{i-1}} k(x)^* & \to & \bigoplus_{x\in X^i} \mathbb{Z} \to 0 \\
 & \downarrow \times n & & \downarrow \times n & & \downarrow \times n \\
\mathcal{C} & \bigoplus_{x\in X^{i-2}} K_2k(x) & \to & \bigoplus_{x\in X^{i-1}} k(x)^* & \to & \bigoplus_{x\in X^i} \mathbb{Z} \to 0 \\
 & \downarrow & & \downarrow & & \downarrow \\
\mathcal{D} & \bigoplus_{x\in X^{i-2}} H^2(k(x),\mu_n^{\otimes 2}) & \to & \bigoplus_{x\in X^{i-1}} H^1(k(x),\mu_n) & \to & \bigoplus_{x\in X^i} H^0(k(x),\mathbb{Z}/n) \to 0 \\
 & \downarrow & & \downarrow & & \downarrow \\
 & 0 & & 0 & & 0
\end{array}$$

Dans ce diagramme, les flèches horizontales du diagramme $\mathcal{C}$ sont les symboles modérés, puis les flèches diviseurs (plus précisément ce sont les flèches obtenues à partir du symbole modéré et de la flèche diviseur après normalisation des variétés considérées, puis somme). Les flèches du complexe $\mathcal{D}$ sont les flèches de résidu en cohomologie galoisienne. Les flèches allant de $\mathcal{C}$ vers $\mathcal{D}$ sont les flèches décrites au § 3.1. Les complexes verticaux sont exacts. Pour la verticale médiane, cela résulte de la théorie de Kummer. Pour la verticale de gauche, c'est le théorème de Merkur'ev/Suslin qui le garantit. Le groupe d'homologie $H_0(\mathcal{C})$ s'identifie à $CH^i(X)$. Une chasse au diagramme n'utilisant que la surjectivité dans le théorème de Merkur'ev/Suslin (on pourrait oublier le coin supérieur gauche du diagramme ci-dessus) donne alors la suite exacte :

$$0 \longrightarrow H_1(\mathcal{C})/n \longrightarrow H_1(\mathcal{D}) \longrightarrow {}_nH_0(\mathcal{C}) \longrightarrow 0$$

soit

(3.1) $$0 \longrightarrow H_1(\mathcal{C})/n \longrightarrow H_1(\mathcal{D}) \longrightarrow {}_nCH^i(X) \longrightarrow 0$$

(cette présentation simplifiée de l'argument initial de Bloch est présentée dans [CSS]. Elle nous avait été signalée par T. Ekedahl).

Problème : comment contrôler les groupes $H_1(\mathcal{C})$ et $H_1(\mathcal{D})$?

Le complexe $\mathcal{C}$ est en fait une partie d'un complexe, le complexe de Gersten :

$$\bigoplus_{x\in X^0} K_ik(x) \longrightarrow \cdots \longrightarrow \bigoplus_{x\in X^{i-j}} K_jk(x) \longrightarrow \cdots \longrightarrow \bigoplus_{x\in X^i} \mathbb{Z} \longrightarrow 0$$

dont les termes sont définis au moyen de la K-théorie supérieure des corps. Lorsque X est *lisse sur un corps*, la *conjecture de Gersten*, établie par Quillen [Q], assure que ce complexe est le complexe des sections globales d'une résolution flasque du faisceau Zariski $\mathcal{K}_i$ sur X, défini par faisceautisation du préfaisceau $U \longrightarrow K_i(H^0(U,\mathcal{O}_X))$, le groupe $K_i(A)$ étant celui associé par Quillen à l'anneau A. Le groupe noté $H_1(\mathcal{C})$ ci-dessus s'identifie donc au groupe de cohomologie (Zariski) $H^{i-1}(X,\mathcal{K}_i)$.

De même, le complexe $\mathcal{D}$ est en fait une partie d'un complexe

$$\bigoplus_{x\in X^0} H^i(k(x),\mu_n^{\otimes i}) - \cdots \longrightarrow \bigoplus_{x\in X^{i-j}} H^{i-j}(k(x),\mu_n^{\otimes i-j}) - \cdots \longrightarrow \bigoplus_{x\in X^i} H^0(k(x),\mathbf{Z}/n) \longrightarrow 0$$

de groupes de cohomologie étale, dont les flèches sont définies à partir d'applications résidus supérieurs. Lorsque X est *lisse*, un analogue de la conjecture de Gersten, établi par Bloch et Ogus [BO], assure que le complexe ci-dessus est le complexe des sections globales d'une résolution flasque, pour la topologie de Zariski sur X, du faisceau Zariski $\mathcal{H}^i(\mu_n^{\otimes i})$ associé au préfaisceau $U \longrightarrow H^i_{et}(U,\mu_n^{\otimes i})$. On a l'énoncé analogue pour les faisceaux $\mathcal{H}^i(\mu_n^{\otimes j})$, i et j quelconques. Le groupe $H_1(\mathcal{D})$ considéré plus haut s'identifie donc au groupe de cohomologie (Zariski) $H^{i-1}(X,\mathcal{H}^i(\mu_n^{\otimes i}))$.

Mettant ensemble ces deux résultats, on voit que la suite exacte (3.1) peut se réécrire, dans le cas lisse :

(3.2) $$0 \longrightarrow H^{i-1}(X,\mathcal{K}_i)/n \longrightarrow H^{i-1}(X,\mathcal{H}^i(\mu_n^{\otimes i})) \longrightarrow {}_nCH^i(X) \longrightarrow 0.$$

Notons que les résultats de Bloch–Ogus montrent également que le groupe $H^i(X,\mathcal{H}^i(\mu_n^{\otimes i}))$ s'identifie au groupe $H_0(\mathcal{D})$, qui n'est autre que le quotient $CH^i(X)/n$.

Le dernier ingrédient dans l'approche de Bloch de la torsion des groupes de Chow est la suite spectrale de passage du local au global (cf. [BO])

(3.3) $$E_2^{pq} = H^p(X,\mathcal{H}^q(\mu_n^{\otimes j})) \Longrightarrow H^n_{et}(X,\mu_n^{\otimes j}).$$

Les termes E_2^{pq} sont concentrés dans le domaine $0 \leq p \leq d = \dim(X)$ comme il sied à la cohomologie de Zariski, et, pour X lisse, dans le cône $p \leq q$. C'est là un résultat moins trivial, résultant de l'existence, dans le cas lisse, des résolutions flasques mentionnées ci-dessus, et qui sont, pour $\mathcal{H}^q(\mu_n^{\otimes j})$, de longueur q. Dans la suite, nous appellerons souvent suite spectrale de Bloch-Ogus la suite spectrale (3.3).

En particulier, il existe des applications naturelles :

(3.4) $$\gamma_i : H^{i-1}(X,\mathcal{H}^i(\mu_n^{\otimes i})) \longrightarrow H^{2i-1}_{et}(X,\mu_n^{\otimes i})$$

et l'on a ainsi le diagramme (X lisse) :

(3.5) $$\begin{array}{ccccccccc} 0 & \longrightarrow & H^{i-1}(X,\mathcal{K}_i)/n & \longrightarrow & H^{i-1}(X,\mathcal{H}^i(\mu_n^{\otimes i})) & \longrightarrow & {}_nCH^i(X) & \longrightarrow & 0 \\ & & & & \downarrow \gamma_i & & & & \\ & & & & H^{2i-1}_{et}(X,\mu_n^{\otimes i}). & & & & \end{array}$$

Faisant parcourir à n les puissances d'un nombre premier l fixé inversible dans k, et passant à la limite inductive, on obtient le diagramme :

(3.6) $$\begin{array}{ccccccccc} 0 & \longrightarrow & H^{i-1}(X,\mathcal{K}_i)\otimes \mathbf{Q}_l/\mathbf{Z}_l & \longrightarrow & H^{i-1}(X,\mathcal{H}^i(\mathbf{Q}_l/\mathbf{Z}_l(i))) & \longrightarrow & CH^i(X)_{l-tors} & \longrightarrow & 0 \\ & & & & \downarrow \gamma_i & & & & \\ & & & & H^{2i-1}_{et}(X,\mathbf{Q}_l/\mathbf{Z}_l(i)). & & & & \end{array}$$

On peut compléter ces diagrammes de la façon suivante. Pour tout entier m inversible dans k, on dispose, sur la k–variété lisse X, d'une application cycle de Grothendieck

$$\rho : CH^i(X) \longrightarrow H^{2i}_{et}(X, \mu_m^{\otimes i})$$

décrite par Deligne dans [SGA4 1/2].

Par ailleurs, de la suite exacte de faisceaux étales sur X :

$$1 \longrightarrow \mu_m^{\otimes i} \longrightarrow \mu_{nm}^{\otimes i} \longrightarrow \mu_n^{\otimes i} \longrightarrow 1$$

on déduit une application bord (*Bockstein*)

$$\beta : H^{2i-1}_{et}(X, \mu_n^{\otimes i}) \longrightarrow H^{2i}_{et}(X, \mu_m^{\otimes i}).$$

Une vérification non triviale, effectuée dans le travail [CSS], assure qu'au signe près, le diagramme

$$(3.7)\qquad \begin{array}{ccc} {}_nCH^i(X) & \xrightarrow{\rho} & H^{2i}_{et}(X, \mu_m^{\otimes i}) \\ \uparrow{\scriptstyle \alpha_i} & & \uparrow{\scriptstyle \beta} \\ H^{i-1}(X, \mathcal{H}^i(\mu_n^{\otimes i})) & \xrightarrow{\gamma_i} & H^{2i-1}_{et}(X, \mu_n^{\otimes i}) \end{array}$$

est commutatif.

Limitant n et m aux puissances de l premier inversible dans k, et passant à la limite inductive sur n et à la limite projective sur m, on obtient le diagramme commutatif au signe près :

$$(3.8)\qquad \begin{array}{ccc} CH^i(X)_{l-tors} & \xrightarrow{\rho} & H^{2i}_{et}(X, \mathbf{Z}_l(i))) \\ \uparrow{\scriptstyle \alpha_i} & & \uparrow{\scriptstyle \beta} \\ H^{i-1}(X, \mathcal{H}^i(\mathbf{Q}_l/\mathbf{Z}_l(i)) & \xrightarrow{\gamma_i} & H^{2i-1}_{et}(X, \mathbf{Q}_l/\mathbf{Z}_l(i)). \end{array}$$

3.3. Cycles de codimension 1 et 2

3.3.1. Cycles de codimension 1

La suite (3.2) se lit ici

$$0 \longrightarrow H^0(X, \mathcal{K}_1)/n \longrightarrow H^1(X, \mathcal{H}^1(\mu_n)) \longrightarrow {}_nCH^1(X) \longrightarrow 0$$

et la suite spectrale de Bloch–Ogus donne un isomorphisme

$$H^1(X, \mathcal{H}^1(\mu_n)) \simeq H^1_{et}(X, \mu_n),$$

si bien que l'on a la suite

$$0 \longrightarrow H^0(X, \mathbf{G}_m)/n \longrightarrow H^1_{et}(X, \mu_n) \longrightarrow {}_nCH^1(X) \longrightarrow 0,$$

dont on peut vérifier qu'elle s'identifie, au signe près, avec la suite de Kummer ([B1], Prop. 3.6).

De la suite spectrale de Bloch–Ogus à coefficients μ_n on tire aussi la suite exacte, essentiellement bien connue :

$$(3.9)\qquad 0 \longrightarrow CH^1(X)/n \longrightarrow H^2_{et}(X,\mu_n) \longrightarrow H^0(X,\mathcal{H}^2(\mu_n)) \longrightarrow 0,$$

où le terme $H^0(X,\mathcal{H}^2(\mu_n))$ peut être identifié à la n-torsion ${}_nH^2_{et}(X,\mathbb{G}_m)$ du groupe de Brauer de X.

3.3.2. Cycles de codimension deux

Dans la suite spectrale de Bloch–Ogus, la flèche γ_2 est toujours injective, et de (3.5) et (3.6) on déduit le résultat important (Bloch+Merkur'ev–Suslin, cf. [CSS]) :

THÉORÈME 3.3.2. — *Pour X lisse sur un corps k, et n inversible dans k, le groupe ${}_nCH^2(X)$ est un sous-quotient du groupe de cohomologie étale $H^3_{et}(X,\mu_n^{\otimes 2})$. Pour l premier à la caractéristique de k, le groupe $CH^2(X)_{l-tors}$ est un sous-quotient du groupe $H^3_{et}(X,\mathbb{Q}_l/\mathbb{Z}_l(2))$.*

Une analyse attentive de l'argument présenté montre que de (3.2) il suffit de retenir la surjection $H^{i-1}(X,\mathcal{H}^i(\mu_n^{\otimes i})) \longrightarrow {}_nCH^i(X) \longrightarrow 0$, laquelle résulte simplement de la surjectivité dans le théorème de Merkur'ev-Suslin (dont l'énoncé n'utilise pas la K-théorie de Quillen).

De la suite spectrale de Bloch–Ogus à coefficients $\mu_n^{\otimes 2}$ on tire la longue suite exacte

$$(3.10)\qquad \begin{aligned} 0 \longrightarrow H^1(X,\mathcal{H}^2(\mu_n^{\otimes 2})) \xrightarrow{\eta} H^3_{et}(X,\mu_n^{\otimes 2}) \longrightarrow H^0(X,\mathcal{H}^3(\mu_n^{\otimes 2})) \longrightarrow \\ CH^2(X)/n \longrightarrow H^4_{et}(X,\mu_n^{\otimes 2}). \end{aligned}$$

Pour usage ultérieur (§ 7) notons tout d'abord qu'en combinant (3.5) et le début de cette suite, on obtient la suite exacte :

$$(3.11)\qquad 0 \longrightarrow H^1(X,\mathcal{K}_2)/n \longrightarrow NH^3_{et}(X,\mu_n^{\otimes 2}) \longrightarrow {}_nCH^2(X) \longrightarrow 0,$$

Par ailleurs, on aimerait obtenir des cas où l'application

$$H^3_{et}(X,\mu_n^{\otimes 2}) \longrightarrow H^0(X,\mathcal{H}^3(\mu_n^{\otimes 2}))$$

est surjective, c'est-à-dire où toute classe de cohomologie sur le corps des fonctions de X, provenant partout localement d'une classe de cohomologie, provient automatiquement d'une classe de cohomologie globale. Ceci permettrait alors de contrôler le quotient $CH^2(X)/n$ via une injection $CH^2(X)/n \longrightarrow H^4_{et}(X,\mu_n^{\otimes 2})$. Malheureusement, comme nous l'a fait observer V. Srinivas, cette dernière application n'est pas toujours injective : il suffit pour s'en rendre compte de considérer la variété lisse X complément dans l'espace projectif complexe $\mathbf{P}^3$ d'une quartique générique (donc de groupe de Picard engendré par les sections hyperplanes). La question de savoir si l'application est injective lorsque X est projective, lisse, et possède un point k-rationnel, est ouverte.

§ 4. Variétés sur un corps séparablement clos

Lorsque le corps de base est séparablement clos, et X connexe lisse de dimension d, la dimension cohomologique des corps $k(x)$, pour x point de X, est au plus d. Ainsi tous les termes avec $q > d$ dans la suite spectrale de Bloch–Ogus s'annulent. Comme cela fut remarqué assez tard (cf. [BV] et [S3]), on en déduit simplement (i.e. sans réduction préliminaire à la dimension 2) les théorèmes de ce paragraphe, dus essentiellement à Roitman et à Bloch.

THÉORÈME 4.1. — *Soit X une variété lisse connexe de dimension d sur un corps séparablement clos k, et n inversible dans k. Alors*

(*i*) *Le groupe ${}_nCH^d(X)$ est un quotient du groupe de cohomologie étale $H^{2d-1}_{et}(X, \mu_n^{\otimes d})$.*

(*ii*) *Le groupe ${}_nCH^d(X)$ est un groupe fini.*

(*iii*) *Si X est une variété affine lisse et $d > 1$, le groupe $CH^d(X)$ est sans torsion première à $car(k)$.*

Démonstration : La suite spectrale à coefficients dans $\mu_n^{\otimes d}$ donne en effet dans ce cas un *isomorphisme*

$$\gamma_d : H^{d-1}(X, \mathcal{H}^d(\mu_n^{\otimes d})) \simeq H^{2d-1}_{et}(X, \mu_n^{\otimes d}),$$

et le diagramme (3.5) donne la suite exacte

$$0 \longrightarrow H^{d-1}(X, \mathcal{K}_d)/n \longrightarrow H^{d-1}(X, \mathcal{H}^d(\mu_n^{\otimes d})) \longrightarrow {}_nCH^d(X) \longrightarrow 0. \tag{4.1}$$

La finitude des groupes de cohomologie étale $H^p_{et}(X, \mu_n^{\otimes j})$ est un phénomène général valable sur toute variété sur un corps algébriquement clos. Enfin, pour toute variété affine X sur un corps séparablement clos, les groupes $H^p_{et}(X, \mu_n^{\otimes j})$ sont nuls pour p plus grand que la dimension de X. □

THÉORÈME 4.2 (théorème de Roitman). — *Soit k un corps séparablement clos, et X une k-variété projective lisse et connexe. L'application d'Albanese :*

$$\mathrm{alb} : A_0(X) \longrightarrow \underline{\mathrm{Alb}}_X(k)$$

induit un isomorphisme sur la torsion première à la caractéristique de k.

Démonstration :

(*i*) Par sections hyperplanes suffisamment générales, on trouve une courbe projective lisse connexe $C \subset X$ telle que l'application induite $\underline{\mathrm{Alb}}_C \longrightarrow \underline{\mathrm{Alb}}_X$ soit un épimorphisme de variétés abéliennes, et plus précisément induise une surjection au niveau des points de l^n-torsion

$$_{l^n}\underline{\mathrm{Alb}}_C(k) \longrightarrow\!\!\!\!\!\rightarrow {}_{l^n}\underline{\mathrm{Alb}}_X(k)$$

pour tout l premier et n entier, $n > 0$ (ceci se ramène à voir que l'application $H^1_{et}(X, \mathbb{Z}/l^n) \longrightarrow H^1_{et}(C, \mathbb{Z}/l^n)$ est injective, ce qui résulte du théorème de connexion de Zariski, voir [B1], ou [B2]).

Dans le diagramme naturellement commutatif

$$\begin{array}{ccc} A_0(C) & \longrightarrow & \underline{\mathrm{Alb}}_C(k) \\ \downarrow & & \downarrow \\ A_0(X) & \longrightarrow & \underline{\mathrm{Alb}}_X(k) \end{array}$$

la flèche horizontale supérieure est un isomorphisme. Ainsi l'application ${}_{l^n}A_0(X) \longrightarrow {}_{l^n}\underline{\mathrm{Alb}}_X(k)$ est une *surjection.*

(*ii*) LEMME. — *Il existe un isomorphisme*

$$H^{2d-1}_{et}(X, \mathbf{Q}_l/\mathbf{Z}_l(d)) \simeq \underline{\mathrm{Alb}}_X(k)_{l-tors}.$$

Démonstration : La dualité de Poincaré assure que l'accouplement

$$H^{2d-1}_{et}(X, \mu_{l^n}^{\otimes d}) \times H^1_{et}(X, \mu_{l^n}) \longrightarrow H^{2d}_{et}(X, \mu_{l^n}^{\otimes d+1}) \simeq \mu_{l^n}$$

est une dualité parfaite. On en déduit un isomorphisme

$$H^{2d-1}_{et}(X, \mu_{l^n}^{\otimes d}) \simeq \mathrm{Hom}({}_{l^n}\mathrm{Pic}(X), \mathbf{Q}_l/\mathbf{Z}_l(1))$$

et en passant à la limite directe :

$$H^{2d-1}_{et}(X, \mathbf{Q}_l/\mathbf{Z}_l(d)) \simeq \mathrm{Hom}(\varprojlim_{n>0} {}_{l^n}\mathrm{Pic}(X), \mathbf{Q}_l/\mathbf{Z}_l(1)) \simeq \mathrm{Hom}(\varprojlim_{n>0} {}_{l^n}\mathrm{Pic}^0(X), \mathbf{Q}_l/\mathbf{Z}_l(1)),$$

le dernier isomorphisme provenant du passage à la limite projective en n dans la suite exacte

$$0 \longrightarrow {}_{l^n}\mathrm{Pic}^0(X) \longrightarrow {}_{l^n}\mathrm{Pic}(X) \longrightarrow {}_{l^n}NS(X) \longrightarrow 0$$

déduite de la suite exacte

$$0 \longrightarrow \mathrm{Pic}^0(X) \longrightarrow \mathrm{Pic}(X) \longrightarrow NS(X) \longrightarrow 0,$$

où le groupe $NS(X)$ est de type fini (voir § 1) et le groupe $\mathrm{Pic}^0(X)$ est le groupe (divisible) des points k-rationnels de la variété de Picard de X (pour un groupe abélien A de type fini, on a toujours $\varprojlim_{n>0} {}_{l^n}A = 0$).

On sait que la variété de Picard est la variété abélienne duale de la variété d'Albanese. Plus précisément, on sait que l'accouplement de Weil

$${}_{l^n}\mathrm{Pic}^0(X) \times {}_{l^n}\underline{\mathrm{Alb}}_X(k) \longrightarrow \mu_{l^n}$$

est une dualité parfaite. Via cet accouplement, l'isomorphisme ci-dessus induit donc un isomorphisme :

$$H^{2d-1}_{et}(X, \mathbf{Q}_l/\mathbf{Z}_l(d)) \simeq \underline{\mathrm{Alb}}_X(k)_{l-tors}.$$

□

(*iii*) Dans le diagramme (3.6), pour $i = d$, on sait que la flèche γ_d est un isomoprhisme (voir la démonstration du théorème 4.1). Compte tenu du lemme ci-dessus, ce diagramme donne donc naissance à une suite exacte :

$$0 \longrightarrow H^{d-1}(X, \mathcal{K}_d) \otimes \mathbf{Q}_l/\mathbf{Z}_l \longrightarrow \underline{\mathrm{Alb}}_X(k)_{l-tors} \longrightarrow CH^d(X)_{l-tors} \longrightarrow 0.$$

De cette suite, et de la divisibilité de $H^{d-1}(X, \mathcal{K}_d) \otimes \mathbf{Q}_l/\mathbf{Z}_l$, on déduit que pour tout n entier, la flèche induite ${}_{l^n}\underline{\mathrm{Alb}}_X(k) \longrightarrow {}_{l^n}CH^d(X)$ est surjective. Mais la torsion de $CH^d(X) = CH_0(X)$ coïncide avec la torsion de $A_0(X)$ (noyau de l'application degré sur $CH_0(X)$. D'après (*i*) on sait donc qu'il y a une application surjective de ${}_{l^n}CH^d(X)$ sur ${}_{l^n}\underline{\mathrm{Alb}}_X(k)$. Mais les deux groupes ${}_{l^n}\underline{\mathrm{Alb}}_X(k)$ et ${}_{l^n}CH^d(X)$ sont finis (c'est clair pour le premier, pour le second, cela résulte de la première surjection, ou du théorème 4.1). En comparant les ordres des groupes, on voit que les deux applications surjectives ci-dessus sont forcément bijectives, ce qui achève la démonstration du théorème de Roitman, et montre également que le groupe $H^{d-1}(X, \mathcal{K}_d) \otimes \mathbf{Q}_l/\mathbf{Z}_l$ est nul. □

Théorème 4.3. — *Soit X une variété projective lisse connexe sur le corps séparablement clos k, et soit l premier, différent de car(k).*

(i) Dans le diagramme (3.6) :

$$0 \longrightarrow H^{i-1}(X,\mathcal{K}_i)\otimes \mathbf{Q}_l/\mathbf{Z}_l \longrightarrow H^{i-1}(X,\mathcal{H}^i(\mathbf{Q}_l/\mathbf{Z}_l(i))) \longrightarrow CH^i(X)_{l-tors} \longrightarrow 0$$
$$\downarrow \gamma_i$$
$$H^{2i-1}_{et}(X,\mathbf{Q}_l/\mathbf{Z}_l(i)),$$

la flèche γ_i se factorise par un homomorphisme

$$\lambda_i : CH^i(X)_{l-tors} \longrightarrow H^{2i-1}_{et}(X,\mathbf{Q}_l/\mathbf{Z}_l(i)).$$

(ii) Cet homomorphisme est bijectif pour $i = 1$ et injectif pour $i = 2$.

(iii) Pour $d = \dim(X)$, l'homomorphisme λ_d est un isomorphisme.

Démonstration : Il nous faut montrer que la flèche composée

$$H^{i-1}(X,\mathcal{K}_i)\otimes \mathbf{Q}_l/\mathbf{Z}_l \longrightarrow H^{i-1}(X,\mathcal{H}^i(\mathbf{Q}_l/\mathbf{Z}_l(i))) \longrightarrow H^{2i-1}_{et}(X,\mathbf{Q}_l/\mathbf{Z}_l(i))$$

est nulle. Pour tout élément $\zeta \in H^{i-1}(X,\mathcal{K}_i)\otimes \mathbf{Q}_l/\mathbf{Z}_l$, il existe un sous-corps F de k, de type fini sur le corps premier, et une F-variété projective et lisse X_0 telle que X provienne de X_0 par changement de base de F à k, et que ζ provienne d'un élément $\zeta_0 \in H^{i-1}(X_0,\mathcal{K}_i)\otimes \mathbf{Q}_l/\mathbf{Z}_l$ (le complexe de Gersten de $\mathcal{K}_i$ sur X étant la limite inductive des complexes de Gersten sur tous les X_0 possibles). Comme le diagramme (3.6) vaut sur un corps quelconque, et qu'il est fonctoriel par changement de corps de base, on a le carré commutatif

$$\begin{array}{ccc} H^{i-1}(X_0,\mathcal{K}_i)\otimes \mathbf{Q}_l/\mathbf{Z}_l & \longrightarrow & H^{2i-1}_{et}(X_0,\mathbf{Q}_l/\mathbf{Z}_l(i)) \\ \downarrow & & \downarrow \\ H^{i-1}(X,\mathcal{K}_i)\otimes \mathbf{Q}_l/\mathbf{Z}_l & \longrightarrow & H^{2i-1}_{et}(X,\mathbf{Q}_l/\mathbf{Z}_l(i)). \end{array}$$

Soit alors F_s la clôture séparable de F dans k puis $G = \mathrm{Gal}(F_s/F)$ et enfin $X_{0,s} = X_0 \times_F F_s$. La flèche

$$H^{2i-1}_{et}(X_0,\mathbf{Q}_l/\mathbf{Z}_l(i)) \longrightarrow H^{2i-1}_{et}(X,\mathbf{Q}_l/\mathbf{Z}_l(i))$$

se factorise par :

$$H^{2i-1}_{et}(X_0,\mathbf{Q}_l/\mathbf{Z}_l(i)) \to H^{2i-1}_{et}(X_{0,s},\mathbf{Q}_l/\mathbf{Z}_l(i))^G \subset H^{2i-1}_{et}(X_{0,s},\mathbf{Q}_l/\mathbf{Z}_l(i)) \to H^{2i-1}_{et}(X,\mathbf{Q}_l/\mathbf{Z}_l(i)),$$

la dernière flèche étant d'ailleurs un isomorphisme.

Mais il résulte des conjectures de Weil généralisées, comme établies par Deligne, que sur un corps F de type fini sur le corps premier, les groupes $H^n_{et}(X_{0,s},\mathbf{Q}_l/\mathbf{Z}_l(i))^G$ sont finis si $n \neq 2i$ (voir [CR1] Thm. 1.5). Ainsi l'image de $H^{i-1}(X_0,\mathcal{K}_i)\otimes\mathbf{Q}_l/\mathbf{Z}_l$ dans le groupe $H^{2i-1}_{et}(X,\mathbf{Q}_l/\mathbf{Z}_l(i))$ est finie. Mais comme le groupe $H^{i-1}(X_0,\mathcal{K}_i)\otimes \mathbf{Q}_l/\mathbf{Z}_l$ est divisible, cette image est forcément nulle, ce qui achève la démonstration du point (i). Que les applications γ_i soient injectives pour $i = 1, 2$ est une propriété générale valable sur tout corps (§ 3), ce qui établit l'injectivité de λ_i dans ces deux cas. Lorsque k est séparablement clos, comme X est propre et intègre, on a $H^0(X,\mathbf{G}_m) = k^*$, donc $H^0(X,\mathbf{G}_m)/l^n = 0$, et la suite de Kummer montre que λ_1 est un isomorphisme. Lorsque k est séparablement clos, γ_d est un isomorphisme (voir la démonstration du théorème 4.1). Il en est donc de même de λ_d. □

Remarque : Nous aurions pu utiliser l'énoncé *(iii)* dans la démonstration du théorème de Roitman, mais il a semblé préférable d'éviter le recours au résultat de Deligne dans la démonstration dudit théorème.

PROPOSITION 4.4. — *Soit X une variété projective lisse connexe de dimension d sur un corps séparablement clos de caractéristique $p \geq 0$. Alors le groupe $H^{d-1}(X, \mathcal{K}_d)$ est une extension du groupe fini*

$$\bigoplus_{l \neq p} H^{2d-1}_{et}(X, \mathbb{Z}_l(d))_{tors}$$

par un groupe divisible par tout entier non nul dans k.

Démonstration : Pour tout l premier, $l \neq car(k)$, et tout entier $m > 0$, on dispose de la suite exacte (4.1) :

$$0 \longrightarrow H^{d-1}(X, \mathcal{K}_d)/l^m \longrightarrow H^{2d-1}(X, \mathcal{H}^d(\mu_{l^m}^{\otimes d})) \longrightarrow {}_{l^m}CH^d(X) \longrightarrow 0 .$$

Par passage à la limite projective, on obtient une suite exacte (les groupes sont tous finis) et on trouve une flèche θ_l de $H^{d-1}(X, \mathcal{K}_d)$ vers $H^{2d-1}_{et}(X, \mathbb{Z}_l(d))$, qu'un argument de *poids* analogue à celui de la démonstration ci-dessus montre avoir son image dans la torsion de $H^{2d-1}_{et}(X, \mathbb{Z}_l(d))$. Procédant comme dans [CR1], Théorème 1.5, on en déduit d'abord que le noyau de la flèche θ_l est l-divisible, puis, comme le module de Tate $T_l(CH^d(X))$ est sans torsion, que l'image de θ_l est exactement le sous-groupe de torsion $H^{2d-1}_{et}(X, \mathbb{Z}_l(d))_{tors}$. Par ailleurs, Gabber a montré que, pour i et j donnés, pour presque tout l, le groupe $H^i_{et}(X, \mathbb{Z}_l(j))$ est sans torsion. Ceci suffit à établir l'énoncé. □

Sources et compléments.

Les applications λ_i furent définies par Bloch ([B1]). Suwa [Sw] les étudie sous le nom d'*applications d'Abel-Jacobi de Bloch*, et donne des conditions nécessaires pour que l'application λ_2 soit surjective (elle ne l'est pas toujours). La construction des applications λ_i donnée ci-dessus est simplifiée par rapport à celle de Bloch, grâce à l'utilisation du théorème de Merkur'ev-Suslin ([MS], 1982).

Les idées intervenant dans les démonstrations des théorèmes 4.2 et 4.3 sont pour l'essentiel dues à Bloch. Dans [B1], Bloch établissait le théorème de Roitman par un argument géométrique asez élaboré, s'appuyant en outre sur les applications λ_i, construites grâce au théorème de Deligne. Dans [B2], Bloch établissait un cas particulier de ce qui devait devenir le théorème de Merkur'ev/Suslin : il établissait la surjectivité de l'application de réciprocité $K_2k/nK_2k \longrightarrow H^2(k, \mu_n^{\otimes 2})$. lorsque k est un corps de fonctions de deux variables sur un corps algébriquement clos. De cela il déduisait le théorème de Roitman pour les surfaces (une petite difficulté étant la commutativité d'un diagramme, difficulté que nous avons contournée grâce à l'argument de comptage à la fin de la démonstration du théorème 4.2) - sans recours au théorème de Deligne. Un argument géométrique permet alors de déduire le théorème de Roitman pour les variétés de dimension supérieure. On voit que la démonstration donnée en 4.2, est très proche de celle de [B2], à ceci près que nous évitons la réduciton au cas de surfaces. Mais le lecteur vérifiera (cf. (3.1) et (3.5)) que, même pour une variété de dimension quelconque, nous n'utilisons que le cas particulier du théorème de Merkur'ev/Suslin établi par Bloch dans [B2].

Que $H^{d-1}(X, \mathcal{K}_d)$ admette une surjection sur le groupe $H^{2d-1}_{et}(X, \mathbb{Z}_l(d))_{tors}$ est une remarque de Bloch [B1], mais la proposition 4.4 ci-dessus est nouvelle. Des théorèmes de structure analogues pour $H^0(X,\mathcal{K}_2)$ et $H^1(X,\mathcal{K}_2)$ apparaissent dans [CR1]. On peut se demander quelle est la généralité de telles présentations, et si l'on dispose de résultats de divisibilité analogues pour les groupes de K–théorie $K_i(X)$ $(i \geq 1)$.

Un corollaire frappant du théorème de Roitman est que la torsion (première à $car(k)$) de $A_0(X)$ ne change pas par extension de corps algébriquement clos. En étendant une méthode de Suslin, F. Lecomte [L] a montré que pour toute variété lisse sur un corps algébriquement clos k, et tout entier n premier à $car(k)$, les groupes ${}_nH^i(X,\mathcal{K}_j)$ et $H^i(X,\mathcal{K}_j)/n$ restent inchangés par extension de corps algébriquement clos.

La plupart des théorèmes discutés ici valent aussi pour la p–torsion ($p = car(k)$), sous une forme convenable (Milne, Gros, Gros–Suwa). Je renvoie le lecteur à l'article de Gros–Suwa [GS] et à sa bibliographie.

Signalons aussi que des versions du théorème de Roitman ont été données pour des variétés singulières (Collino, Levine, Srinivas (voir [Sr] et sa bibliographie), Barbieri-Viale [BV], Saito [S3]).

§ 5. Variétés sur les corps finis

THÉORÈME 5.1 [CSS]. — *Soit X une variété lisse sur un corps fini F. Alors pour tout entier n premier à la caractéristique de F, le sous-groupe de n-torsion ${}_nCH^2(X)$ est fini.*

Démonstration : C'est une conséquence immédiate du théorème 3.3.2, selon lequel sur un corps quelconque, le groupe ${}_nCH^2(X)$ est un sous-quotient du groupe de cohomologie étale $H^3_{et}(X, \mu_n^{\otimes 2})$. En effet, la suite spectrale de Hochschild-Serre en cohomologie étale

$$H^p(F, H^q_{et}(\overline{X}, \mu_n^{\otimes j})) \Longrightarrow H^n_{et}(X, \mu_n^{\otimes j})$$

(où $\overline{X} = X \times_F \overline{F}$) assure que tous les groupes $H^n_{et}(X, \mu_n^{\otimes j})$ sont finis, puisqu'il en est ainsi des groupes $H^q_{et}(\overline{X}, \mu_n^{\otimes j})$ (cohomologie étale sur un corps séparablement clos) et que les groupes de cohomologie galoisienne d'un corps fini à valeurs dans un module fini sont des groupes finis. □

(Le même résultat, avec la même démonstration, vaut sur un corps local, p-adique ou réel. Voir 8.1).

THÉORÈME 5.2 [CSS]. — *Soit X une variété projective lisse et géométriquement connexe sur un corps fini F. Alors le sous-groupe de torsion de $CH^2(X)$ est fini.*

Esquisse de démonstration : D'après le théorème (3.3.2), le groupe $CH^2(X)_{l-tors}$ est un sous-quotient du groupe de cohomologie étale $H^3_{et}(X, \mathbf{Q}_l/\mathbf{Z}_l(2))$. Mais on déduit du théorème de Deligne (conjectures de Weil, avec coefficients tordus), par une suite spectrale de Hochschild-Serre, que pour une variété projective et lisse X sur un corps fini F, si $j \neq 2i$, $2i+1$, alors les groupes $H^j_{et}(X, \mathbf{Q}_l(i))$ sont nuls, et les groupes $H^j_{et}(X, \mathbf{Q}_l/\mathbf{Z}_l(i))$ sont finis ([CSS], Théorème 2 p. 780). Ainsi $H^3_{et}(X, \mathbf{Q}_l/\mathbf{Z}_l(2))$ est un groupe fini, et $CH^2(X)_{l-tors}$ est donc un groupe fini.

Par ailleurs, on peut montrer que pour presque tout premier l, le groupe $H^3_{et}(X, \mathbf{Q}_l/\mathbf{Z}_l(2))$ est en fait nul. Ainsi, la torsion première à $p = car(F)$ de $CH^2(X)$ est finie. La partie p-primaire requiert des arguments plus délicats, qui sont développés dans [CSS], et sont essentiellement dus à O. Gabber. □

Remarque 5.2.1 : L'utilisation du diagramme commutatif (3.8) (assez délicat à établir) permet même de montrer qu'on dispose d'une application injective, induite par l'application cycle :

$$\rho : CH^2(X)_{l-tors} \longrightarrow H^4_{et}(X, \mathbf{Z}_l(2))_{tors}.$$

C'est sur cette application injective qu'est fondée l'approche de [CSS] du corps de classes non ramifié de Kato–Saito pour les surfaces (théorème 5.3 ci-après).

On note $\pi_1^{ab}(X)$ le plus grand quotient abélien du groupe fondamental de X. Une autre façon de le définir est :

$$\pi_1^{ab}(X) = \mathrm{Hom}(H^1_{et}(X, \mathbf{Q}/\mathbf{Z}), \mathbf{Q}/\mathbf{Z}).$$

Le groupe fondamental abélien géométrique $\pi_1^{ab,geom}(X)$ est par définition le noyau de l'application

$$\pi_1^{ab}(X) \longrightarrow \pi_1^{ab}(\mathrm{Spec}(F)) = \widehat{\mathbf{Z}}$$

induite par le morphisme structural (l' égalité provenant du fait que F est un corps fini). C'est aussi le dual (à valeurs dans $\mathbf{Q}/\mathbf{Z}$) du groupe

$$\mathrm{Coker}[H^1(F, \mathbf{Q}/\mathbf{Z}) \longrightarrow H^1_{et}(X, \mathbf{Q}/\mathbf{Z})]$$

la flèche étant définie par le morphisme structural.

Les conjectures de Weil (le résultat de Weil) impliquent que le groupe $\pi_1^{ab,geom}(X)$ est un groupe fini.

THÉORÈME 5.3 (Kato–Saito [KS]). — *Soit X une variété projective lisse et géométriquement connexe sur un corps fini F, et $d = \dim(X)$. Alors le groupe $A_0(X)$ des 0–cycles de degré zéro modulo l'équivalence rationnelle est un groupe fini, isomorphe au groupe $\pi_1^{ab,geom}(X)$, groupe fondamental abélien géométrique de X.*

Esquisse de démonstration ([CR2]) : On commence par observer que le groupe $A_0(X)$ est un groupe de torsion (en se ramenant au cas des courbes par un argument de type Bertini). On définit classiquement (Lang) un homomorphisme *surjectif* (c'est un théorème de type Tchebotarev) et fonctoriel

$$\theta : A_0(X) \twoheadrightarrow \pi_1^{ab,geom}(X).$$

Nous voulons montrer que cette application est une bijection. Un argument de type Lefschetz permet de ramener cette dernière assertion au cas des surfaces ($\dim(X) = 2$). Soit l premier, $l \neq car(F)$. Le diagramme (3.6) s'écrit ici :

$$\begin{array}{ccccccc} 0 \longrightarrow & H^1(X, \mathcal{K}_2) \otimes \mathbf{Q}_l/\mathbf{Z}_l & \longrightarrow & H^1_{zar}(X, \mathcal{H}^2(\mathbf{Q}_l/\mathbf{Z}_l(2))) & \longrightarrow & A_0(X)_{l-tors} & \longrightarrow 0 \\ & & & \downarrow \gamma_2 & & & \\ & & & H^3_{et}(X, \mathbf{Q}_l/\mathbf{Z}_l(2)), & & & \end{array}$$

la flèche γ_2 étant une injection. D'après Deligne (cf. démonstration du théorème 5.2), le groupe $H^3_{et}(X, \mathbf{Q}_l/\mathbf{Z}_l(2))$ est fini. Comme le groupe $H^1(X, \mathcal{K}_2) \otimes \mathbf{Q}_l/\mathbf{Z}_l$ est divisible, on conclut que l'injection γ_2 se factorise par une injection $A_0(X)_{l-tors} \longrightarrow H^3_{et}(X, \mathbf{Q}_l/\mathbf{Z}_l(2))$. Ceci montre déjà que $A_0(X)_{l-tors}$ est fini, et nul pour presque tout l (résultat conjecturé par Parshin, établi par Kato–Saito). En utilisant des arguments de dualité (Poincaré au niveau de la clôture algébrique de F, plus dualité pour la cohomologie des corps finis) on identifie le groupe fini $H^3_{et}(X, \mathbf{Q}_l/\mathbf{Z}_l(2))$ à la partie l–primaire du groupe $\pi_1^{ab,geom}(X)$. On a donc une *injection*

$$\sigma : A_0(X)_{l-tors} \hookrightarrow \pi_1^{ab,geom}(X)_{l-tors}.$$

Il serait pénible de comparer cette injection à l'application θ de Lang. Mais nous n'avons pas besoin de le faire. Comme θ est surjective et σ injective, et que les groupes concernés sont finis, les deux applications σ et θ doivent être des bijections.

Le cas de la torsion p–primaire peut être traité par des méthodes p–adiques convenables [CR2]. □

Remarque 5.4 : En utilisant le théorème ci-dessus et les diagrammes (3.5), (3.6) et (3.8), on peut aller plus loin dans l'analyse de la suite spectrale de Bloch-Ogus pour une variété projective et lisse sur un corps fini. Le lecteur intéressé pourra consulter [K] et [C2].

§ 6. Variétés sur les corps de nombres, I

Pour une variété X lisse sur un corps de nombres k (ou plus généralement sur un corps de type fini sur le corps premier), on peut se demander si les groupes de Chow $CH^i(X)$ sont des groupes de type fini (c'est une variante d'une conjecture de H. Bass). On a vu au chapitre 1 qu'il en est bien ainsi lorsque $i = 1$. On ne dispose que de très peu de résultats lorsque $i \geq 2$. Des conjectures très ambitieuses ont été énoncées ces dernières années sur les dimensions, supposée finies, des groupes $CH^i(X) \otimes \mathbf{Q}$, lorsque X est projective et lisse sur un corps de nombres (par exemple, Bloch et Beilinson conjecturent un isomorphisme $A_0(X) \otimes \mathbf{Q} \simeq \underline{\mathrm{Alb}}_X(k) \otimes \mathbf{Q}$), mais la finitude même de ces dimensions n'a été établie dans aucun cas non trivial.

En 1981, Bloch [B4] obtint via la K-théorie un résultat de **finitude** pour le groupe (de torsion) $A_0(X)$ de toute une classe de surfaces rationnelles définies sur un corps de nombres. Depuis, une série de travaux à permis d'étendre ce résultat de finitude de la **torsion** à d'autres variétés. Dans ce chapitre, je décrirai le résultat le plus général obtenu à ce jour :

THÉORÈME 6.1 (Colliot-Thélène/Raskind [CR3]; Salberger [Sb2]). — *Soit X une variété projective, lisse, géométriquement connexe sur un corps de nombres k. Supposons le groupe $H^2(X, \mathcal{O}_X) = 0$. Alors le sous-groupe de torsion $CH^2(X)_{tors}$ est fini.*

Attributions : Je renvoie à [CR3] pour le détail des résultats particuliers précédemment obtenus. Disons simplement que sur un corps de nombres, tous les résultats obtenus jusqu'à une date récente concernaient des variétés pour lequel le groupe de Néron–Severi (géométrique) est sans torsion. K. Coombes [Cb] fut le premier à établir un résultat de finitude pour certaines *surfaces* possédant de la torsion dans leur groupe de Néron–Severi (par exemple les surfaces d'Enriques). La méthode de Coombes [Cb] a son originalité, et j'engage le lecteur à se reporter à l'article [Cb]. Sous la forme générale ci-dessus, le théorème, obtenu fin 1989, résulte d'une combinaison de travaux de Raskind et de l'auteur, et d'une idée de Salberger (théorème 6.7 ci-dessous). Ce dernier obtint peu de temps après nous une démonstration du théorème indépendante de la nôtre, démonstration dont je ne pus prendre complète connaissance qu'en Août 1991 (voir cependant § 9). Comme on le verra au § 7, en 1990/91, S. Saito ([S4]), s'appuyant en partie sur des indications de Salberger et de l'auteur, proposa ensuite une troisième démonstration du théorème.

Notre démonstration requiert un certain nombre de résultats préliminaires. Elle utilise un modèle projectif et lisse de la variété X au-dessus d'un ouvert de l'anneau des entiers de k, et la notion de groupe de Chow dans un contexte plus général que celui décrit au chapitre 1 (voir par exemple [F], chap. 20).

Le résultat suivant, établi dans [CR3], avait été annoncé par Gillet en 1985.

THÉORÈME 6.2. — *Soit k un corps de nombres, R son anneau des entiers, $f \in R$, $f \neq 0$ et A l'anneau localisé R_f. Soit $\mathbf{X}$ un A-schéma lisse, et soit n un entier naturel inversible dans A. Alors le sous-groupe de n-torsion ${}_nCH^2(\mathbf{X})$ est fini.*

Démonstration (esquisse) :

a) Pour le schéma $\mathbf{X}$ *lisse* sur un anneau de Dedekind A, soit $\pi : \mathbf{X} \longrightarrow \mathrm{Spec}(A)$, on dispose (Gillet, non publié) de l'analogue de la conjecture de Gersten en théorie de Bloch-Ogus pour les faisceaux $\mathcal{H}^i(\mathbf{X}, \mu_n^{\otimes j})$. L'argument formel à base de complexes développé au § 3 au-dessus d'un corps (cf. (3.2)) donne une *surjection* :

$$H^1_{Zar}(\mathbf{X}, \mathcal{H}^2(\mathbf{X}, \mu_n^{\otimes 2})) \twoheadrightarrow {}_nCH^2(\mathbf{X}).$$

De même, la suite spectrale de Bloch–Ogus et la conjecture de Gersten pour les faisceaux $\mathcal{H}^i(\mathbf{X}, \mu_n^{\otimes j})$ donnent une *injection*

$$H^1_{Zar}(\mathbf{X}, \mathcal{H}^2(\mathbf{X}, \mu_n^{\otimes 2})) \hookrightarrow H^3_{et}(\mathbf{X}, \mu_n^{\otimes 2}).$$

b) De façon générale, pour $\mathbf{X}$ de type fini sur $A = R_f$, avec R l'anneau des entiers d'un corps de nombres, les groupes $H^i_{et}(\mathbf{X}, \mu_n^{\otimes j})$ sont finis, comme on voit en utilisant la suite spectrale

$$H^p_{et}(\mathrm{Spec}(A), R^q\pi_*\mu_n^{\otimes j}) \Longrightarrow H^*_{et}(\mathbf{X}, \mu_n^{\otimes j})$$

la constructibilité du faisceau $R^q\pi_*\mu_n^{\otimes j}$ (valable pour π de type fini quelconque et A de Dedekind, voir [SGA4 1/2], Théorèmes de finitude), et les théorèmes de finitude pour la cohomologie étale des faisceaux constructibles sur un ouvert d'un anneau d'entiers de corps de nombres ([M2], II, 7.1). □

Je commencerai par établir des cas particuliers du théorème 6.1 de façon à bien mettre en évidence les différentes techniques concourant au résultat général. Ainsi, la démonstration du théorème suivant (qui remonte à Juillet 1989) va illustrer la **méthode de localisation**.

THÉORÈME 6.3 (Colliot-Thélène/Raskind). — *Soit X une variété projective, lisse, géométriquement connexe sur un corps de nombres k. Supposons les groupes $H^2(X, \mathcal{O}_X)$ et $H^1(X, \mathcal{O}_X)$ nuls. Soit $n > 0$ un entier naturel. Alors le sous-groupe de n-torsion ${}_nCH^2(X)$ est fini.*

Démonstration :

a) On sait que la dimension de l'espace vectoriel $H^1(X, \mathcal{O}_X)$ est égale à la dimension de la variété de Picard de X. L'hypothèse $H^1(X, \mathcal{O}_X) = 0$ assure donc que le groupe de Picard $\mathrm{Pic}(\overline{X})$ coïncide avec le groupe de Néron–Severi $NS(\overline{X})$ (Théorème 1.2). Comme ce dernier est un groupe de type fini, on voit qu'il existe une extension finie de corps L/k telle que $\mathrm{Pic}(X_L) = \mathrm{Pic}(\overline{X})$. Soit $m = [L : k]$.

b) Soit R l'anneau des entiers de k. On peut trouver un localisé $A = R_f$ de R (avec $f \in R$, $f \neq 0$) et un A-schéma projectif et lisse $\mathbf{X}$ de fibre générique X/k. Quitte à restreindre A, i.e. à inverser un nouvel f, on peut supposer que m et n sont inversibles dans A, que la clôture intégrale B de A dans L est finie étale sur A, et que le groupe des classes $\mathrm{Pic}(B)$ est nul. Enfin, quitte à restreindre un peu plus A, et grâce à un théorème de semi-continuité de Grothendieck, l'hypothèse $H^2(X, \mathcal{O}_X) = 0$ permet de supposer $H^2(\mathbf{X}_p, \mathcal{O}_{\mathbf{X}_p}) = 0$ pour chacune des fibres $\mathbf{X}_p$ au-dessus d'un point fermé $p \in \mathrm{Spec}(A)$, le même énoncé valant alors aussi pour les fibres du schéma $\mathbf{X}_B / \mathrm{Spec}(B)$ aux points fermés $q \in \mathrm{Spec}(B)$.

Encore d'après Grothendieck, l'égalité $H^2(\mathbf{X}_q, \mathcal{O}_{\mathbf{X}_q}) = 0$ implique qu'au niveau du complété de l'anneau local de B en un point fermé q, l'application de restriction $\mathrm{Pic}(\mathbf{X}_{\widehat{B}_q}) \longrightarrow \mathrm{Pic}(\mathbf{X}_q)$ est surjective (c'est ici un point crucial de notre démonstration). Enfin, l'égalité $\mathrm{Pic}(X_L) = \mathrm{Pic}(\overline{X})$ assure qu'au niveau de B, le groupe de Picard est *constant* localement. En particulier, l'application naturelle $\mathrm{Pic}(\mathbf{X}_{B_q}) \longrightarrow \mathrm{Pic}(\mathbf{X}_{\widehat{B}_q})$ est surjective, si bien qu'il en est ainsi aussi de l'application $\mathrm{Pic}(\mathbf{X}_{\widehat{B}_q}) \longrightarrow \mathrm{Pic}(\mathbf{X}_q)$.

c) On dispose de la suite de localisation :

$$\bigoplus_{q \in \mathrm{Spec}(B)^{(1)}} \mathrm{Pic}(\mathbf{X}_q) \longrightarrow CH^2(\mathbf{X}_B) \longrightarrow CH^2(X_L) \longrightarrow 0.$$

De façon purement formelle, pour tout point fermé $q \in \mathrm{Spec}(B)^{(1)}$ définissant un diviseur principal sur $\mathrm{Spec}(B)$, i.e. défini par un élément $\pi \in B$, et toute classe $\mathcal{L}$ dans $\mathrm{Pic}(\mathbf{X}_B)$, l'application composée

$$\operatorname{Pic}(\mathbf{X}_B) \longrightarrow \operatorname{Pic}(\mathbf{X}_q) \longrightarrow CH^2(\mathbf{X}_B)$$

est *nulle.* Représentant en effet $\mathcal{L}$ par un diviseur D ne contenant pas le diviseur principal $\mathbf{X}_q \subset \mathbf{X}_B$, on a l'égalité

$$(D_{|\mathbf{X}_q})_{|\mathbf{X}_B} = \operatorname{div}(\pi_{|D})$$

(la flèche composée associe à D la classe du cycle de codimension deux intersection du diviseur D et du diviseur *principal* $\mathbf{X}_q$. Cette classe est donc nulle.)

Dans les conditions ci-dessus, l'anneau B est principal et chaque application $\operatorname{Pic}(\mathbf{X}_B) \longrightarrow \operatorname{Pic}(\mathbf{X}_q)$ surjective. On conclut donc que la flèche de gauche dans la suite de localisation est nulle, et donc que la flèche de restriction

$$CH^2(\mathbf{X}_B) \longrightarrow CH^2(X_L)$$

est un isomorphisme.

c) On dispose du diagramme commutatif de flèches de restriction :

$$\begin{array}{ccc} CH^2(\mathbf{X}_B) & \longrightarrow & CH^2(X_L) \\ \uparrow & & \uparrow \\ CH^2(\mathbf{X}) & \longrightarrow & CH^2(X), \end{array}$$

où les flèches horizontales sont surjectives. D'après ce qui précède, la flèche supérieure est un isomorphisme. Enfin, le noyau de la flèche $CH^2(\mathbf{X}) \longrightarrow CH^2(\mathbf{X}_B)$ est, par un argument de norme, annulé par le degré $m = [B : A] = [L : k]$. Il est donc contenu dans le groupe de m-torsion ${}_mCH^2(\mathbf{X})$, groupe fini d'après le théorème 6.2. D'apré ce même théorème, le groupe de n-torsion ${}_nCH^2(\mathbf{X}_B)$ est aussi fini. Une chasse au diagramme que je laisserai au lecteur assure alors la finitude du groupe de n-torsion ${}_nCH^2(X)$. □

Remarque 6.3.1 : En fait, sous les hypothèses du théorème 6.3, on peut conclure à la finitude de tout le sous-groupe de torsion $CH^2(X)_{tors}$. Il suffit en effet de combiner le précédent théorème avec le résultat plus général suivant, dont la démonstration illustre la **méthode galoisienne.**

THÉORÈME 6.4. — *Soit X une variété projective, lisse, géométriquement connexe sur un corps k de type fini sur le corps premier. Supposons les groupes $H^2(X, \mathcal{O}_X)$ et $H^1(X, \mathcal{O}_X)$ nuls. Alors le groupe de torsion $CH^2(X)_{tors}$ est d'exposant fini.*

Démonstration : Soit $\overline{k}$ une clôture algébrique de k, puis $\overline{X} = X \times_k \overline{k}$, enfin $G = \operatorname{Gal}(\overline{k}/k)$.

a) D'après le théorème 3.3.2, on dispose d'une inclusion naturelle

$$CH^2(\overline{X})_{tors} \hookrightarrow H^3_{et}(\overline{X}, \mathbf{Q}/\mathbf{Z}(2)).$$

Passant aux points fixes sous l'action de Galois, on a l'inclusion

$$[CH^2(\overline{X})_{tors}]^G \hookrightarrow H^3_{et}(\overline{X}, \mathbf{Q}/\mathbf{Z}(2))^G.$$

Mais l'hypothèse que k est un corps de fini sur le corps premier et une réduction au cas des corps finis assure, via le théorème de Deligne sur la conjecture de Weil, que le groupe de droite est un groupe fini (voir [CR1], Theorem 1.5).

b) Comme la flèche $CH^2(X) \longrightarrow CH^2(\overline{X})$ envoie clairement le premier groupe dans les invariants $CH^2(\overline{X})^G$, et que son noyau est de torsion (argument de transfert), pour établir le théorème, il suffit donc de montrer que le groupe

$$\mathrm{Ker}[CH^2(X) \longrightarrow CH^2(\overline{X})]$$

est d'exposant fini. Or ceci vaut pour toute variété projective et lisse sur un corps k de caractéristique zéro, sous les hypothèses $H^1(X, \mathcal{O}_X) = 0$ et $H^2(X, \mathcal{O}_X) = 0$. Ceci est établi dans [CR1], dont je présente maintenant brièvement les principaux arguments.

On emploie pour cela la méthode galoisienne, qui fut utilisée pour la première fois par S. Bloch [B4], et est systématisée dans [CR1]. Elle consiste à faire de la cohomologie galoisienne sur le complexe des sections globales de la résolution de Quillen du faisceau $\mathcal{K}_2$ sur $\overline{X}$, ce qui donne lieu à une suite exacte :

$$\cdots \longrightarrow H^1(G, K_2(\overline{k}(X)/H^0(\overline{X}, \mathcal{K}_2)) \longrightarrow \mathrm{Ker}[CH^2(X) \longrightarrow CH^2(\overline{X})] \longrightarrow H^1(G, H^1(\overline{X}, \mathcal{K}_2)).$$

Le groupe de gauche est d'exposant fini sous la simple hypothèse $H^1(X, \mathcal{O}_X) = 0$, et le groupe de droite est d'exposant fini sous l'hypothèse $H^2(X, \mathcal{O}_X) = 0$. Pour établir ces résultats, [CR1] utilise la version cohomologique du théorème 90 pour K_2 ([C1]), à savoir

$$H^1(G, K_2\overline{k}(X)/K_2\overline{k}) = 0$$

sous la seule hypothèse $X(k) \neq \emptyset$ – sans hypothèse sur les groupes de cohomologie cohérente – ce qui permet de ramener l'étude du groupe de gauche à celle du groupe

$$\mathrm{Ker}[H^2(G, H^0(\overline{X}, \mathcal{K}_2)) \longrightarrow H^2(G, K_2\overline{k}(X))],$$

et analyse la structure des groupes $H^0(\overline{X}, \mathcal{K}_2)$ et $H^1(\overline{X}, \mathcal{K}_2)$, en s'appuyant sur plusieurs des résultats fondamentaux de Merkur'ev et Suslin. Sous l'hypothèse $H^1(X, \mathcal{O}_X) = 0$, on montre que le groupe $H^2(G, H^0(\overline{X}, \mathcal{K}_2))$ lui-même est d'exposant fini. Je renvoie à [CR1] et au rapport [R1] pour plus de détails. □

Une version plus complète de la suite de la localisation d'une part, un résultat fondamental de Bloch développé par Kato–Saito et Somekawa d'autre part, permettent d'éliminer la restriction $H^1(X, \mathcal{O}_X) = 0$ dans le théorème 6.3.

THÉORÈME 6.5. — *Soit X une variété projective, lisse, géométriquement connexe sur un corps de nombres k. Supposons $H^2(X, \mathcal{O}_X) = 0$. Soit $n > 0$ un entier naturel. Alors le sous-groupe de n-torsion ${}_nCH^2(X)$ est fini.*

Démonstration :

a) Il existe une extension finie de corps L/k telle que $X(L) \neq \emptyset$ et que l'application naturelle de groupes de type fini $NS(X_L) \longrightarrow NS(\overline{X})$ soit un isomorphisme. Soit $m = [L : k]$. Comme dans la démonstration de 6.3, on peut trouver un localisé $A = R_f$ de l'anneau R des entiers de k (avec $f \in R$, $f \neq 0$) et un A-schéma projectif et lisse $\mathbf{X}$ de fibre générique X/k, de telle sorte que m et n soient inversibles dans A, que la clôture intégrale B de A dans L soit finie étale sur A, que le groupe des classes $\mathrm{Pic}(B)$ soit nul (i.e. B est principal), et que (via le théorème de semi-continuité de Grothendieck) $H^2(\mathbf{X}_p, \mathcal{O}_{\mathbf{X}_p}) = 0$ pour chacune des fibres $\mathbf{X}_p$ au-dessus d'un point fermé $p \in \mathrm{Spec}(A)$, le même énoncé valant alors aussi pour les fibres du schéma $\mathbf{X}_B/\mathrm{Spec}(B)$ aux points fermés $q \in \mathrm{Spec}(B)$.

b) La suite de localisation utilisée en 6.3 peut s'étendre à gauche en une suite exacte (qui peut d'ailleurs être prolongée à gauche [R2], Prop. 1.2) :

$$H^1(X_L, \mathcal{K}_2) \longrightarrow \bigoplus_{q \in \mathrm{Spec}(B)^{(1)}} \mathrm{Pic}(\mathbf{X}_q) \longrightarrow CH^2(\mathbf{X}_B) \longrightarrow CH^2(X_L) \longrightarrow 0.$$

Le cup–produit définit une application

$$\mathrm{Pic}(X_L) \otimes L^* \longrightarrow H^1(X_L, \mathcal{K}_2),$$

(facile à décrire via la résolution de Gersten–Quillen du faisceau $\mathcal{K}_2$). On dispose donc de l'application composée ρ :

$$\mathrm{Pic}(\mathbf{X}_B) \otimes L^* \longrightarrow \mathrm{Pic}(X_L) \otimes L^* \longrightarrow H^1(X_L, \mathcal{K}_2) \longrightarrow \bigoplus_{q \in \mathrm{Spec}(B)^{(1)}} \mathrm{Pic}(\mathbf{X}_q),$$

dont on vérifie aisément qu'elle envoie la classe de $\mathcal{L} \otimes \alpha \in \mathrm{Pic}(\mathbf{X}_B) \otimes L^*$ sur la famille $(\mathcal{L}_{|\mathbf{X}_q})^{v_q(\alpha)}$. Un argument similaire à celui développé en 6.3, et reposant sur le théorème de relèvement de Grothendieck, montre qu'en composant ρ avec la projection

$$\bigoplus_{q \in \mathrm{Spec}(B)^{(1)}} \mathrm{Pic}(\mathbf{X}_q) \longrightarrow \bigoplus_{q \in \mathrm{Spec}(B)^{(1)}} NS(\mathbf{X}_q)$$

on obtient une surjection.

Ainsi la flèche composée

$$H^1(X_L, \mathcal{K}_2) \longrightarrow \bigoplus_{q \in \mathrm{Spec}(B)^{(1)}} \mathrm{Pic}(\mathbf{X}_q) \longrightarrow \bigoplus_{q \in \mathrm{Spec}(B)^{(1)}} NS(\mathbf{X}_q)$$

est surjective (l'hypothèse $H^2(X, \mathcal{O}_X) = 0$ a été ici utilisée de façon cruciale). Mais pour terminer la démonstration comme dans 6.3, on a besoin de contrôler le conoyau de la flèche

$$H^1(X_L, \mathcal{K}_2) \longrightarrow \bigoplus_{q \in \mathrm{Spec}(B)^{(1)}} \mathrm{Pic}(\mathbf{X}_q),$$

c'est-à-dire qu'on voudrait aussi attraper les éléments du groupe

$$\bigoplus_{q \in \mathrm{Spec}(B)^{(1)}} \mathrm{Pic}^0(\mathbf{X}_q)$$

comme *bords* d'éléments de $H^1(X_L, \underline{K}_2)$. C'est ici qu'intervient le résultat de Bloch, Kato–Saito et Somekawa, que je rappelle au point suivant.

c) Soit R l'anneau des entiers d'un corps p–adique K (extension finie de $\mathbf{Q}_p$), de valuation $v : K^* \longrightarrow \mathbf{Z}$, de corps résiduel $\mathbf{F}$. Soit J un R-schéma abélien. L'identification $J(R) = J(K)$ permet de définir une application de spécialisation $J(K) \longrightarrow J(\mathbf{F})$, soit $x \longrightarrow \overline{x}$. On définit alors une application

$$J(K) \otimes K^* \longrightarrow J(\mathbf{F})$$

en associant à $x \otimes \lambda$ l'élément $(v(\lambda) \cdot \overline{x})$. Si M/K est une extension finie de corps, on définit une application

$$\rho_{M/K} : J(M) \otimes M^* \longrightarrow J(\mathbf{F})$$

comme l'application composée

$$J(M)\otimes M^* \longrightarrow J(\mathbb{F}_M) \xrightarrow{N} J(\mathbb{F}),$$

où N désigne la norme pour l'extension de corps résiduels $\mathbb{F}_M/\mathbb{F}$.

Si maintenant K est un corps de nombres, $R = A_f$ un localisé de son anneau A des entiers ($f \in A$, $f \neq 0$), et J un R-schéma abélien, on peut pour chaque point fermé $p \in \operatorname{Spec}(R)$, de corps résiduel $\mathbb{F}_p$ (ici p est un point fermé, non un nombre premier!) et toute extension finie M/K, définir une application

$$\rho_{M/K,p} : J(M)\otimes M^* \longrightarrow J(\mathbb{F}_p)$$

par addition des applications ρ_{M_q/K_p} correspondant aux diverses extensions de la place p à une place q de la clôture intégrale de R dans M. Le résultat-clé que nous utiliserons, dont l'idée est due à Bloch [B5] et les développements ultérieurs à Kato/Saito [KS] puis Somekawa [So] est :

THÉORÈME. — *Dans la situation ci-dessus, le conoyau de l'application*

$$\bigoplus_{M/K}\bigoplus_{p}\rho_{M/K,p} : \bigoplus_{M/K}(J(M)\otimes M^*) \longrightarrow \bigoplus_{p} J(\mathbb{F}_p),$$

où M/K parcourt les extensions finies de K, et p parcourt les idéaux premiers de R, est un groupe fini.

d) En utilisant la nullité des $H^2(\mathbf{X}_q, \mathcal{O}_{\mathbf{X}_q}) = 0$, et quitte à restreindre un peu plus A, on peut assurer (là encore grâce à un résultat de Grothendieck) que le foncteur de Picard $\underline{\operatorname{Pic}}_{\mathbf{X}_B/B}$ est représentable par un B-schéma en groupes localement lisse. Sa composante neutre, soit $J = \underline{\operatorname{Pic}}^o_{\mathbf{X}_B/B}$, est un B-schéma abélien.

Considérons alors le diagramme :

$$\begin{array}{ccc}
 & \nearrow & \bigoplus_{q\in \operatorname{Spec}(B)^{(1)}} NS(\mathbf{X}_q) \\
 & & \uparrow \\
H^1(X_L,\mathcal{K}_2) & \longrightarrow & \bigoplus_{q\in \operatorname{Spec}(B)^{(1)}} \operatorname{Pic}(\mathbf{X}_q) \\
\uparrow & & \uparrow \\
\bigoplus_{M/L}(J(M)\otimes M^*) & \longrightarrow & \bigoplus_{q\in \operatorname{Spec}(B)^{(1)}} J(\mathbb{F}_q)
\end{array}$$

défini comme suit. Les corps M parcourent les extensions finies de L, une par classe d'isomorphisme. La flèche horizontale médiane provient de la suite de localisation. La flèche verticale de gauche est définie par les homomorphismes composés

$$J(M)\otimes M^* \longrightarrow \operatorname{Pic}(X_M)\otimes M^* \xrightarrow{\text{cup}} H^1(X_M,\mathcal{K}_2) \xrightarrow{N_{M/L}} H^1(X_L,\mathcal{K}_2).$$

La flèche oblique, par définition, fait commuter le triangle. La flèche horizontale du bas est celle décrite au point c) ci-dessus. On peut vérifier que le carré est commutatif.

D'après b), la flèche oblique est surjective, et d'après c) la flèche horizontale inférieure a un conoyau fini. Il en résulte donc que la flèche horizontale médiane a un conoyau fini.

e) De la suite de localisation (point b) ci-dessus) on conclut que la flèche

$$CH^2(\mathbf{X}_B) \longrightarrow CH^2(X_L)$$

est surjective à noyau fini. Le théorème 6.2 et une chasse au diagramme analogue à celle effectuée à la fin de la démonstration de 6.3 permettent alors de conclure à la finitude de ${}_nCH^2(X)$. □

Pour établir le théorème 6.1 en toute généralité, à partir du théorème 6.5, il reste à éliminer l'hypothèse $H^1(X, \mathcal{O}_X) = 0$ dans le théorème 6.4, autrement dit à établir :

THÉORÈME 6.6 (Salberger). — *Soit X une variété projective, lisse, géométriquement connexe sur un corps de nombres k. Supposons le groupe $H^2(X, \mathcal{O}_X)$ nul. Alors le groupe de torsion $CH^2(X)_{tors}$ est d'exposant fini.*

Suivant la méthode de notre théorème 6.4, on voit que ceci résulte de l'énoncé :

THÉORÈME 6.7 (Salberger). — *Soit X une variété projective, lisse, géométriquement connexe sur un corps de nombres k. Le groupe*

$$\operatorname{Ker} \rho_X : H^2(G, H^0(\overline{X}, \mathcal{K}_2)) \longrightarrow H^2(G, K_2\overline{k}(X))$$

est d'exposant fini.

Démonstration (esquisse) : L'idée–clé de Salberger est de ramener cet énoncé au cas des courbes, par sections hyperplanes lisses successives (possible, d'après Bertini). Soit $Y \subset X$ une telle section. On montre que la flèche de restriction $H^2(G, H^0(\overline{X}, \mathcal{K}_2)) \longrightarrow H^2(G, H^0(\overline{Y}, \mathcal{K}_2))$ induit une flèche $\operatorname{Ker}(\rho_X) \longrightarrow \operatorname{Ker}(\rho_Y)$. Par ailleurs, on dispose ([CT-R 1985], fondé sur des résultats de Suslin) de la suite exacte :

$$0 \longrightarrow H^1_{et}(\overline{X}, \mathbf{Q}/\mathbf{Z}(2)) \longrightarrow H^0(\overline{X}, \mathcal{K}_2) \longrightarrow H^0(\overline{X}, \mathcal{K}_2) \otimes \mathbf{Q} \longrightarrow 0,$$

donc d'isomorphismes

$$H^2(G, H^1_{et}(\overline{X}, \mathbf{Q}/\mathbf{Z}(2))) \simeq H^2(G, H^0(\overline{X}, \underline{K}_2)).$$

La restriction aux sous–groupes divisibles maximaux des G–modules $H^1_{et}(\overline{X}, \mathbf{Q}/\mathbf{Z}(2))$ et $H^1_{et}(\overline{Y}, \mathbf{Q}/\mathbf{Z}(2))$ de la flèche de restriction, soit

$$H^1_{et}(\overline{X}, \mathbf{Q}/\mathbf{Z}(2))^o \longrightarrow H^1_{et}(\overline{Y}, \mathbf{Q}/\mathbf{Z}(2))^o$$

est une application G–équivariante qui admet une presque rétraction, c'est-à-dire qu'il existe une flèche G–équivariante

$$H^1_{et}(\overline{Y}, \mathbf{Q}/\mathbf{Z}(2))^o \longrightarrow H^1_{et}(\overline{X}, \mathbf{Q}/\mathbf{Z}(2))^o$$

qui composée avec la précédente est la multiplication par un entier strictement positif. Dans [CR3] nous établissons ce point en utilisant le théorème de complète réductibilité de Poincaré. Le sous–groupe divisible maximal de $H^1_{et}(\overline{X}, \mathbf{Q}/\mathbf{Z}(1))^o$ est le groupe des points de torsion de la variété de Picard de X, et on applique le dit théorème à l'homomorphisme injectif de variétés abéliennes

$$\underline{\mathrm{Pic}}^0_{X/k} \longrightarrow \underline{\mathrm{Pic}}^0_{Y/k}.$$

Pour établir le théorème, il suffit donc de considérer le cas des courbes, où le théorème est dû à Raskind [R3], lequel s'appuie sur des résultats de Jannsen (principe local-global en théorie du corps de classes des courbes sur un corps de nombres [J]) et de S. Saito (corps de classes pour les courbes sur un corps local). Dans le cas des courbes, on a en fait plus : sous la seule hypothèse $X(k) \neq \emptyset$, le groupe $\operatorname{Ker}(\rho_X)$ est nul. □

§ 7. Variétés sur les corps de nombres, II

Dans ce chapitre, je décris l'approche de Shuji Saito [S4] des théorèmes de finitude pour la torsion décrits au chapitre précédent, approche qui permet parfois d'aller plus loin (voir le théorème 7.6 ci-dessous). Comme indiqué plus haut, l'article de Saito incorpore certaines suggestions de Salberger et de moi-même. Un des aspects intéressants de la méthode de Saito est qu'elle garantit, sous certaines hypothèses, que les applications *cycles*, à valeurs dans la cohomologie étale, sont *injectives* sur la torsion du groupe de Chow.

La présentation choisie met en relief certains points un peu cachés dans l'article [S4].

Soit X une variété lisse intègre sur un corps k de caractéristique zéro. Pour tout entier $n > 0$, on dispose de la suite exacte (3.11) :

$$0 \longrightarrow H^1(X,\mathcal{K}_2)/n \longrightarrow NH^3_{et}(X,\mu_n^{\otimes 2}) \longrightarrow {}_nCH^2(X) \longrightarrow 0,$$

où par définition,

$$NH^3_{et}(X,\mu_n^{\otimes 2}) = \mathrm{Ker}[H^3_{et}(X,\mu_n^{\otimes 2}) \longrightarrow H^3(k(X),\mu_n^{\otimes 2})].$$

En passant à la limite inductive sur tous les entiers $n > 0$, on a la suite exacte

$$0 \longrightarrow H^1(X,\mathcal{K}_2)\otimes \mathbf{Q}/\mathbf{Z} \longrightarrow NH^3_{et}(X,\mathbf{Q}/\mathbf{Z}(2)) \longrightarrow CH^2(X)_{tors} \longrightarrow 0,$$

où

$$NH^3_{et}(X,\mathbf{Q}/\mathbf{Z}(2)) = \mathrm{Ker}[H^3_{et}(X,\mathbf{Q}/\mathbf{Z}(2)) \longrightarrow H^3(k(X),\mathbf{Q}/\mathbf{Z}(2))].$$

Notons

$$D = H^0(X,\mathcal{H}^3(\mathbf{Q}/\mathbf{Z}(2)) \subset H^3(k(X),\mathbf{Q}/\mathbf{Z}(2))$$

puis

$$\Theta = \mathrm{Coker}[H^1(X,\mathcal{K}_2)\otimes \mathbf{Q}/\mathbf{Z} \longrightarrow H^3_{et}(X,\mathbf{Q}/\mathbf{Z}(2))].$$

On dispose donc du diagramme trivialement commutatif de suites exactes :

$$(7.2)\qquad \begin{array}{ccccccccc} & & & & 0 & & 0 & & \\ & & & & \downarrow & & \downarrow & & \\ 0 & \longrightarrow & H^1(X,\mathcal{K}_2)\otimes\mathbf{Q}/\mathbf{Z} & \longrightarrow & NH^3_{et}(X,\mathbf{Q}/\mathbf{Z}(2)) & \longrightarrow & CH^2(X)_{tors} & \longrightarrow & 0 \\ & & \downarrow = & & \downarrow & & \downarrow & & \\ 0 & \longrightarrow & H^1(X,\mathcal{K}_2)\otimes\mathbf{Q}/\mathbf{Z} & \longrightarrow & H^3_{et}(X,\mathbf{Q}/\mathbf{Z}(2)) & \longrightarrow & \Theta & \longrightarrow & 0 \\ & & & & \downarrow & \swarrow \Psi & & & \\ & & & & D & & & & \end{array}$$

où la flèche Ψ est la flèche induite par la flèche verticale du bas.

Le théorème suivant dégage le contenu algébrique de l'approche de Saito. Nous verrons plus bas des conditions arithmétiques et géométriques garantissant que l'hypothèse (H) est satisfaite.

THÉORÈME 7.1. — *Soit X une variété lisse intègre sur un corps k de caractéristique zéro. Faisons l'hypothèse*

(H) *Le groupe* $\Theta = \mathrm{Coker}[H^1(X,\mathcal{K}_2)\otimes\mathbf{Q}/\mathbf{Z} \longrightarrow H^3_{et}(X,\mathbf{Q}/\mathbf{Z}(2))]$ *est d'exposant fini, annulé par l'entier $N > 0$.*

Alors $CH^2(X)_{tors}$ est d'exposant fini, divisant N, et, pour tout entier $n > 0$ multiple de N, l'application composée

$$CH^2(X)_{tors} \longrightarrow CH^2(X)/n \longrightarrow H^4_{et}(X, \mu_n^{\otimes 2})$$

de la projection naturelle et de l'application cycle est injective.

Démonstration :

a) Le premier énoncé, à savoir que sous l'hypothèse (H), le groupe $CH^2(X)_{tors}$ est annulé par N, se lit immédiatement sur le diagramme (7.2) : de fait, $CH^2(X)_{tors}$ est un sous-groupe de Θ.

b) LEMME ([S4], Prop. 2.4). — *Pour toute k-variété lisse intègre X, et pour tout entier $n > 0$, on a une suite exacte*

$${}_n\Theta \longrightarrow {}_nD \longrightarrow CH^2(X)/n \longrightarrow H^4_{et}(X, \mu_n^{\otimes 2}).$$

Démonstration : On considère le diagramme commutatif suivant :

(7.3)

$$\begin{array}{ccccccc}
 & & & & 0 & & \\
 & & & & \downarrow & & \\
 & & H^3_{et}(X,\mu_n^{\otimes 2}) & \rightarrow & H^0(X,\mathcal{H}^3(\mu_n^{\otimes 2})) & \rightarrow CH^2(X)/n \rightarrow & H^4_{et}(X,\mu_n^{\otimes 2}) \\
 & & \downarrow & & \downarrow & & \\
H^1(X,\mathcal{K}_2)\otimes \mathbf{Q}/\mathbf{Z} & \rightarrow & H^3_{et}(X,\mathbf{Q}/\mathbf{Z}(2)) & \rightarrow & H^0(X,\mathcal{H}^3(\mathbf{Q}/\mathbf{Z}(2))) & & \\
 & & \downarrow \times n & & \downarrow \times n & & \\
 & & H^3_{et}(X,\mathbf{Q}/\mathbf{Z}(2)) & \rightarrow & H^0(X,\mathcal{H}^3(\mathbf{Q}/\mathbf{Z}(2))). & &
\end{array}$$

Dans ce diagramme, la suite horizontale supérieure est la suite exacte provenant de la théorie de Bloch–Ogus (voir § 3), la suite médiane est un complexe (début du présent paragraphe), la suite verticale de gauche est exacte : c'est la suite exacte de Kummer déduite de la suite exacte de faisceaux étales

$$1 \longrightarrow \mu_n^{\otimes 2} \longrightarrow \mathbf{Q}/\mathbf{Z}(2) \xrightarrow{\times n} \mathbf{Q}/\mathbf{Z}(2) \longrightarrow 1.$$

Montrons que la suite verticale de droite est exacte. On dispose du diagramme commutatif

$$\begin{array}{ccccccc}
 & & 0 & & 0 & & 0 \\
 & & \downarrow & & \downarrow & & \downarrow \\
0 & \longrightarrow & H^0(X,\mathcal{H}^3(\mu_n^{\otimes 2})) & \longrightarrow & H^3_{et}(k(X),\mu_n^{\otimes 2}) & \longrightarrow & \bigoplus_{x\in X^1} H^2_{et}(k(x),\mu_n) \\
 & & \downarrow & & \downarrow & & \downarrow \\
0 & \longrightarrow & H^0(X,\mathcal{H}^3(\mathbf{Q}/\mathbf{Z}(2))) & \longrightarrow & H^3_{et}(k(X),\mathbf{Q}/\mathbf{Z}(2)) & \longrightarrow & \bigoplus_{x\in X^1} H^2_{et}(k(x),\mathbf{Q}/\mathbf{Z}(1)) \\
 & & \downarrow & & \downarrow \times n & & \downarrow \times n \\
0 & \longrightarrow & H^0(X,\mathcal{H}^3(\mathbf{Q}/\mathbf{Z}(2))) & \longrightarrow & H^3_{et}(k(X),\mathbf{Q}/\mathbf{Z}(2)) & \longrightarrow & \bigoplus_{x\in X^1} H^2_{et}(k(x),\mathbf{Q}/\mathbf{Z}(1))
\end{array}$$

où les suites horizontales sont exactes. Les suites verticales proviennent de suites de Kummer. Elles sont donc exactes sauf peut-être en leur terme initial. Que $H^2_{et}(k(x),\mu_n)$ s'injecte dans $H^2_{et}(k(x),\mathbf{Q}/\mathbf{Z}(1))$ est une conséquence du théorème 90 de Hilbert. Que $H^3_{et}(k(X),\mu_n^{\otimes 2})$ s'injecte dans $H^3_{et}(k(x),\mathbf{Q}/\mathbf{Z}(2))$ est hautement non trivial : c'est une conséquence du théorème principal de Merkur'ev–Suslin [MS].

Des propriétés d'exactitude du diagramme ci-dessus résulte alors l'exactitude de la suite verticale de droite dans le diagramme (7.3). De ce diagramme on déduit alors une suite exacte

$$H^3_{et}(X,\mu_n^{\otimes 2}) \longrightarrow {}_nD \longrightarrow CH^2(X)/n \longrightarrow H^4_{et}(X,\mu_n^{\otimes 2}),$$

puis, comme la première flèche se factorise par ${}_n\Theta$ (voir (7.2)), la suite exacte

$${}_n\Theta \longrightarrow {}_nD \longrightarrow CH^2(X)/n \longrightarrow H^4_{et}(X,\mu_n^{\otimes 2})$$

annoncée. □

c) On considère le diagramme

$$\begin{array}{ccccccc} & & & & CH^2(X)_{tors} & & \\ & & & & \downarrow & & \\ {}_n\Theta & \longrightarrow & {}_nD & \longrightarrow & CH^2(X)/n & \longrightarrow & H^4_{et}(X,\mu_n^{\otimes 2}) \\ \downarrow & & \downarrow & & \downarrow & & \\ \Theta & \longrightarrow & D & \longrightarrow & CH^2(X)\otimes\mathbf{Q}/\mathbf{Z}, & & \end{array}$$

où la suite horizontale médiane est la suite exacte décrite ci-dessus, la suite horizontale inférieure est obtenue par passage à la limite inductive sur les suites précédentes, et ρ est l'application composée. Observons :

– La suite verticale est un complexe.

– Si N divise n, donc n annule Θ, alors ${}_n\Theta \longrightarrow \Theta$ est surjectif.

– Si N divise n, alors, comme nous l'avons déjà remarqué, n annule $CH^2(X)_{tors}$. Il en résulte que l'application $CH^2(X)_{tors} \longrightarrow CH^2(X)/n$ est injective.

Fixons $n > 0$ multiple de N, et soit $z_0 \in CH^2(X)_{tors}$ tel que $\rho(z_0) = 0$. La nullité de z_0, et donc le théorème, résultent maintenant d'une chasse au diagramme que nous représenterons symboliquement, les éléments z_i étant numérotés suivant leur ordre d'apparition :

$$\begin{array}{ccccccc} & & & & z_0 & & \\ & & & & \downarrow \rho & & \\ z_5 & \longrightarrow & z_6\ z_2 & \longrightarrow & z_1 & \longrightarrow & 0 \\ \downarrow & & \downarrow\ \downarrow & & \downarrow & & \\ z_4 & \longrightarrow & z_3 & \longrightarrow & 0 & & \end{array}$$

De l'injection ${}_nD \subset D$, on conclut $z_6 = z_2$, et donc $z_1 = 0$, soit finalement $z_0 = 0$. □

Remarque 7.1.1 : La méthode permettant d'obtenir l'injectivité dans le théorème ci-dessus est aparue pour la première fois dans un travail en préparation de P. Salberger. Dans ce travail, Salberger s'intéresse à l'application de réciprocité sur le groupe $SK_1(X)$ d'une surface X projective et lisse sur un corps p-adique, et sous certaines hypothèses, établit une propriété d'injectivité. C'est Saito qui eut l'idée d'utiliser la même technique dans l'étude des cycles de torsion.

Le théorème 7.1 admet la variante suivante, utile lorsque le corps de base k est de dimension cohomologique plus grande que 2.

THÉORÈME 7.2. — *Soit X une variété lisse intègre sur un corps k de caractéristique zéro. Faisons l'hypothèse*

(H') Le groupe $\Theta' = \operatorname{Coker}[H^3_{et}(k,\mathbf{Q}/\mathbf{Z}(2)) \oplus (H^1(X,\mathcal{K}_2)\otimes \mathbf{Q}/\mathbf{Z}) \longrightarrow H^3_{et}(X,\mathbf{Q}/\mathbf{Z}(2))]$ *est d'exposant fini, annulé par l'entier* $N > 0$.

Supposons de plus $X(k) \neq \emptyset$. Alors $CH^2(X)_{tors}$ est d'exposant fini, divisant N, et, pour tout entier $n > 0$ multiple de N, l'application composée

$$CH^2(X)_{tors} \longrightarrow CH^2(X)/n \longrightarrow H^4_{et}(X,\mu_n^{\otimes 2})$$

de la projection naturelle et de l'application cycle est injective.

Démonstration (esquisse) : On choisit un point $P \in X(k)$. Ce point permet de réaliser une section de la flèche de changement de base $H^3_{et}(k,\mathbf{Q}/\mathbf{Z}(2)) \longrightarrow H^3_{et}(X,\mathbf{Q}/\mathbf{Z}(2))$ utilisée dans la définition de Θ', et donc d'écrire $\Theta = \Theta' \oplus H^3_{et}(k,\mathbf{Q}/\mathbf{Z}(2))$. Via le même point P, on peut aussi décomposer $D = H^0(X,\mathcal{H}^3(\mathbf{Q}/\mathbf{Z}(2)))$ en $D = D' \oplus H^3_{et}(k,\mathbf{Q}/\mathbf{Z}(2))$. On obtient alors la suite exacte

$${}_n\Theta' \longrightarrow {}_nD' \longrightarrow CH^2(X)/n \longrightarrow H^4_{et}(X,\mu_n^{\otimes 2}),$$

et la démonstration se poursuit comme celle du théorème 7.1. □

Lorsque X est un variété projective lisse et géométriquement intègre sur un corps k de caractéristique zéro, on a ([CR1]) :

$$H^1(\overline{X},\mathcal{K}_2)\otimes \mathbf{Q}/\mathbf{Z} = 0.$$

On voit donc que la flèche

$$H^3_{et}(k,\mathbf{Q}/\mathbf{Z}(2)) \oplus (H^1(X,\mathcal{K}_2)\otimes \mathbf{Q}/\mathbf{Z}) \longrightarrow H^3_{et}(X,\mathbf{Q}/\mathbf{Z}(2)),$$

dont le conoyau est Θ', se factorise alors en une flèche :

$$H^3_{et}(k,\mathbf{Q}/\mathbf{Z}(2)) \oplus (H^1(X,\mathcal{K}_2)\otimes \mathbf{Q}/\mathbf{Z}) \longrightarrow \operatorname{Ker}[H^3_{et}(X,\mathbf{Q}/\mathbf{Z}(2)) \longrightarrow H^3_{et}(\overline{X},\mathbf{Q}/\mathbf{Z}(2))].$$

D'exposés de Salberger (voir § 9) et de l'article de Saito on peut encore extraire le résultat suivant, valable sans hypothèse arithmétique sur k :

THÉORÈME 7.3. — *Soit X une variété projective lisse et géométriquement intègre sur un corps k de caractéristique zéro. Supposons $H^1(X,\mathcal{O}_X) = 0$ et $H^2(X,\mathcal{O}_X) = 0$. Alors le conoyau de l'application*

$$H^3_{et}(k,\mathbf{Q}/\mathbf{Z}(2)) \oplus (H^1(X,\mathcal{K}_2)\otimes \mathbf{Q}/\mathbf{Z}) \longrightarrow \operatorname{Ker}[H^3_{et}(X,\mathbf{Q}/\mathbf{Z}(2)) \longrightarrow H^3_{et}(\overline{X},\mathbf{Q}/\mathbf{Z}(2))]$$

est d'exposant fini.

Démonstration : La suite spectrale de Hochschild–Serre en cohomologie étale

$$H^p(k, H^p_{et}(\overline{X}, \mathbf{Q}/\mathbf{Z}(2))) \Longrightarrow H^n_{et}(X, \mathbf{Q}/\mathbf{Z}(2))$$

donne lieu à une filtration $F^0 \subset F^1 \subset F^2$ sur le groupe

$$F^2 = \mathrm{Ker}[H^3_{et}(X, \mathbf{Q}/\mathbf{Z}(2)) \longrightarrow H^3_{et}(\overline{X}, \mathbf{Q}/\mathbf{Z}(2))].$$

On a une surjection $H^3_{et}(k, \mathbf{Q}/\mathbf{Z}(2)) \longrightarrow F^0$ qui composée avec $F^0 \subset F^2$ est la flèche naturelle $H^3_{et}(k, \mathbf{Q}/\mathbf{Z}(2)) \longrightarrow H^3_{et}(X, \mathbf{Q}/\mathbf{Z}(2))$. Le quotient F^1/F^0 est un sous-quotient du groupe $H^2(k, H^1_{et}(\overline{X}, \mathbf{Q}/\mathbf{Z}(2)))$. Mais sous l'hypothèse $H^1(X, \mathcal{O}_X) = 0$, le groupe $H^1_{et}(\overline{X}, \mathbf{Q}/\mathbf{Z}(2))$ est un groupe fini, et donc F^1/F^0 est un groupe d'exposant fini. Le quotient F^2/F^1 est un sous-groupe de $H^1(k, H^2_{et}(\overline{X}, \mathbf{Q}/\mathbf{Z}(2)))$. Pour établir le théorème, il suffira donc de montrer que le groupe

$$\mathrm{Coker}[H^1(X, \mathcal{K}_2) \otimes \mathbf{Q}/\mathbf{Z} \longrightarrow H^1(k, H^2_{et}(\overline{X}, \mathbf{Q}/\mathbf{Z}(2)))]$$

est un groupe d'exposant fini dès que $H^2(X, \mathcal{O}_X) = 0$.

Puisqu'on veut seulement montrer que ce groupe est d'exposant fini, un argument de transfert permet de remplacer le corps k par une extension finie, que je noterai encore k, et de supposer que $X(k) \neq \emptyset$ et que le groupe de Galois $G = \mathrm{Gal}(\overline{k}/k)$ agit trivialement sur le groupe de Néron-Severi $NS(\overline{X})$ (on aurait pu faire cet argument de transfert dès le début de la démonstration). Ainsi $NS(X) = NS(\overline{X})$. On considère alors le diagramme (où t = torsion du groupe considéré) :

$$\begin{array}{ccccc}
\mathrm{Pic}(X) \otimes k^* \otimes \mathbf{Q}/\mathbf{Z} & \longrightarrow & H^1(X, \mathcal{K}_2) \otimes \mathbf{Q}/\mathbf{Z} & \longrightarrow & H^1(k, H^2_{et}(\overline{X}, \mathbf{Q}/\mathbf{Z}(2))) \\
\downarrow = & & & & \uparrow \varphi \\
\mathrm{Pic}(X) \otimes H^1(k, \mathbf{Q}/\mathbf{Z}(1)) & & & & \\
\downarrow & & & & \\
(NS(X)/t) \otimes H^1(k, \mathbf{Q}/\mathbf{Z}(1)) & \simeq & H^0(k, (NS(\overline{X})/t) \otimes H^1(k, \mathbf{Q}/\mathbf{Z}(1))) & \rightarrow & H^1(k, (NS(\overline{X})/t) \otimes \mathbf{Q}/\mathbf{Z}(1))
\end{array}$$

où la première flèche horizontale est donnée par le cup–produit, et où φ est induite par la flèche

$$NS(\overline{X}) \otimes \mathbf{Q}/\mathbf{Z}(1) \longrightarrow H^2_{et}(\overline{X}, \mathbf{Q}/\mathbf{Z}(2))$$

déduite, par torsion par $\mathbf{Q}/\mathbf{Z}(1)$, de la flèche

$$\delta : NS(\overline{X}) \otimes \mathbf{Q}/\mathbf{Z} = \mathrm{Pic}(\overline{X}) \otimes \mathbf{Q}/\mathbf{Z} \longrightarrow H^2_{et}(\overline{X}, \mathbf{Q}/\mathbf{Z}(1))$$

issue de la suite de Kummer. Selon Saito, le diagramme ci–dessus commute. Comme nous sommes en caractéristique zéro, l'hypothèse $H^2(X, \mathcal{O}_X) = 0$ est, par la théorie de Hodge, équivalente au fait que δ a un conoyau fini. Ainsi $\mathrm{Coker}(\varphi)$ est d'exposant fini, et le théorème résulte alors du diagramme ci–dessus. □

Remarque 7.3.1 : Pour établir l'énoncé, on s'est seulement servi de l'image de $\mathrm{Pic}(X) \otimes k^*$ dans $H^1(X, \mathcal{K}_2)$ (après extension du corps de base). Comme l'observe Salberger, il serait très désirable de savoir fabriquer des éléments plus intéressants dans $H^1(X, \mathcal{K}_2)$.

Remarque 7.3.2 : De l'égalité

$$H^1(\overline{X}, \mathcal{K}_2) \otimes \mathbf{Q}/\mathbf{Z} = 0,$$

du diagramme (7.2), et du théorème 7.3, on déduit que pour X une variété projective lisse et géométriquement intègre sur un corps k de caractéristique zéro, avec $H^1(X, \mathcal{O}_X) = 0$ et

$H^2(X,\mathcal{O}_X)=0$, le groupe $\mathrm{Ker}[CH^2(X)\longrightarrow CH^2(\overline{X})]$ est d'exposant fini (résultat de [CR1] rappelé dans la démonstration du théorème 6.4).

Le théorème 7.3 a l'application arithmétique immédiate :

COROLLAIRE 7.4. — *Soit k un corps de type fini sur le corps $\mathbb{Q}$ des rationnels, et soit X une variété projective lisse et géométriquement intègre sur k. Supposons $H^1(X,\mathcal{O}_X)=0$ et $H^2(X,\mathcal{O}_X)=0$. Alors le conoyau de l'application*

$$H^3_{et}(k,\mathbb{Q}/\mathbb{Z}(2))\oplus(H^1(X,\mathcal{K}_2)\otimes\mathbb{Q}/\mathbb{Z})\longrightarrow H^3_{et}(X,\mathbb{Q}/\mathbb{Z}(2))$$

est d'exposant fini.

Démonstration : De fait, l'image de l'application

$$H^3_{et}(X,\mathbb{Q}/\mathbb{Z}(2))\longrightarrow H^3_{et}(\overline{X},\mathbb{Q}/\mathbb{Z}(2))$$

est dans le sous-groupe $H^3_{et}(\overline{X},\mathbb{Q}/\mathbb{Z}(2))^G$ des invariants sous le groupe de Galois $G=\mathrm{Gal}(\overline{k}/k)$. Comme on l'a déjà rappelé dans la démonstration du théorème 6.4, l'hypothèse que k est un corps de type fini sur le corps premier et une réduction au cas des corps finis assure, via le théorème de Deligne sur la conjecture de Weil, que ce groupe, comme d'ailleurs tout groupe $H^i_{et}(\overline{X},\mathbb{Q}/\mathbb{Z}(j))^G$ pour $i\neq 2j$, est un groupe fini ([CR1], Theorem 1.5). L'énoncé résulte alors directement du théorème 7.3. □

Le corollaire 7.4 assure que, sous ses hypothèses, la condition (H') du théorème 7.2 est satisfaite.

Or, comme annoncé en 1.9.2, on a le théorème général suivant :

THÉORÈME 7.5 ([S4]). — *Soit X une variété lisse (non nécessairement propre) sur un corps k de type fini sur $\mathbb{Q}$. Alors, pour tout entier $n>0$, et pour tout entier naturel i, l'image de l'application cycle*

$$CH^i(X)/n\longrightarrow H^{2i}_{et}(X,\mu_n^{\otimes i})$$

est un groupe fini.

Démonstration (résumé) : On peut trouver un anneau R de corps de fractions k, de type fini et lisse sur $\mathbb{Z}$, avec n inversible dans R, et un schéma $\mathbf{X}$ lisse sur $\mathrm{Spec}(R)$ tel que $\mathbf{X}\times_R k=X$. En utilisant la lissité de $p:\mathbf{X}\longrightarrow\mathrm{Spec}(\mathbb{Z})$, on obtient que les faisceaux étales $R^jp_*\mu_n^{\otimes i}$ sont constructibles sur $\mathrm{Spec}(\mathbb{Z})$, donc à cohomologie étale finie, ce qui, via la suite spectrale de Leray, assure la finitude des groupes $H^r_{et}(\mathbf{X},\mu_n^{\otimes s})$ (voir 6.2). L'application $CH^i(\mathbf{X})\longrightarrow CH^i(X)$ est surjective, mais l'on ne sait pas encore définir une application cycle allant de $CH^i(\mathbf{X})$ dans $H^{2i}_{et}(\mathbf{X},\mu_n^{\otimes i})$ (et qui ferait commuter le diagramme évident, assurant ainsi la finitude voulue). Le point délicat de la démonstration de Saito consiste à montrer qu'à tout le moins l'image de l'application cycle $CH^i(X)\longrightarrow H^{2i}_{et}(X,\mu_n^{\otimes i})$ est contenue dans l'image de $H^{2i}_{et}(\mathbf{X},\mu_n^{\otimes i})$ dans $H^{2i}_{et}(X,\mu_n^{\otimes i})$, ce qui assure sa finitude. □

La combinaison de 7.2, 7.4 et 7.5 donne donc le théorème de finitude :

THÉORÈME 7.6. — *Soit k un corps de type fini sur le corps $\mathbb{Q}$ des rationnels, et soit X une variété projective lisse et géométriquement intègre sur k. Supposons $H^1(X,\mathcal{O}_X)=0$ et $H^2(X,\mathcal{O}_X)=0$, et $X(k)\neq\emptyset$. Alors le sous-groupe de torsion $CH^2(X)_{tors}$ de $CH^2(X)$ est un groupe fini.* □

Remarque 7.6.1 : Ce théorème généralise l'énoncé obtenu en 6.3.1 sur un corps de nombres – à la restriction $X(k) \neq \emptyset$ près, qu'on n'avait pas imposée dans 6.3.1. Sur un corps k de type fini sur $\mathbf{Q}$, un seul cas avait été jusqu'alors traité, celui des surfaces rationnelles ([C1]), là encore avec la restriction $X(k) \neq \emptyset$. De fait, il existe une surface rationnelle X sur un corps $k = \mathbf{Q}_p(T)$ de fonctions rationnelles en une variable sur un corps p-adique $\mathbf{Q}_p$, avec $X(k) = \emptyset$ et $CH^2(X)_{tors} \neq 0$ non détectable par les applications cycles (Sansuc et l'auteur, J. of Algebra **84** (1985); réinterprétation par Salberger, non publiée).

Par sa méthode, mais en s'appuyant d'une part sur le résultat de Salberger (Théorème 6.6) bornant la torsion, d'autre part sur des résultats de Jannsen proches de ceux utilisés dans la démonstration du théorème 6.7, Saito a réussi à obtenir le théorème de finitude le plus général obtenu au chapitre précédent :

THÉORÈME 7.7 (= Théorème 6.1). — *Soit X une variété projective, lisse, géométriquement connexe sur un corps de nombres k. Supposons le groupe $H^2(X, \mathcal{O}_X) = 0$. Alors le sous-groupe de torsion $CH^2(X)_{tors}$ est fini.*

Démonstration : D'après 6.6, le groupe $CH^2(X)_{tors}$ est d'exposant fini. Pour presque tout premier l, la partie l-primaire $CH^2(X)_{l-tors}$ de $CH^2(X)_{tors}$ est donc nulle, et pour établir le théorème, il suffit d'établir la finitude de $CH^2(X)_{l-tors}$ pour tout premier l. Comme au théorème 7.3, on considère la suite spectrale en cohomologie étale

$$H^p(k, H^q_{et}(\overline{X}, \mathbf{Q}_l/\mathbf{Z}_l(2))) \Longrightarrow H^n_{et}(X, \mathbf{Q}_l/\mathbf{Z}_l(2)),$$

qui donne lieu à une filtration $F^0 \subset F^1 \subset F^2 \subset F^3$ sur le groupe $H^3_{et}(X, \mathbf{Q}_l/\mathbf{Z}_l(2))$.

Il sera très commode de travailler dans la catégorie de Serre $\mathcal{C}$ quotient de la catégorie des groupes abéliens par celle des groupes abéliens finis.

Le groupe F^0, quotient de $H^3(k, \mathbf{Q}_l/\mathbf{Z}_l(2))$, est fini, donc nul dans $\mathcal{C}$, car k est un corps de nombres. Le quotient F^3/F^2 est un sous-groupe de $H^3_{et}(\overline{X}, \mathbf{Q}_l/\mathbf{Z}_l(2))^G$, groupe fini (voir la démonstration de 7.4), donc nul dans $\mathcal{C}$. On a une injection

$$F^2/F^1 \hookrightarrow H^1(k, H^2_{et}(\overline{X}, \mathbf{Q}_l/\mathbf{Z}_l(2)).$$

Dans $\mathcal{C}$, on a une surjection

$$H^2(k, H^1_{et}(\overline{X}, \mathbf{Q}_l/\mathbf{Z}_l(2))) \twoheadrightarrow F^1/F^0.$$

Sous l'hypothèse que k est un corps de nombres, Jannsen ([J], Cor. 7 (b) p. 355) a établi l'existence d'un entier $r \geq 0$ (dépendant a priori de l – et qu'on peut choisir nul pour presque tout l) et d'une surjection

$$(\mathbf{Q}_l/\mathbf{Z}_l)^r \twoheadrightarrow H^2(k, H^1_{et}(\overline{X}, \mathbf{Q}_l/\mathbf{Z}_l(2))^o)$$

(M^o désigne le sous-groupe divisible maximal d'un groupe M). On observera que le quotient $H^1_{et}(\overline{X}, \mathbf{Q}_l/\mathbf{Z}_l(2))/H^1_{et}(\overline{X}, \mathbf{Q}_l/\mathbf{Z}_l(2))^o$ est un groupe fini, et donc que l'application

$$H^2(k, H^1_{et}(\overline{X}, \mathbf{Q}_l/\mathbf{Z}_l(2))^o) \longrightarrow H^2(k, H^1_{et}(\overline{X}, \mathbf{Q}_l/\mathbf{Z}_l(2)))$$

a un conoyau d'exposant fini.

On a $F^2 = \mathrm{Ker}[H^3_{et}(X, \mathbf{Q}_l/\mathbf{Z}_l(2)) \longrightarrow H^3_{et}(\overline{X}, \mathbf{Q}_l/\mathbf{Z}_l(2))]$). Comme on l'a établi dans la démonstration du théorème 7.3, l'hypothèse $H^2(X, \mathcal{O}_X) = 0$ implique, sur un corps k de caractéristique zéro, que l'application composée

$$H^1(X, \mathcal{K}_2) \otimes \mathbf{Q}_l/\mathbf{Z}_l \longrightarrow F^2 \longrightarrow H^1(k, H^2_{et}(\overline{X}, \mathbf{Q}_l/\mathbf{Z}_l(2)))$$

a un conoyau d'exposant fini.

On dispose donc dans $\mathcal{C}$ de l'application composée

$$(\mathbf{Q}_l/\mathbf{Z}_l)^r \longrightarrow H^2(k, H^1_{et}(\overline{X}, \mathbf{Q}_l/\mathbf{Z}_l(2))^\circ) \longrightarrow H^2(k, H^1_{et}(\overline{X}, \mathbf{Q}_l/\mathbf{Z}_l(2)) \longrightarrow H^3_{et}(X, \mathbf{Q}_l/\mathbf{Z}_l(2)).$$

Notons ici $\Theta = \mathrm{Coker}[H^1(X,\mathcal{K}_2)\otimes \mathbf{Q}_l/\mathbf{Z}_l \longrightarrow H^3_{et}(X, \mathbf{Q}_l/\mathbf{Z}_l(2))]$. Du diagramme

$$\begin{array}{ccccc} & & 0 & & \\ & & \uparrow & & \\ & & \Theta & & \\ & & \uparrow & & \\ (\mathbf{Q}_l/\mathbf{Z}_l)^r & \longrightarrow & H^3_{et}(X, \mathbf{Q}_l/\mathbf{Z}_l(2)) & \longrightarrow & H^1(k, H^2_{et}(\overline{X}, \mathbf{Q}_l/\mathbf{Z}_l(2))) \\ & & \uparrow & & \\ & & H^1(X,\mathcal{K}_2)\otimes \mathbf{Q}_l/\mathbf{Z}_l & & \end{array}$$

et des considérations précédentes, on voit que sous nos hypothèses, le groupe

$$\Theta'' = \mathrm{Coker}[(\mathbf{Q}_l/\mathbf{Z}_l)^r \longrightarrow \Theta]$$

est un groupe d'exposant fini.

Soit (voir le diagramme (7.2)) :

$$D'' = \mathrm{Coker}[(\mathbf{Q}_l/\mathbf{Z}_l)^r \longrightarrow H^0(X, \mathcal{H}^3(\mathbf{Q}_l/\mathbf{Z}_l(2)))].$$

Tout quotient d'un groupe $(\mathbf{Q}_l/\mathbf{Z}_l)^a$ est de la forme $(\mathbf{Q}_l/\mathbf{Z}_l)^b$ (avec $b \leq a$). On en déduit donc que dans la catégorie $\mathcal{C}$, pour tout entier $n > 0$, les flèches naturelles induites sur la n–torsion ${}_n\Theta \longrightarrow {}_n\Theta''$ et ${}_nD \longrightarrow {}_nD''$ sont des isomorphismes, et qu'il existe un entier $n > 0$ (puissance de l) tel que l'inclusion ${}_n\Theta'' \longrightarrow \Theta''$ soit un isomorphisme. Comme en c) de la démonstration du théorème 7.1, on considère dans $\mathcal{C}$ le diagramme commutatif

$$\begin{array}{ccccccc} & & & & CH^2(X)_{l-tors} & & \\ & & & & \downarrow & & \\ {}_n\Theta'' & \longrightarrow & {}_nD'' & \longrightarrow & CH^2(X)/n & \longrightarrow & H^4_{et}(X, \mu_n^{\otimes 2}) \\ \downarrow & & \downarrow & & \downarrow & & \\ \Theta'' & \longrightarrow & D'' & \longrightarrow & CH^2(X)\otimes \mathbf{Q}_l/\mathbf{Z}_l\,, & & \end{array}$$

où la suite horizontale est exacte (via Merkur'ev–Suslin), la suite verticale est un complexe, et la flèche ${}_n\Theta'' \longrightarrow \Theta''$ un isomorphisme. D'après le théorème 6.6 (Salberger), $CH^2(X)(l)$ est d'exposant fini. Quitte à remplacer n par une puissance plus élevée de l, annulant $CH^2(X)_{l-tors}$, on peut donc assurer que l'application $CH^2(X)_{l-tors} \longrightarrow CH^2(X)/n$ est injective.

La même chasse au diagramme qu'au théorème 7.1 montre alors que l'application

$$CH^2(X)_{l-tors} \longrightarrow H^4_{et}(X, \mu_n^{\otimes 2})$$

est une injection dans $\mathcal{C}$. Ainsi, dans la catégorie des groupes abéliens, l'application

$$CH^2(X)_{l-tors} \longrightarrow H^4_{et}(X, \mu_n^{\otimes 2})$$

a un noyau fini. Comme l'image de cette application est finie (Théorème 7.5), on conclut que le groupe $CH^2(X)_{l-tors}$ est lui-même fini. $\square$

Remarque 7.7.1 : Dans l'approche ci-dessus, on établit la finitude de $CH^2(X)_{tors}$ sans montrer d'abord, comme on l'avait fait au paragraphe 6, que pour tout entier $n > 0$ le groupe ${}_nCH^2(X)$ est fini.

§ 8. Variétés sur les corps locaux

Par *corps local*, on entendra ici une extension finie du corps p-adique $\mathbb{Q}_p$. On a le résultat général suivant (cf. [CSS] Cor. 2 p. 773) :

THÉORÈME 8.1. — *Soit k un corps local, et X une k-variété lisse. Alors pour tout entier $n > 0$, le sous-groupe de n-torsion ${}_nCH^2(X)$ est un groupe fini.*

Démonstration : C'est une conséquence immédiate du théorème 3.3.2 : le groupe ${}_nCH^2(X)$ est un sous-quotient du groupe de cohomologie étale $H^3_{et}(X, \mu_n^{\otimes 2})$. Or pour une variété X sur un corps local, et tout $i \geq 0$ et j entiers, le groupe $H^i_{et}(X, \mu_n^{\otimes j})$ est fini, comme on le voit en utilisant la suite spectrale de Hochschild–Serre, la finitude des groupes de cohomologie $H^q_{et}(\overline{X}, \mu_n^{\otimes j})$ ($q \geq 0$) et celle des groupes $H^p(\mathrm{Gal}(\overline{k}/k), F)$ ($p \geq 0$) pour k local et F un $\mathrm{Gal}(\overline{k}/k)$–module fini. □

Le théorème suivant avait été obtenu sous des hypothèses plus restrictives dans [CR1]. C'est l'énoncé 8.3 ci–dessous qui permet d'aller un peu plus loin, en éliminant les hypothèses du type : $H^1(X, \mathcal{O}_X) = 0$, ou X (ou $\underline{\mathrm{Alb}}_X$) a bonne réduction.

THÉORÈME 8.2. — *Soient k un corps local, extension finie de $\mathbb{Q}_p$, et X une k-variété projective, lisse et géométriquement intègre. Supposons $H^2(X, \mathcal{O}_X) = 0$. Alors :*

(*a*) *Le groupe*

$$\ker[CH^2(X) \longrightarrow CH^2(\overline{X})]$$

est fini.

(*b*) *Si de plus l'image du groupe $NH^3_{et}(X, \mathbb{Q}/\mathbb{Z}(2))$ dans le groupe $H^3_{et}(\overline{X}, \mathbb{Q}/\mathbb{Z}(2))$ est fini, par exemple si $H^3_{et}(\overline{X}, \mathbb{Q}/\mathbb{Z}(2))^G$ est fini, alors le groupe $CH^2(X)_{tors}$ est fini.*

(*c*) ([CR1], 3.16). *Si X a potentiellement bonne réduction sur k, la torsion première à p de $CH^2(X)$ est un groupe fini.*

(*d*) ([CR3, 6.1) *Si X est une surface, le groupe $CH^2(X)_{tors}$ est fini.*

Démonstration : Tout d'abord, d'après le théorème 4.2, le groupe $CH^2(\overline{X})_{tors}$ est, de façon G–équivariante, un sous–groupe de $H^3_{et}(\overline{X}, \mathbb{Q}/\mathbb{Z}(2))$, et un coup d'oeil au diagramme (3.6) montre que l'image de $CH^2(X)_{tors}$ dans $CH^2(\overline{X})_{tors} \subset H^3_{et}(\overline{X}, \mathbb{Q}/\mathbb{Z}(2))$ s'identifie à l'image de

$$NH^3_{et}(X, \mathbb{Q}/\mathbb{Z}(2)) = \mathrm{Ker}[H^3_{et}(X, \mathbb{Q}/\mathbb{Z}(2)) \longrightarrow H^3_{et}(k(X), \mathbb{Q}/\mathbb{Z}(2))]$$

dans $H^3_{et}(\overline{X}, \mathbb{Q}/\mathbb{Z}(2))$. L'énoncé (*b*) résulte donc de (*a*).

Si X a bonne réduction potentielle, des théorèmes de changement de base en cohomologie étale et du théorème de Deligne (conjecture de Weil) résulte la finitude des groupes $H^3_{et}(\overline{X}, \mathbb{Q}_l/\mathbb{Z}_l(2))^G$ pour chaque l premier $\neq p$, et leur nullité pour presque tout l ([CR1], 1.5.1). Ainsi (*c*) résulte de (*b*).

Lorsque X est une surface, le G–module $H^3_{et}(\overline{X}, \mathbb{Q}/\mathbb{Z}(2))$ s'identifie au groupe $\underline{\mathrm{Alb}}_X(\overline{k})_{tors}$ des points de torsion de la variété d'Albanese $\underline{\mathrm{Alb}}_X$ de X. Ainsi $H^3_{et}(\overline{X}, \mathbb{Q}/\mathbb{Z}(2))^G$ s'identifie au groupe $\underline{\mathrm{Alb}}_X(k)_{tors}$, et ce dernier groupe est fini (voir § 1, Prop. 1.7). Ainsi (*d*) résulte de (*b*).

Etablissons (*a*). Comme on l'a rappelé dans la démonstration du théorème 6.4, on dispose d'une suite exacte

$$H^1(G, K_2\overline{k}(X)/H^0(\overline{X}, \mathcal{K}_2)) \longrightarrow \mathrm{Ker}[CH^2(X) \longrightarrow CH^2(\overline{X})] \longrightarrow H^1(G, H^1(\overline{X}, \mathcal{K}_2)).$$

Sous l'hypothèse $H^2(X, \mathcal{O}_X) = 0$ et k corps local, on sait que le groupe $H^1(G, H^1(\overline{X}, \mathcal{K}_2))$ est fini ([CR1] Prop. 3.9). En utilisant la nullité de $H^1(G, K_2\overline{k}(X)/K_2\overline{k})$ pour une variété sur un corps local ([C1]), et l'unique divisibilité du groupe $K_2\overline{k}$, on voit que le groupe de gauche s'identifie au noyau

$$\operatorname{Ker}\rho_X : H^2(G, H^0(\overline{X}, \mathcal{K}_2)) \longrightarrow H^2(G, K_2\overline{k}(X)).$$

Or le théorème 8.3 ci-dessous dit que ce groupe est fini. Ainsi le groupe médian dans la suite exacte ci-dessus est fini. $\square$

Remarque 8.2.1 : On peut se demander si l'hypothèse de (*b*) n'est pas toujours satisfaite (cf. [J] Remark 5 p. 349).

Il reste donc à établir le théorème suivant ([CR3], Prop. 4.2), analogue sur un corps local du théorème 6.7 (Salberger) :

THÉORÈME 8.3. — *Soit X une variété projective, lisse, géométriquement connexe sur un corps local k. Le groupe*

$$H^1(G, K_2\overline{k}(X)/H^0(\overline{X}, \mathcal{K}_2)) = \operatorname{Ker}\rho_X : H^2(G, H^0(\overline{X}, \mathcal{K}_2)) \longrightarrow H^2(G, K_2\overline{k}(X))$$

est un groupe fini.

Démonstration : Elle est entièrement analogue à celle du théorème 6.7, dont nous reprenons les notations. Supposons $\dim(X) > 1$. Soit $Y \subset X$ une section hyperplane lisse. La flèche de restriction $H^2(G, H^0(\overline{X}, \mathcal{K}_2)) \longrightarrow H^2(G, H^0(\overline{Y}, \mathcal{K}_2))$ induit une flèche $\operatorname{Ker}(\rho_X) \longrightarrow \operatorname{Ker}(\rho_Y)$. Le noyau de $H^2(G, H^0(\overline{X}, \mathcal{K}_2)) \longrightarrow H^2(G, H^0(\overline{Y}, \mathcal{K}_2))$ s'identifie à celui de

$$H^2(G, H^1_{et}(\overline{X}, \mathbf{Q}/\mathbf{Z}(2))) \longrightarrow H^2(G, H^1_{et}(\overline{Y}, \mathbf{Q}/\mathbf{Z}(2))).$$

Comme la G–cohomologie d'un G–module fini est finie (k est local), à groupes finis près, ce noyau s'identifie à celui de

$$H^2(G, H^1_{et}(\overline{X}, \mathbf{Q}/\mathbf{Z}(2))^o) \longrightarrow H^2(G, H^1_{et}(\overline{Y}, \mathbf{Q}/\mathbf{Z}(2))^o).$$

La flèche de restriction

$$H^1_{et}(\overline{X}, \mathbf{Q}/\mathbf{Z}(2))^o \longrightarrow H^1_{et}(\overline{Y}, \mathbf{Q}/\mathbf{Z}(2))^o$$

est une application G–équivariante qui admet une presque rétraction, c'est-à-dire qu'il existe une flèche G–équivariante

$$H^1_{et}(\overline{Y}, \mathbf{Q}/\mathbf{Z}(2))^o \longrightarrow H^1_{et}(\overline{X}, \mathbf{Q}/\mathbf{Z}(2))^o$$

qui composée avec la précédente est la multiplication par un entier $m > 0$. Ainsi le noyau de

$$H^2(G, H^1_{et}(\overline{X}, \mathbf{Q}/\mathbf{Z}(2))^o) \longrightarrow H^2(G, H^1_{et}(\overline{Y}, \mathbf{Q}/\mathbf{Z}(2))^o)$$

est contenu dans le groupe de m–torsion de $H^2(G, H^1_{et}(\overline{X}, \mathbf{Q}/\mathbf{Z}(2))^o)$, et ce groupe est un quotient de $H^2(G, {}_m(H^1_{et}(\overline{X}, \mathbf{Q}/\mathbf{Z}(2))^o))$, groupe fini puisque k est local. On voit donc que la flèche $\operatorname{Ker}(\rho_X) \longrightarrow \operatorname{Ker}(\rho_Y)$ a un noyau fini. On est donc ramené à établir le théorème pour une courbe C sur un corps local. Ceci est fait par Raskind dans [R3], dont nous rappelons brièvement les arguments. Comme on a $CH^2(C) = 0$ pour une courbe C, la méthode galoisienne montre que le groupe $H^1(G, K_2\overline{k}(C)/H^0(\overline{C}, \mathcal{K}_2))$ s'identifie au conoyau de la flèche

$$H^1(C, \mathcal{K}_2) \longrightarrow H^1(\overline{C}, \mathcal{K}_2)^G.$$

Notons $V(C) = \mathrm{Ker}[H^1(C,\mathcal{K}_2) \longrightarrow k^*]$ et $V(\overline{C}) = \mathrm{Ker}[H^1(\overline{C},\mathcal{K}_2) \longrightarrow \overline{k}^*]$, les flèches provenant des applications de réciprocité usuelles. Comme l'explique Raskind (op. cit. 3.5 et lemme suivant) un théorème de S. Saito en théorie du corps de classes supérieur (sur un corps local) assure que l'application $V(C) \longrightarrow V(\overline{C})^G$ est surjective. En utilisant l'unique divisibilité de $V(\overline{C})$ ([R3], Lemma 1.1) et la G–cohomologie de la suite exacte

$$0 \longrightarrow V(\overline{C}) \longrightarrow H^1(\overline{C},\mathcal{K}_2) \longrightarrow \overline{k}^* \longrightarrow 0,$$

on voit que le conoyau de $H^1(C,\mathcal{K}_2) \longrightarrow H^1(\overline{C},\mathcal{K}_2)^G$ s'identifie à celui de

$$H^1(C,\mathcal{K}_2) \longrightarrow k^*,$$

i.e. à celui de la flèche

$$N : \bigoplus_{P \in C^{(0)}} k(P)^* \longrightarrow k^*,$$

où les flèches $N_P : k(P)^* \longrightarrow k^*$ ne sont autres que les applications normes pour tout point fermé P. La flèche N est surjective si C possède un k–point. Dans ce cas, on a donc même $H^1(G, K_2\overline{k}(C)/H^0(\overline{C},\mathcal{K}_2)) = 0$. En l'absence de point rationnel, il est cependant clair que l'image de N dans k^* est d'indice fini, car c'est déjà le cas pour chaque N_P, k étant un corps local. On peut se demander si N n'est pas toujours surjective, même en l'absence de point k–rationnel. $\square$

Pour certaines surfaces, l'approche de Saito permet de contrôler le groupe $CH^2(X)_{tors} = CH_0(X)_{tors}$ au moyen d'applications cycles, et de façon peut–être plus frappante encore, au moyen du groupe de Brauer. Le théorème suivant s'applique en particulier aux surfaces avec $H^2(X,\mathcal{O}_X) = 0$ et $H^1(X,\mathcal{O}_X) = 0$ (cette dernière condition assurant la trivialité de la variété d'Albanese), en particulier aux surfaces d'Enriques et aux surfaces rationnelles. Seul ce dernier cas était connu ([C1]).

THÉORÈME 8.4 (Saito, [S2]). — *Soit k un corps local, et soit X une surface projective lisse et géométriquement intègre sur k. Supposons $H^2(X,\mathcal{O}_X) = 0$ et supposons que la variété d'Albanese $\underline{\mathrm{Alb}}_X$ de X ait potentielle bonne réduction. Alors :*

(*a*) *Il existe un entier $N > 0$ tel que pour tout entier n multiple de N, l'application cycle*

$$CH_0(X)_{tors} \longrightarrow H^4_{et}(X, \mu_n^{\otimes 2})$$

soit injective.

(*b*) *L'accouplement naturel*

$$CH_0(X) \times Br(X) \longrightarrow Br(k) = \mathbf{Q}/\mathbf{Z}$$

induit une injection

$$CH_0(X)_{tors} \longrightarrow \mathrm{Hom}(Br(X), \mathbf{Q}/\mathbf{Z}).$$

Démonstration : Pour établir (*a*), il suffit, d'après le théorème 7.1, d'établir que le conoyau de la flèche

$$H^1(X,\mathcal{K}_2) \otimes \mathbf{Q}/\mathbf{Z} \longrightarrow H^3_{et}(X, \mathbf{Q}/\mathbf{Z}(2))$$

est d'exposant fini. Pour cela, on reprend la démonstration du théorème 7.3. On analyse le groupe $H^3_{et}(X, \mathbf{Q}/\mathbf{Z}(2))$ au moyen de la suite spectrale de Hochschild–Serre. Le groupe $H^3(k, \mathbf{Q}/\mathbf{Z}(2))$ est nul (k est local, donc de dimension cohomologique $cd(k) = 2$). Le groupe $H^3_{et}(\overline{X}, \mathbf{Q}/\mathbf{Z}(2))^G$ est fini, car X est une surface sur un corps local (voir la démonstration de 8.2 (*d*) ci–dessus). Procédant comme en 7.3, on voit qu'il suffit ici de montrer que le groupe

$H^2(k, H^1_{et}(\overline{X}, \mathbb{Q}/\mathbb{Z}(2)))$ est d'exposant fini (en 7.3, l'hypothèse $H^1(X, \mathcal{O}_X) = 0$ garantit la nullité du sous-groupe divisible maximal de $H^1_{et}(\overline{X}, \mathbb{Q}/\mathbb{Z}(2))$, ce qui assure à peu de frais que le groupe $H^2(k, H^1_{et}(\overline{X}, \mathbb{Q}/\mathbb{Z}(2)))$ est d'exposant fini).

Or le groupe $H^2(k, H^1_{et}(\overline{X}, \mathbb{Q}/\mathbb{Z}(2)))$ est ici fini, car l'hypothèse de bonne réduction de la variété d'Albanese, donc de la variété de Picard, sur une extension finie K de k assure que le groupe $H^2(k, H^1_{et}(\overline{X}, \mathbb{Q}/\mathbb{Z}(2))^o)$ est nul (cf. [CR1], p. 190/191), donc que le groupe $H^2(k, H^1_{et}(\overline{X}, \mathbb{Q}/\mathbb{Z}(2))^o)$ est fini (en fait, nul, par exemple parce que $cd(k) \leq 2$ implique que ce groupe est divisible). Ceci établit (*a*).

Passons à (*b*). La combinaison de la dualité en théorie du corps de classes local et de la dualité de Poincaré pour une variété projective et lisse sur un corps séparablement clos donne lieu pour la surface X à un accouplement non dégénéré de groupes abéliens finis

$$H^4_{et}(X, \mu_n^{\otimes 2}) \times H^2_{et}(X, \mu_n) \longrightarrow H^6_{et}(X, \mu_n^{\otimes 3}) = \mathbb{Z}/n,$$

(voir [S1], 2.9; pour une dualité analogue, mais sur un corps fini, voir [CSS] démonstration de (i), p. 790/791), et la suite de Kummer donne lieu à une suite exacte

$$0 \longrightarrow \mathrm{Pic}(X)/n \longrightarrow H^2_{et}(X, \mu_n) \longrightarrow {}_nBr(X) \longrightarrow 0$$

et l'on vérifie que l'accouplement

$$CH^2(X)/n \times \mathrm{Pic}(X)/n \longrightarrow H^6_{et}(X, \mu_n^{\otimes 3}) = \mathbb{Z}/n$$

induit par l'accouplement ci-dessus, l'application cycle et la flèche déduite de la suite de Kummer, est nul (l'application cycle envoie un point fermé P dans la cohomologie à support dans ce point, et tout diviseur sur X admet un représentant à support étranger à P). Choisissant alors n multiple de N comme en (*a*), on déduit alors de (*a*) une injection :

$$CH_0(X)_{tors} \hookrightarrow \mathrm{Hom}({}_nBr(X), \mathbb{Z}/n)$$

et donc a fortiori une injection $CH_0(X)_{tors} \hookrightarrow \mathrm{Hom}(Br(X), \mathbb{Q}/\mathbb{Z})$ comme annoncé. Un petit travail permet de vérifier que cette application est bien déduite de l'accouplement naturel

$$CH_0(X) \times Br(X) \longrightarrow Br(k) = \mathbb{Q}/\mathbb{Z}$$

(pour un calcul analogue, voir [CSS] démonstration de (ii), p. 791/792). □

Remarque 8.4.1 : Pour éliminer l'hypothèse (désagréable) de bonne réduction, il suffirait d'établir que l'image de $H^3_{et}(X, \mathbb{Q}/\mathbb{Z}(2))$ dans $H^3_{et}(k(X), \mathbb{Q}/\mathbb{Z}(2))$ est finie, ou du moins d'exposant fini. Sans hypothèse de bonne réduction pour la surface X, on sait en effet que $CH^2(X)_{tors}$ est fini (Théorème 8.2). Si l'image de $H^3_{et}(X, \mathbb{Q}/\mathbb{Z}(2))$ dans $H^3_{et}(k(X), \mathbb{Q}/\mathbb{Z}(2))$ est finie, le diagramme (7.2) montre que le groupe Θ est d'exposant fini, ce qui permet alors d'appliquer le théorème 7.1.

Remarque 8.4.2 : Sous les hypothèses de 8.4, et sous l'hypothèse supplémentaire, sans doute superfétatoire (conjecture de Bloch sur les surfaces avec $H^2(X, \mathcal{O}_X) = 0$), que l'application d'Albanese $A_0(\overline{X}) \longrightarrow \underline{\mathrm{Alb}}_X(\overline{k})$ est un isomorphisme, Saito [S2] démontre qu'en fait l'accouplement

$$A_0(X) \times Br(X) \longrightarrow \mathbb{Q}/\mathbb{Z}$$

est non dégénéré à gauche. L'idée est de contrôler l'image de $A_0(X)$ dans $\underline{\mathrm{Alb}}_X(k)$ au moyen de la dualité de Tate pour les variétés abéliennes sur les corps locaux, qui donne un isomorphisme :

$$\underline{\mathrm{Alb}}_X(k) \simeq \mathrm{Hom}(H^1(k, \underline{\mathrm{Pic}}^0_X), \mathbb{Q}/\mathbb{Z}).$$

L'homomorphisme composé naturel

$$H^1(k, \underline{\mathrm{Pic}}^0_X) \longrightarrow H^1(k, \underline{\mathrm{Pic}}_X) \longrightarrow Br(X),$$

donne alors lieu à un diagramme dont on vérifie la commutativité :

$$\begin{array}{ccc} A_0(X) & \longrightarrow & \underline{\mathrm{Alb}}_X(k) \\ \downarrow & & \downarrow \simeq \\ \mathrm{Hom}(Br(X), \mathbb{Q}/\mathbb{Z}) & \longrightarrow & \mathrm{Hom}(H^1(k, \underline{\mathrm{Pic}}^0_X), \mathbb{Q}/\mathbb{Z}). \end{array}$$

Ainsi le noyau de la flèche verticale de gauche est inclus dans le noyau de la flèche horizontale supérieure, qui par l'hypothèse supplémentaire est un groupe de torsion. Le théorème précédent assure que la restriction à $A_0(X)_{tors}$ de la flèche verticale de gauche est injective.

THÉORÈME 8.5. — *Soit X une surface projective et lisse, géométriquement intègre sur un corps local k. Supposons*

(i) $H^2(X, \mathcal{O}_X) = 0$, et

(ii) L'application d'Albanese $A_0(\overline{X}) \longrightarrow \underline{\mathrm{Alb}}_X(\overline{k})$ est un isomorphisme (conséquence de (i) selon une conjecture de Bloch).

Alors le groupe $A_0(X)$ est une extension d'un sous-groupe ouvert de $\underline{\mathrm{Alb}}_X(k)$ par un groupe fini. En particulier, le quotient $A_0(X)/n$ est fini pour tout entier $n > 0$, et le quotient $A_0(X)/l$ est nul pour presque tout premier l.

Démonstration : D'après (*ii*), le noyau de la flèche $A_0(X) \longrightarrow \underline{\mathrm{Alb}}_X(k)$ est un groupe de torsion, donc fini sous l'hypothèse (*i*) d'après le théorème 8.2 (*d*). Etudions l'image de $A_0(X)$ dans $\underline{\mathrm{Alb}}_X(k)$. Comme ce dernier groupe est un groupe analytique p-adique compact commutatif, ses sous-groupes ouverts ne sont autres que ses sous-groupes d'indice fini. Soit C une courbe section hyperplane lisse de X. On a un épimorphisme de variétés abéliennes $\underline{\mathrm{Alb}}_C \longrightarrow \underline{\mathrm{Alb}}_X$, induisant un homomorphisme $\underline{\mathrm{Alb}}_C(k) \longrightarrow \underline{\mathrm{Alb}}_X(k)$ d'image ouverte, donc d'indice fini, dans $\underline{\mathrm{Alb}}_X(k)$. L'application $A_0(C) \longrightarrow \underline{\mathrm{Alb}}_C$ est un isomorphisme si $C(k) \neq \emptyset$, dans le cas général son image est un sous-groupe d'indice fini de $\underline{\mathrm{Alb}}_C(k)$ (k est local). Ainsi l'image de $A_0(X) \longrightarrow \underline{\mathrm{Alb}}_X(k)$ contient un sous-groupe d'indice fini de $\underline{\mathrm{Alb}}_X(k)$, c'est donc un sous-groupe d'indice fini, donc ouvert. Le groupe analytique p-adique compact $\underline{\mathrm{Alb}}_X(k)$ est une extension d'un groupe fini par un sous-groupe isomorphe à une somme directe finie d'exemplaires de $\mathbb{Z}_p$, ce qui implique alors la dernière partie de l'énoncé. □

Pour terminer, et sans démonstration, je citerai un résultat obtenu par la méthode de localisation.

THÉORÈME 8.6 ([CR3], 6.3). — *Soit k un corps local k, extension finie de $\mathbb{Q}_p$, soit R son anneau des entiers et $\mathbb{F}$ son corps résiduel. Soit $\mathbf{X}$ un R-schéma projectif et lisse à fibres géométriques intègres, de fibre générique X/k et de fibre spéciale $Y/\mathbb{F}$. Supposons $H^2(Y, \mathcal{O}_Y) = 0$. Alors, pour tout premier $l \neq p$, les sous-groupes de torsion l-primaire $CH^2(X)_{l-tors}$ et $CH^2(Y)_{l-tors}$ sont des groupes finis naturellement isomorphes.* □

§ 9. Variétés sur les corps de nombres, III

Aux paragraphes précédents, j'ai donné des démonstrations, dues pour l'essentiel à Salberger, au moins dans le cas global, des théorèmes bornant l'exposant de la torsion de $CH^2(X)$ lorsque X satisfait $H^2(X, \mathcal{O}_X) = 0$ (Théorème 6.6 sur un corps global; théorème 8.2 (*b*) sur un corps local – avec quelques hypothèses parasites). Dans ce paragraphe, je commence par décrire la méthode originale de Salberger permettant d' établir ce résultat de torsion bornée (théorème 9.1 et début de la démonstration du théorème 9.2). La démonstration présentée ici a été reconstituée à partir d'exposés de Salberger en 1990.

J'expose ensuite comment Salberger (communication personnelle) peut modifier la technique de localisation exposée au § 6 de façon à établir le théorème de finitude 6.1, ici baptisé 9.2, sans recourir, comme Raskind et moi-même l'avions fait, au résultat fin de Bloch, Kato-Saito et Somekawa (démonstration du théorème 6.5, point c)).

La récente prépublication [Sb2] contient de nombreuses autres idées et démonstrations intéressantes, et j'engage le lecteur à la consulter.

THÉORÈME 9.1. — *Soit k un corps local (non archimédien) ou global de caractéristique zéro, et soit X une k-variété projective et lisse géométriquement intègre. Supposons $H^2(X, \mathcal{O}_X) = 0$. Alors le conoyau de l'application*

$$H^1(X, \mathcal{K}_2) \otimes \mathbf{Q}/\mathbf{Z} \longrightarrow \operatorname{Ker}[H^3_{et}(X, \mathbf{Q}/\mathbf{Z}(2)) \longrightarrow H^3_{et}(\overline{X}, \mathbf{Q}/\mathbf{Z}(2))^G \oplus H^3_{et}(k(X), \mathbf{Q}/\mathbf{Z}(2))]$$

est d'exposant fini.

Démonstration : On procède comme au théorème 7.3. La suite spectrale de Hochschild–Serre en cohomologie étale

$$H^p(k, H^q_{et}(\overline{X}, \mathbf{Q}/\mathbf{Z}(2))) \Longrightarrow H^n_{et}(X, \mathbf{Q}/\mathbf{Z}(2))$$

donne lieu à une filtration $F^0_X \subset F^1_X \subset F^2_X$ sur le groupe $\operatorname{Ker}[H^3_{et}(X, \mathbf{Q}/\mathbf{Z}(2)) \longrightarrow H^3_{et}(\overline{X}, \mathbf{Q}/\mathbf{Z}(2))^G]$. Notons

$$N_r H^3_{et}(X, \mathbf{Q}/\mathbf{Z}(2)) = \operatorname{Ker}[H^3_{et}(X, \mathbf{Q}/\mathbf{Z}(2)) \longrightarrow H^3_{et}(\overline{X}, \mathbf{Q}/\mathbf{Z}(2))^G \oplus H^3_{et}(k(X), \mathbf{Q}/\mathbf{Z}(2))],$$

et considérons la filtration induite sur ce sous-groupe, soit

$$F^0_{X,r} \subset F^1_{X,r} \subset F^2_{X,r}.$$

Le groupe $F^0_{X,r}$ est un sous-groupe de F^0_X, lui-même quotient de $H^3(k, \mathbf{Q}/\mathbf{Z}(2))$. Sur un corps local, ce dernier groupe est fini. Sur un corps de nombres, il coïncide avec $(\mathbf{Z}/2)^s$, où s est le nombre de complétions réelles de k (en fait, sur k quelconque, si $X(k) \neq \emptyset$, alors $F^0_{X,r} = 0$, comme on voit par un argument de spécialisation).

Comme au théorème 7.3, l'hypothèse $H^2(X, \mathcal{O}_X) = 0$ implique que l'application

$$H^1(X, \mathcal{K}_2) \otimes \mathbf{Q}/\mathbf{Z} \longrightarrow F^2_X / F^1_X$$

a un conoyau d'exposant fini, a fortiori en est-il de même pour l'application

$$H^1(X, \mathcal{K}_2) \otimes \mathbf{Q}/\mathbf{Z} \longrightarrow F^2_{X,r} / F^1_{X,r}.$$

Pour établir le théorème, il suffit donc de montrer que le groupe $F^1_{X,r}$ est d'exposant fini. Pour cela on utilise la technique des sections hyperplanes déjà employée dans 6.7 et 8.3. Soit $C \subset X$ une k-courbe projective lisse géométriquement intègre obtenue par sections hyperplanes successives.

La suite spectrale de Hochschild–Serre est fonctorielle contravariante par morphismes quelconques. On obtient donc un diagramme commutatif :

$$\begin{array}{ccccccccc} 0 & \longrightarrow & F^0_X & \longrightarrow & F^1_X & \longrightarrow & E^{21}_{\infty,X} & \longrightarrow & 0 \\ & & \downarrow & & \downarrow & & \downarrow & & \\ 0 & \longrightarrow & F^0_C & \longrightarrow & F^1_C & \longrightarrow & E^{21}_{\infty,C} & \longrightarrow & 0 \ , \end{array}$$

où les groupes F^0_X et F^0_C sont finis. Par ailleurs, pour k local ou global, le groupe $H^4(k, \mathbf{Q}/\mathbf{Z}(2))$ est nul. De la fonctorialité de la suite spectrale on tire donc un diagramme commutatif de suites exactes

$$\begin{array}{ccccccc} & & H^2(k, H^1_{et}(\overline{X}, \mathbf{Q}/\mathbf{Z}(2))) & \longrightarrow & E^{21}_{\infty,X} & \longrightarrow & 0 \\ & & \downarrow & & \downarrow & & \\ H^0(\overline{C}, \mathbf{Q}/\mathbf{Z}(2)))^G & \longrightarrow & H^2(k, H^1_{et}(\overline{C}, \mathbf{Q}/\mathbf{Z}(2))) & \longrightarrow & E^{21}_{\infty,C} & \longrightarrow & 0 \end{array}$$

Dans ce diagramme, le groupe $H^0(\overline{C}, \mathbf{Q}/\mathbf{Z}(2)))^G = \mathbf{Q}/\mathbf{Z}(1)^G = \mu(k)$ est le groupe des racines de l'unité dans k ; c'est donc un groupe fini. Les arguments développés en 6.7 (cas global) et 8.3 (cas local) montrent que le noyau de la flèche verticale de gauche est d'exposant fini. Une chasse aux diagrammes montre alors que le noyau de la flèche $F^1_X \longrightarrow F^1_C$ est aussi d'exposant fini.

Soit $Y \subset X$ est une section hyperplane lisse géométriquement intègre. Si R désigne l'anneau local de X au point générique de Y, tout élément de $NH^3_{et}(X, \mathbf{Q}/\mathbf{Z}(2))$ a, par la théorie de Bloch–Ogus [Bl–O], une image nulle dans $H^3_{et}(R, \mathbf{Q}/\mathbf{Z}(2))$, donc, par passage au corps résiduel $k(Y)$ de R, une image nulle dans $H^3_{et}(k(Y), \mathbf{Q}/\mathbf{Z}(2))$. Par induction, on voit donc que la flèche de restriction $H^3_{et}(X, \mathbf{Q}/\mathbf{Z}(2)) \longrightarrow H^3_{et}(C, \mathbf{Q}/\mathbf{Z}(2))$ envoie le sous-groupe $NH^3_{et}(X, \mathbf{Q}/\mathbf{Z}(2))$ dans $NH^3_{et}(C, \mathbf{Q}/\mathbf{Z}(2))$, et aussi le groupe $N_rH^3_{et}(X, \mathbf{Q}/\mathbf{Z}(2))$ dans $N_rH^3_{et}(C, \mathbf{Q}/\mathbf{Z}(2))$.

En particulier, l'application $F^1_{X,r} \longrightarrow F^1_C$ a une image contenue dans $F^1_{C,r} = \mathrm{Ker}[F^1_C \longrightarrow H^3(k(C), \mathbf{Q}/\mathbf{Z}(2))]$.

Comparant les suites spectrales de Hochschild–Serre pour C et pour le corps des fonctions $k(C)$, on obtient un diagramme commutatif

$$\begin{array}{ccccccccc} 0 & \longrightarrow & F^0_C & \longrightarrow & F^1_C & \longrightarrow & H^2(k, H^1_{et}(\overline{C}, \mathbf{Q}/\mathbf{Z}(2)))/im(\mu(k) & \longrightarrow & 0 \\ & & \downarrow & & \downarrow & & \downarrow & & \\ 0 & \longrightarrow & F^0_{k(C)} & \longrightarrow & F^1_{k(C)} & \longrightarrow & H^2(k, H^1(\overline{k}(C), \mathbf{Q}/\mathbf{Z}(2))) & \longrightarrow & 0 \ , \end{array}$$

où F^0_C et $F^0_{k(C)}$ sont des groupes finis quotients de $H^3(k, \mathbf{Q}/\mathbf{Z}(2))$. Ainsi, à groupe fini près, $F^1_{C,r}$ coïncide avec

$$\mathrm{Ker}[H^2(k, H^1_{et}(\overline{C}, \mathbf{Q}/\mathbf{Z}(2))) \longrightarrow H^2(k, H^1(\overline{k}(C), \mathbf{Q}/\mathbf{Z}(2)))].$$

Ce groupe n'est autre que le groupe

$$\mathrm{Ker}[H^2(G, H^0(\overline{C}, \mathcal{K}_2)) \longrightarrow H^2(G, K_2\overline{k}(C))]$$

dont on a déjà dit (6.7 et 8.3) qu'il est d'exposant fini pour une courbe sur un corps local ou global - et nul lorsque cette courbe possède un point rationnel ([R3]).

Comme l'application $F^1_{X,r} \longrightarrow F^1_{C,r}$ a un noyau d'exposant fini, on conclut que $F^1_{X,r}$ est d'exposant fini, ce qui achève la démonstration.

(Comme on le vérifiera aisément, dans le cas local on peut partout dans la démonstration remplacer *d'exposant fini* par *fini.*) □

THÉORÈME 9.2 (= Théorème 6.1). — *Soit X une variété projective, lisse, géométriquement connexe sur un corps de nombres k. Supposons le groupe $H^2(X,\mathcal{O}_X)=0$. Alors le sous-groupe de torsion $CH^2(X)_{tors}$ est fini.*

Démonstration : Montrons d'abord que le le groupe $CH^2(X)_{tors}$ est un groupe d'exposant fini. Comparant la suite exacte

$$0 \longrightarrow H^1(X,\mathcal{K}_2)\otimes \mathbf{Q}/\mathbf{Z} \longrightarrow NH^3_{et}(X,\mathbf{Q}/\mathbf{Z}(2)) \longrightarrow CH^2(X)_{tors} \longrightarrow 0,$$

et la même suite au niveau de la clôture algébrique

$$0 \longrightarrow H^1(\overline{X},\mathcal{K}_2)\otimes \mathbf{Q}/\mathbf{Z} \longrightarrow NH^3_{et}(\overline{X},\mathbf{Q}/\mathbf{Z}(2)) \longrightarrow CH^2(\overline{X})_{tors} \longrightarrow 0,$$

et utilisant la nullité de $H^1(\overline{X},\mathcal{K}_2)\otimes \mathbf{Q}/\mathbf{Z}$ ([CR1]) ainsi que la finitude de $H^3_{et}(\overline{X},\mathbf{Q}/\mathbf{Z}(2))^G$ (cf. démonstration du théorème 6.4), on déduit du théorème 9.1 ci-dessus que le groupe $CH^2(X)_{tors}$ est un groupe d'exposant fini.

Ainsi $CH^2(X)_{l-tors}$ est nul pour presque tout premier l, et pour établir la finitude de $CH^2(X)_{tors}$, il suffit, pour un premier l donné arbitraire, d'établir la finitude de $CH^2(X)_{l-tors}$.

Reprenons alors la démonstration du théorème 6.5, dont je ne répéterai pas ici les notations. Reportons-nous au point b) de cette démonstration. On dispose de la suite de localisation

$$H^1(X_L,\mathcal{K}_2) \longrightarrow \bigoplus_{q\in \mathrm{Spec}(B)^{(1)}} \mathrm{Pic}(\mathbf{X}_q) \longrightarrow CH^2(\mathbf{X}_B) \longrightarrow CH^2(X_L) \longrightarrow 0,$$

et on a établi que l'application composée

$$H^1(X_L,\mathcal{K}_2) \longrightarrow \bigoplus_{q\in \mathrm{Spec}(B)^{(1)}} \mathrm{Pic}(\mathbf{X}_q) \longrightarrow \bigoplus_{q\in \mathrm{Spec}(B)^{(1)}} NS(\mathbf{X}_q)$$

est surjective. A fortiori en est-il de même de l'application composée induite

$$H^1(X_L,\mathcal{K}_2)\otimes\mathbf{Q} \longrightarrow \bigoplus_{q\in \mathrm{Spec}(B)^{(1)}} \mathrm{Pic}(\mathbf{X}_q)\otimes\mathbf{Q} \longrightarrow \bigoplus_{q\in \mathrm{Spec}(B)^{(1)}} NS(\mathbf{X}_q)\otimes\mathbf{Q}.$$

Mais la flèche de droite est ici un isomorphisme, car les groupes

$$J(\mathbf{F}_q) = Ker[\mathrm{Pic}(\mathbf{X}_q) \longrightarrow NS(\mathbf{X}_q)]$$

sont de torsion (et même finis) , et donc $J(\mathbf{F}_q)\otimes\mathbf{Q}=0$. Ainsi la flèche

$$H^1(X_L,\mathcal{K}_2)\otimes\mathbf{Q} \longrightarrow \bigoplus_{q\in \mathrm{Spec}(B)^{(1)}} \mathrm{Pic}(\mathbf{X}_q)\otimes\mathbf{Q}$$

est-elle surjective. De cette surjectivité et de la suite de localisation rappelée ci-dessus on déduit alors par un argument formel que l'application de restriction à la fibre générique sur les groupes de Chow, application bien sûr surjective, induit encore une application *surjective* $CH^2(\mathbf{X}_B)_{tors} \longrightarrow CH^2(X_L)_{tors}$ sur les groupes de torsion.

Plaçons-nous maintenant dans la situation du théorème 6.2, dont je reprends les notations. En faisant varier l'entier n parmi les puissances d'un nombre premier l, on obtient un diagramme

$$\begin{array}{ccc} H^1(\mathbf{X}, \mathcal{H}^2(\mathbf{Q}_l/\mathbf{Z}_l(2))) & \longrightarrow CH^2(\mathbf{X})_{l-tors} & \longrightarrow 0 \\ \downarrow \gamma_2 & & \\ H^3_{et}(\mathbf{X}, \mathbf{Q}_l/\mathbf{Z}_l(2)) & & \end{array}$$

où la flèche γ_2 est injective et la flèche horizontale surjective (les limites inductives filtrantes respectent surjections et injections).

Le groupe de torsion l-primaire $H^3_{et}(\mathbf{X}, \mathbf{Q}_l/\mathbf{Z}_l(2))$, limite directe des groupes finis $H^3_{et}(\mathbf{X}, \mu_{l^n}^{\otimes 2})$, est un groupe de cotype fini, i.e. extension d'un groupe fini l-primaire par un groupe du type $(\mathbf{Q}_l/\mathbf{Z}_l)^m$, soit encore somme directe d'un groupe fini l-primaire et d'une somme finie de groupes $(\mathbf{Q}_l/\mathbf{Z}_l)$. Ceci vaut plus généralement pour tout groupe $H^i_{et}(\mathbf{X}, \mathbf{Q}_l/\mathbf{Z}_l(j))$. Pour le voir, on utilise les arguments classiques suivants (cf. [M1], p. 163 à 166 et [CSS], p. 773/774 et 780/781). On considère la longue suite exacte de cohomologie associée à la suite exacte de faisceaux étales

$$1 \longrightarrow \mu_{l^n}^{\otimes j} \longrightarrow \mu_{l^{n+m}}^{\otimes j} \longrightarrow \mu_{l^m}^{\otimes j} \longrightarrow 1$$

sur $\mathbf{X}$. On passe à la limite projective en n sur cette suite. On obtient encore une suite exacte, car les groupes $H^i_{et}(\mathbf{X}, \mu_{l^n}^{\otimes j})$ sont finis (voir la démonstration du théorème 6.2 b)). On passe ensuite à la limite inductive sur m dans la longue suite exact obtenue. On obtient ainsi des suites exactes courtes

$$0 \longrightarrow H^i(\mathbf{X}, \mathbf{Z}_l(j)) \longrightarrow H^i(\mathbf{X}, \mathbf{Q}_l(j)) \longrightarrow H^i(\mathbf{X}, \mathbf{Q}_l/\mathbf{Z}_l(j)) \longrightarrow H^{i+1}(\mathbf{X}, \mathbf{Z}_l(j))_{l-tors} \longrightarrow 0$$

où $H^i(\mathbf{X}, \mathbf{Q}_l(j)) = H^i(\mathbf{X}, \mathbf{Z}_l(j)) \otimes_{\mathbf{Z}_l} \mathbf{Q}_l$. On observe que pour tout i et j entiers, la limite *projective* des groupes $H^i_{et}(\mathbf{X}, \mu_{l^n}^{\otimes j})$ est un $\mathbf{Z}_l$-module de type fini. De la suite exacte ci-dessus résulte donc bien que $H^i(\mathbf{X}, \mathbf{Q}_l/\mathbf{Z}_l(j))$ est une extension d'un groupe fini l-primaire par une somme directe de groupes $\mathbf{Q}_l/\mathbf{Z}_l$, et donc en fait une somme directe d'un groupe fini l-primaire et de groupes $\mathbf{Q}_l/\mathbf{Z}_l$.

En utilisant la dualité entre les $\mathbf{Z}_l$-modules compacts et modules de torsion l-primaires discrets, on voit que cette propriété se transmet à tout sous-quotient. Ainsi le groupe $CH^2(\mathbf{X})_{l-tors}$, sous-quotient de $H^3_{et}(\mathbf{X}, \mathbf{Q}_l/\mathbf{Z}_l(2))$, est-il de cotype fini, et il en est de même du groupe $CH^2(X)_{l-tors}$, dont nous avons établi qu'il est un quotient du précédent.

Mais, d'après le début de la démonstration, le groupe $CH^2(X)_{l-tors}$ est d'exposant fini. C'est donc un groupe fini. □

BIBLIOGRAPHIE

[BV] L. Barbieri-viale. — Cohomology theories and algebraic cycles on singular varieties, Contemp. Math. **126** (1992) 197–217; Theorie coomologiche e cicli algebrici sulle varieta' singolari, Tesi di dottorato, Universita' di Genova (1991).

[B1] S. Bloch. — Torsion algebraic cycles and a theorem of Roitman, Comp. Math. **39** (1979) 107–127.

[B2] S. Bloch. — *Lectures on algebraic cycles*, Duke Univ. Math. Ser. 4, Durham, N.C. 1980.

[B3] S. Bloch. — Torsion algebraic cycles, K_2 and Brauer groups of function fields, in L.N.M. **844** (ed. M. Kervaire et M. Ojanguren) Springer 1981.

[B4] S. Bloch. — On the Chow groups of certain rational surfaces, Ann. Sc. Ec. Norm. Sup. **14** (1981) 41–59.

[B5] S. Bloch. — Algebraic K–theory and class field theory for arithmetic surfaces, Ann. of Math. **114** (1981) 229–265.

[BO] S. Bloch and A. Ogus. — Gersten's conjecture and the homology of schemes, Ann. Sc. E.N.S. **7** (1974) 181–202.

[BS] S. Bloch and V. Srinivas. — Remarks on correspondences and algebraic cycles, Am. J. Math. **105** (1983) 1235–1253.

[C1] J.-L. Colliot-Thélène. — Hilbert's theorem 90 for K_2, with application to the Chow group of rational surfaces, Invent. Math. **35** (1983) 1–20.

[C2] J.-L. Colliot-Thélène. — On the reciprocity sequence in higher class field theory of function fields, Prépublication d'Orsay **92-52** (1992).

[CR1] J.-L. Colliot-Thélène and W. Raskind. — K_2–cohomology and the second Chow group, Math. Ann. **270** (1985) 165–199.

[CR2] J.-L. Colliot-Thélène and W. Raskind. — On the reciprocity law for surfaces over finite fields, J. Fac. Sc. Univ. Tokyo, Sect. IA **33** (1986) 283–294.

[CR3] J.-L. Colliot-Thélène et W. Raskind. — Groupe de Chow de codimension deux des variétés définies sur un corps de nombres : un théorème de finitude pour la torsion, Invent. Math. **105** (1991) 221–245.

[CS] J.-L. Colliot-Thélène and J.-J. Sansuc. — On the Chow groups of certain rational surfaces : a sequel to a paper of S. Bloch, Duke Math. J. **48** (1981) 421–447.

[CSS] J.-L. Colliot-Thélène, J.-J. Sansuc et C. Soulé. — Torsion dans le groupe de Chow de codimension deux, Duke Math. J. **50** (1983) 763–801.

[Cb] K. Coombes. — The arithmetic of zero–cycles on surfaces with geometric genus and irregularity zero, Math. Ann. **291** (1991) 429–452.

[F] W. Fulton. — *Intersection Theory*, Ergeb. der Math. und ihr. Grenzgeb. 3. Folge, Bd. **2**, Springer 1984.

[G] M. Gros. — 0–cycles de degré zéro sur les surfaces fibrées en coniques, J. reine und angew. Math. **373** (1987) 166–184.

[GS] M. GROS et N. SUWA. — Application d'Abel–Jacobi p–adique et cycles algébriques, Duke Math. J. **57** (1988) 579–613.

[J] U. JANNSEN. — On the l-adic cohomology of varieties over number fields and its Galois cohomology, in : *Galois groups over* $\mathbb{Q}$ (ed. Y. Ihara, K. Ribert, J.–P. Serre) Springer–Verlag 1987.

[K] K. KATO. — A Hasse principle for two dimensional global fields, J. für die reine und angew. Math. **366** (1986) 142–181.

[KS] K. KATO and S. SAITO. — Unramified class field theory of arithmetical surfaces, Ann. of Math. **118** (1983) 241–275.

[L] F. LECOMTE. — Rigidité des groupes de Chow, Duke Math. J. **53** (1986) 405-426.

[MS] A.S. MERKUR'EV and A.A. SUSLIN. — K–cohomology of Severi–Brauer varieties and norm residue homomorphism, Izv. Akad. Nauk SSSR **46** (1982) 1011–1146 = Math. USSR Izv. **21** (1983) 307–341.

[M1] J.S. MILNE. — *Etale cohomology*, Princeton University Press, **33**, Princeton, N.J. 1980.

[M2] J.S. MILNE. — *Arithmetic Duality Theorems*, Perspect. Math. **1**, Academic Press, Boston 1986.

[Q] D. QUILLEN. — Higher algebraic K–theory I, in *Algebraic K–theory* I (H. Bass, ed.) LNM **341**, Springer–Verlag 1973.

[R1] W. RASKIND. — Algebraic K–theory, étale cohomology and torsion algebraic cycles, in *Algebraic K–theory and Algebraic Number Theory*, Honolulu 1987, Contemporary Mathematics **83** (1989) 311–341.

[R2] W. RASKIND. — Torsion algebraic cycles on varieties over local fields, in : *Algebraic K–theory : Connections with Geometry and Topology*, Lake Louise 1987, (J.F. Jardine and V.P. Snaith, ed.) Kluwer Academic Publishers 1989.

[R3] W. RASKIND. — On K_1 of curves over global fields, Math. Ann. **288** (1990) 179–193.

[S1] S. SAITO. — A global duality theorem for varieties over global fields, in : *Algebraic K–theory : Connections with Geometry and Topology*, Lake Louise 1987, (J.F. Jardine and V.P. Snaith, ed.) Kluwer Academic Publishers 1989.

[S2] S. SAITO. — A conjecture of Bloch and Brauer groups of surfaces over p–adic fields, preprint, May 1990.

[S3] S. SAITO. — Torsion zero–cycles and étale homology of singular schemes, Duke Math. J. **64** (1991) 71–83.

[S4] S. SAITO. — Cycle map on torsion algebraic cycles of codimension two, Invent. Math. **106** (1991) 443–460.

[Sb1] P. SALBERGER. — Zero–cycles on rational surfaces over number fields, Invent. Math. **91** (1988) 505–524.

[Sb2] P. SALBERGER. — Chow groups of codimension two and l-adic realizations of motivic cohomology, typescript, September 1991.

[Sc] C. SCHOEN. — Some examples of torsion in the Griffiths group, Math. Ann. **293** (1992) 651–679.

[So] M. SOMEKAWA. — On Milnor K–groups attached to semi–abelian varieties, Journal of K–theory **4** (1990) 105–119.

[S] C. SOULÉ. — K_2 et le groupe de Brauer [d'après A.S. Merkurjev et A.A. Suslin], Séminaire Bourbaki, 1982/83, n° **601**.

[Sr] V. SRINIVAS. — Rational equivalence of 0-cycles on normal varieties over **C**, Proceedings of Symposia in Pure Mathematics **46**, Part 2, (1987) 475-482.

[Su] A.A. SUSLIN. — Torsion in K_2 of fields, Journal of K-theory **1** (1987) 5–29.

[Sw] N. SUWA. — Sur l'image de l'application d'Abel–Jacobi de Bloch, Bull. Soc. Math. France **116** (1988) 69–101.

[SGA4 1/2] *Cohomologie étale*, Séminaire de Géométrie Algébrique du Bois–Marie SGA 4 1/2, par P. Deligne, Springer L.N.M. **569** (1977).

Lectures on the approach to Iwasawa theory for Hasse-Weil L-functions via B_{dR}.

Part I.

By Kazuya Kato.

As the night sky, mathematics has two hemispheres; the archimedean hemisphere and the non-archimedean hemisphere. For some reasons, the latter hemisphere is usually under the horizon of our world, and the study of it is historically always behind the study of the former.

The arithmetic of special values of various zeta functions has been one of the main themes in number theory. By the recent development of the theory of p-adic periods (the p-adic Hodge theory) by Tate, Fontaine-Messing, Faltings, ..., the non-archimedean hemisphere about zeta values can be seen now better than before ([BK], [FP_1], [FP_2], [Ka_2], ...). The aim of this paper is;

—— to review the known relationship between zeta values and the theory of p-adic periods, and

—— to discuss that we can see an arm of a big galaxy, the galaxy of "p-adic zeta elements", in the non-archimedean hemisphere of zeta values, but that the total shape of this galaxy is still hiden under the horizon. Precisely speaking, we expect the following (Chap. I, §3): As all automorphic representations have zeta functions with values in $\mathbf{C}$, all Galois representations of number fields with coefficients in any p-adic ring Λ (which can be a finite ring) have p-adic zeta elements which are canonical bases of some invertible Λ-modules. We have the picture in which the p-adic side is conjectural:

archimedean side	p-adic side
automorphic representations	p-adic Galois representations
zeta functions	p-adic zeta elements
ε-factors	ε-elements
....	....

The harmony between the two hemispheres should be called the generalized Iwasawa theory or equally the generalized Deligne-Beilinson conjectures on zeta values. For Riemann zeta function and Dirichlet L-functions, Iwasawa theory is the best theory at present for the arithmetic of zeta values. How nice it would be if we can construct the Iwasawa theory of Hasse-Weil L-functions.

Where is the homeland of zeta values to which the true reasons of celestial phenomena of zeta values are attributed? How can we find a galaxy train [Mi] to approach it, which runs through the galaxy of p-adic zeta elements and whose engine is the theory of p-adic periods? I imagine that one coach of the train has the name "the explicit reciprocity law of p-adic Galois representations" (cf. the introduction of Chap. II, §2).

The structure of this paper (Part I + Part II) is as follows.

Contents.

Chapter III. Global subjects.

§1. Explicit reciprocity law and zeta values.

§2.*) Zeta values of motives.

§3.*) p-adic L-functions, duality.

Part I (resp. II) consists of the sections without (resp. with) the sign *).

The contents of each chapter are as follows.

In Chap. I, after we review some known results and conjectures on zeta values (§1, §2), we formulate our generalized Iwasawa main conjecture in §3. Our conjecture (3.2.2) states that "p-adic zeta elements" (non-archimedean analogues of zeta values in $\mathbb{C}$) of p-adic Galois representations exist, and that for p-adic Galois representations coming from motives, they have a certain relationship (condition 3.2.2 (v)) with zeta values in $\mathbb{C}$. (Though some important cases of the condition 3.2.2 (v) are discussed in Chap. I §3, the full description of the condition 3.2.2 (v) is given in Chap. III §2 by using the theory of Fontaine and Perrin-Riou [FP_1][FP_2].) This conjecture 3.2.2 generalizes the conjectures on zeta values given in [BK][FP_1][FP_2][Ka_2].

We do not solve the conjecture. Instead we consider related interesting questions; the relationship between zeta values and explicit reciprocity laws, the relationship between p-adic zeta elements and p-adic zeta functions, etc. These are done in Chap. III.

In Chapter II, we study the relationship between the arithmetic of a p-adic field and the ring B_{dR} of Fontaine. In §1 we review the ring B_{dR} and the related theories of p-adic Galois representaions and p-adic periods. §2 and §3 are theoretically hearts of this paper. In

§2 we prove a generalized explicit reciprocity law for Lubin-Tate groups. In §3 we formulate a conjecture called the local main conjecture, which is a local analogue of our generalized Iwasawa main conjecture. This conjecture says that a Galois representation of a p-adic local field with coefficients in a p-adic ring Λ has a p-adic ε-element, which is a basis of a certain Λ-module, and is an analogue of the ε-factor of an ℓ ($\neq$ p)-adic Galois representation.

In Chapter III §1, we consider the relationship between the explicit reciprocity law of Chap. II §2, and values of partial Riemann zeta functions and of Hecke L-series of quadratic imaginary fields, generalizing an old consideration about it by Coates-Wiles. In §2, we review the theory of Fontaine and Perrin-Riou [FP_1] [FP_2] and complete the description of the condition (v) in Chap. I, 3.2.2. In Chap. III §3, we consider the relationship between p-adic zeta elements and p-adic zeta functions. We show that a conjecture of Coates and Perrin-Riou on p-adic L-functions [CP] follows from our generalized Iwasawa conjecture. We also check the compatibility of our conjectures with the duality of motives.

The author was trying to make this paper into the lecture notes style which is good to beginners. Chap. I §1 and Chap. III §1.1 are written carefully according to this intention. He put some funny comments in Chap. I, 3.4.11 (about mouths), Chap. III, 1.2.8 (about a crane [Ki]), etc. (and put the above unusual beginning of the paper), since he likes joyful lectures. However he is afraid that, by the limitation of his ability and energy, other parts became not so kind and not so joyful (it is assumed in the rest of Chap. I (resp. in Chap. II) that the reader knows etale (resp. crystalline) cohomology theory well).

The author is thankful to S. Bloch who introduced him this fantastic field of zeta values and whose idea of Tamagawa numbers of motives ([BK]) was the base of all his studies on zeta values. He is thankful to K. Kimura, N. Kurokawa, S. Mochizuki, C. Nakayama, T. Saito and T. Uzawa for advice.

In this paper, for an abelian group A and an integer n, A/n denotes A/nA, and ${}_nA$ denotes $\mathrm{Ker}(n : A \longrightarrow A)$.

Chapter I. Motivations, the generalized Iwasawa main conjecture.

§1. Introduction, Riemann zeta function.

In §1.1 and §1.2, I explain what is interesting about zeta values (from my point of view) assuming a young beginner reads this paper. §1.3 is a review on values of partial Riemann zeta functions.

§1.1. The three phases of understandings of zeta values.

I think there are three phases in our understanding of values of zeta and L-functions. I describe them by taking Riemann zeta function $\zeta(s) = \sum_{n\geq 1} n^{-s}$ as an example.

1.1.1. <u>Rationality of values (the first phase).</u>

We first find that zeta values have some "rationality property".

Euler proved

(1) $\zeta(2) = \pi^2/6, \quad \zeta(4) = \pi^4/90, \quad \ldots$

(2) $\zeta(r)\pi^{-r} \in \mathbf{Q}$ <u>for even positive integers</u> r.

(Cf. Weil [We_2] to see how Euler was excited and happy in these discoveries.)

The Riemann zeta function converges absolutely when $\mathrm{Re}(s) > 1$, and is extended to a holomorphic function on $\mathbf{C} \setminus \{1\}$ with simple pole at $s = 1$, by analytic continuation. It satisfies the functional equation

$$\Lambda(s) = \Lambda(1-s) \quad \text{where} \quad \Lambda(s) = \Gamma(s/2)\pi^{-s/2}\zeta(s).$$

We obtain from this and the above (2);

(3) <u>Let</u> r <u>be an integer</u> ≤ 0. <u>Then</u>

(i) $\zeta(r) \in \mathbf{Q}$.

(ii) $\zeta(r) = 0$ <u>if and only if</u> $r < 0$ <u>and</u> r <u>is even</u>.

1.1.2. <u>p-adic properties of values (the second phase).</u>

We next find that zeta values have some p-adic properties. For Riemann zeta function, this was found first by Kummer. The theory of p-adic Riemann zeta functions of Kubota-Leopoldt says :

<u>For a prime number</u> p, <u>the function</u>

$$\{s \in \mathbf{Z} \,;\, s \leq 0\} \longrightarrow \mathbf{Q} \quad ; \quad s \longmapsto (1-p^{-s})\zeta(s)$$

<u>is extended uniquely to a continuous function</u>

$$(\varprojlim_n (\mathbf{Z}/p^{n-1}(p-1))) \setminus \{1\} \longrightarrow \mathbf{Q}_p .$$

(The set on the left hand side contains $\{s \in \mathbf{Z} \,;\, s \leq 0\}$ as a dense subset.)

In these phase 1 and phase 2, we find that zeta values are not just numbers in the real-complex analytic world, but are objects in some world which is beyond both the real-complex analytic world and p-adic world.

1.1.3. Arithmetic significances of values (the third phase).

In the third phase, we find that zeta values have profound relationship with important groups in arithmetic such as ideal class groups, Tate-Shafarevich groups, ...

Kummer is the first person also in this direction. Let $\alpha_N = \exp(2\pi i/N)$ for $N \geq 1$, and let C_N be the ideal class group of $\mathbf{Q}(\alpha_N)$. In his attempt to prove Fermat's last theorem, Kummer obtained the Kummer criterion:

Let p be an odd prime number. Then, p divides the order of C_p if and only if p divides the numerator of the rational number $\zeta(r)$ for some odd integer $r < 0$.

(A result of Kummer on Fermat's last theorem was the following (Cf. [Wa] Chap. 1, Chap. 9): If p is an odd prime number and p does not divide the order of C_p, then Fermat's equation $x^p + y^p = z^p$ has no solution such that $x, y, z \in \mathbf{Z}$, $x \neq 0$, $y \neq 0$, $z \neq 0$.)

A refinement of Kummer's criterion is a conjecture of Lichtenbaum proved by Mazur and Wiles [MW]:

Let p be an odd prime number, let $r \geq 2$ be an even integer, and let $\zeta(1-r)\{p\}$ be the p-power part of the rational number $\zeta(1-r)$. Then,

$$\zeta(1-r)\{p\} = \#(H^2(\mathbf{Z}[\tfrac{1}{p}], \mathbf{Z}_p(r)))\ \#(H^1(\mathbf{Z}[\tfrac{1}{p}], \mathbf{Z}_p(r)))^{-1}.$$

Here H^* are etale cohomology groups, (r) means the Tate twist, and # means the order of a set (these etale cohomology groups are finite [So_2]; furthermore, $H^q = 0$ for $q \neq 1, 2$). This formula is related to Kummer's criterion because the group $H^2(\mathbf{Z}[\frac{1}{p}], \mathbf{Z}_p(r))/p$ is embedded into C_p/p for any r.

The classical Iwasawa theory of cyclotomic fields is the generalization of Kummer criterion and also of the conjecture of Lichtenbaum, and it concerns with the $Gal(\mathbb{Q}(\alpha_N)/\mathbb{Q})$-module structures of C_N, not only with the orders of C_N (cf. §3.3.9).

§1.2. Hasse-Weil L-functions.

1.2.1. The relationship between Hasse-Weil L-functions of varieties X over number fields (cf. 2.2.3) and arithmetic properties of X is an attractive theme in number theory. Various results and conjectures have been known concerning "phases 1 and 2" for the values of Hasse-Weil L-functions at integer points. However we do not know much about "phase 3". In this §1.2, I explain it is expected that the theory of classical periods (resp. p-adic periods) is related to the "phases 1" (resp. "phases 2 and 3") for Hasse-Weil L-functions.

1.2.2. I introduce the theory of classical periods ((1) below) and the theory of p-adic periods ((2) below).

For a smooth proper variety X over a field K, the de Rham cohomology groups $H^m_{dR}(X/K)$ $(m \in \mathbb{Z})$ are defined by using differential forms on X, to be the hyper-cohomology groups $H^m(X, \Omega^{\cdot}_{X/K})$ where $\Omega^{\cdot}_{X/K}$ is the de Rham complex

$$[\mathcal{O}_X \xrightarrow{d} \Omega^1_{X/K} \xrightarrow{d} \cdots\cdots].$$

For a smooth scheme X over $\mathbb{C}$, there exists a canonical isomorphism

(1) $\quad H^m_{dR}(X/\mathbb{C}) \cong H^m(X(\mathbb{C}), \mathbb{C})$,

a mysterious relationship between differential forms and the Betti cohomology. Here $X(\mathbb{C})$ is the set of $\mathbb{C}$-rational points of X endowed with the classical topology.

On the other hand, for a smooth scheme X over a complete discrete valuation field K with perfect residue field k such that char(K) = 0 and char(k) = p > 0, there exists a canonical isomorphism obtained rather recently

(2) $$H^m_{dR}(X/K) \otimes_K B_{dR} \cong H^m_{et}(X \otimes_K \bar{K}, \mathbf{Q}_p) \otimes_{\mathbf{Q}_p} B_{dR},$$

a mysterious relationship between differential forms and p-adic etale cohomology. Here B_{dR} is a big field (big physically and big also in the history of mathematics) defined by Fontaine [Fo_1] (see Ch. II §1 for the definition of B_{dR}). The existence of the isomorphism (2) was conjectured by Fontaine in [Fo_1] (called the de Rham conjecture by him), proved under certain assumptions by him and Messing [Fo_1][FM], and proved in general by Faltings [Fa].

1.2.3. I explain how the classical isomorphism 1.2.2 (1) is related to values of Hasse-Weil L-functions. For example, consider the elliptic curve E over $\mathbf{Q}$

$$E = \{(x, y) ; y^2 = x^3 + 1\} \cup \{\infty\}.$$

The isomorphism 1.2.2 (1) for $X = E \otimes \mathbf{C}$ with m = 1

$$H^1_{dR}(X/\mathbf{C}) \xrightarrow{\cong} H^1(X(\mathbf{C}), \mathbf{C}) = \mathrm{Hom}(H_1(X(\mathbf{C}), \mathbf{Q}), \mathbf{C})$$

sends a differential form $\omega \in H^0(X, \Omega^1_{X/\mathbf{C}}) \subset H^1_{dR}(X/\mathbf{C})$ to its cohomology class

$$H_1(X(\mathbf{C}), \mathbf{Q}) \longrightarrow \mathbf{C} \;\; ; \;\; \gamma \longmapsto \int_\gamma \omega .$$

Thus the isomorphism 1.2.2 (1) is related to period integrals such as $\int_\gamma \omega$, and called the period isomorphism. By the analogy, the isomorphism 1.2.2 (2) is called the p-adic period isomorphism.

There is a mysterious relationship between a value of the Hasse-Weil L-function $L(H^1(E), s)$ (cf. 2.2.3) and a period integral: If we put

$$\omega = \frac{dx}{y} \in H^0(E, \Omega^1_{E/Q}) \subset H^0(E, \Omega^1_{E/Q}) \otimes C = H^0(X, \Omega^1_{X/C})$$

γ = the loop $E(\mathbb{R})$ (the orientation of γ is from lower to upper on the (x, y)-plane) $\in H_1(X(\mathbb{C}), Q)$ (note we took a "rational differential" and a rational homology class), then

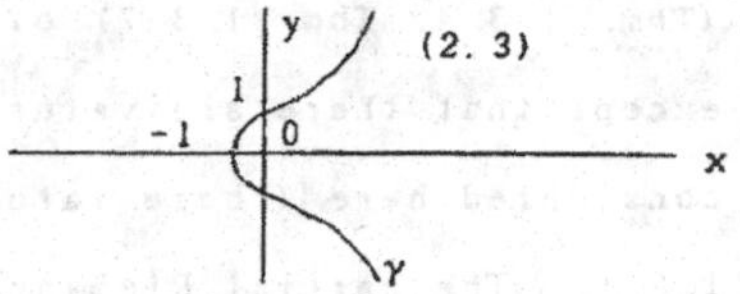

(2.3)

(3) $\quad L(H^1(E), 1) = 12^{-1} \int_\gamma \omega.$

(This equation (3) is not at all an obvious one, as the equation $\zeta(2) = \pi^2/6$ was not.)

Such "rationality property" is generalized to a conjecture of Deligne [De_3], and to a conjecture of Beilinson [Be_1][Ra] (cf. §2).

1.2.4. A well known principle in number theory is that, in the study of objects over a number field, local theories over local fields play important roles. I expect that for the Iwasawa theory of Hasse-Weil L-functions (in which we fix a prime number p and consider the relationship between zeta values and some important p-adic groups like the p-primary part of ideal class groups), the theory of p-adic periods over p-adic local fields 1.2.2 (2) is the most important local theory. Shortly speaking, we expect the following: zeta values are related to differential forms via the theory of period integrals, and then related to the etale cohomology theory (which is a deep arithmetic theory related to ideal class groups, Tate shfarevich groups, etc.) via the theory of p-adic periods.

1.2.5. The theory of p-adic periods is still new and there have been not so many studies of zeta values via B_{dR}. However great progress often happens at the place where two subjects with big distance meet, such as B_{dR} and zeta functions.

§1. 3. <u>Review on some values of Riemann zeta function.</u>

In this section, we review the phase 1 (Thm. 1. 3. 2) and the phase 2 (Thm. 1. 3. 4, Thm. 1. 3. 7) of Riemann zeta function in the sense of §1, except that there are values like $\zeta(3)$, $\zeta(5)$, etc. which are not considered here (these values are related to K-theory; see §2. 2).

1. 3. 1. The <u>partial Riemann zeta functions</u> $\zeta_{a(N)}(s)$ $(a, N \in \mathbb{Z}, N \geq 1)$ are defined by

$$\zeta_{a(N)}(s) = \sum_{\substack{n\geq 1 \\ n\equiv a \bmod N}} n^{-s}.$$

The function $\zeta_{a(N)}(s)$ absolutely converges when $\mathrm{Re}(s) > 1$. It is extended to a holomorphic function on $\mathbb{C} \setminus \{1\}$ which has a simple pole with residue $1/N$ at $s = 1$.

For $a \in \mathbb{Z}/N$, $\zeta_{a(N)}(s)$ means $\zeta_{b(N)}(s)$ with b any integer such that $b \bmod N = a$.

<u>Theorem</u> 1. 3. 2. <u>Let</u> $N \geq 1$ <u>be an integer and let</u> $a \in \mathbb{Z}/N$.

(1) $\zeta_{a(N)}(r) \in \mathbb{Q}$ <u>for any integer</u> $r \leq 0$.

(2) <u>Let</u> $\alpha = \exp(2\pi i a/N)$. <u>Then for any</u> $r \geq 1$,

$$(2\pi i)^{-r}(\zeta_{a(N)} + (-1)^r\zeta_{-a(N)})(r) \in \mathbb{Q}(\alpha),$$

<u>and an automorphism</u> σ <u>of</u> $\mathbb{Q}(\alpha)$ <u>sends this element to</u>

$$(2\pi i)^{-r}(\zeta_{ac(N)} + (-1)^r\zeta_{-ac(N)})(r) \quad \text{if } \sigma(\alpha) = \alpha^c.$$

The case $N = 1$ of Thm. 1. 3. 2 shows $\zeta(r) \in \mathbb{Q}\pi^r$ for even integers $r \geq 2$ and $\zeta(r) \in \mathbb{Q}$ for integers $r \leq 0$.

1. 3. 3. We fix some notations. Let R be a commutative ring over $\mathbb{Q}$, α a root of 1 in R, and let $r \leq 0$ be an integer.

(1) Define $\zeta(\alpha, r) \in R$ by

$$\zeta(\alpha, r) = \sum_{a\in\mathbb{Z}/N} \zeta_{a(N)}(r)\alpha^a$$

where N is any integer ≥ 1 such that $\alpha^N = 1$ (note $\zeta_{a(N)}(r) \in \mathbb{Q}$ by

1.3.2 (1)). Then $\zeta(\alpha, r)$ is independent of the choice of N.

(2) For a prime p, let

$$\zeta^{(p)}(\alpha, r) = \zeta(\alpha, r) - p^{-r}\zeta(\alpha^p, r) \in R .$$

(3) For an integer $c \geq 1$, let

$$\zeta_c(\alpha, r) = \zeta(\alpha, r) - c^{1-r}\zeta(\alpha^c, r) \in R .$$

(4) For p and c as in (2) and (3), respectively, let

$$\zeta_c^{(p)}(\alpha, r) = \zeta^{(p)}(\alpha, r) - c^{1-r}\zeta^{(p)}(\alpha^c, r) = \zeta_c(\alpha, r) - p^{-r}\zeta_c(\alpha^p, r) .$$

<u>Theorem</u> 1.3.4. <u>Let</u> p <u>be a prime number</u>, <u>let</u> $N \in \mathbb{Z}$, $N \geq 1$, <u>and let</u> c <u>be an integer which is prime to</u> pN. <u>In</u> (1) (3) (4), <u>let</u> R <u>be a ring over</u> $\mathbb{Q}_p$ <u>and let</u> $\alpha \in R$ <u>be an</u> N-th <u>root of</u> 1.

(1) <u>For any integer</u> $r \leq 0$, <u>we have</u>

$$\zeta_c(\alpha, r) \in \mathbb{Z}_{(p)}[\alpha] \quad \underline{\text{in}} \quad R.$$

(Here $\mathbb{Z}_{(p)} = \{\frac{n}{m} ;\ m, n \in \mathbb{Z},\ (p, m) = 1\} \subset \mathbb{Q}$.)

(2) <u>Let</u> $r, r' \in \mathbb{Z}$, $r' \leq r \leq 0$. <u>Let</u> $n \geq 0$ <u>and assume</u> $p^n | N$. <u>Then the additive map</u>

$$(\mathbb{Z}/p^n)[t]/(t^N - 1) \longrightarrow (\mathbb{Z}/p^n)[t]/(t^N - 1) \quad ; \quad t^a \longmapsto a^{r-r'}t^a$$

<u>sends</u> $\zeta_c(t, r) \bmod p^n$ <u>to</u> $\zeta_c(t, r') \bmod p^n$.

(3) <u>If</u> $r, r' \in \mathbb{Z}$, $r, r' \leq 0$ and $r \equiv r' \bmod (p-1)p^{n-1}$, <u>we have</u>

$$\zeta_c^{(p)}(\alpha, r) \equiv \zeta_c^{(p)}(\alpha, r') \mod p^n\mathbb{Z}_{(p)}[\alpha].$$

(4) <u>The function</u>

$$\{r \in \mathbb{Z} ;\ r \leq 0\} \longrightarrow \mathbb{Z}_{(p)}[\alpha] \quad ; \quad r \longmapsto \zeta_c^{(p)}(\alpha, r)$$

(resp. $\{r \in \mathbb{Z} ;\ r \leq 0\} \longrightarrow \mathbb{Q}[\alpha] \quad ; \quad r \longmapsto \zeta^{(p)}(\alpha, r)$)

<u>extends uniquely to a continuous function</u>

$$\varprojlim \mathbb{Z}/(p-1)p^n\mathbb{Z} \longrightarrow \mathbb{Z}_p[\alpha]$$

(resp. $(\varprojlim_n \mathbb{Z}/(p-1)p^n\mathbb{Z}) \setminus \{1\} \longrightarrow \mathbb{Q}_p[\alpha]$).

1.3.5. (1) Let R be a commutative ring. For a group G, we denote by

R[G] the group ring of G over R. For a pro-finite group G, we denote by R[[G]] the ring $\varprojlim_H R[G/H]$ where H ranges over all open subgroups of G.

(2) For a commutative ring R, let Q(R) be the total quotient ring of R, that is, $Q(R) = S^{-1}R$ where S is the set of all non-zero-divisors of R.

1.3.6. For an integer $r \leq 0$ and for an integer $N \geq 1$, the <u>Stickelberger element</u> $\theta_N(r) \in Q[(\mathbb{Z}/N)^\times]$ is defined by

$$\theta_N(r) = \sum_{a \in (\mathbb{Z}/N)^\times} \zeta_{a(N)}(r)\,[a]^{-1}.$$

Here we denoted the element a of $(\mathbb{Z}/N)^\times$ regarded as an element of the group ring, by [a] to avoid a confusion.

For a prime number p, let

$$\theta_{Np^\infty}(r) = \varprojlim_{n \geq 1} \theta_{Np^n}(r) \in Q[[(\mathbb{Z}/Np^\infty)^\times]]$$

where $(\mathbb{Z}/Np^\infty)^\times = \varprojlim_n (\mathbb{Z}/Np^n)^\times$.

Let $\chi_{cyclo} : (\mathbb{Z}/Np^\infty)^\times \longrightarrow \varprojlim_n (\mathbb{Z}/p^n)^\times = (\mathbb{Z}_p)^\times \subset \mathbb{Z}_p$ be the canonical map.

The following theorem is deduced easily from Thm. 1.3.4.

<u>Theorem</u> 1.3.7. <u>Let</u> p <u>be a prime number</u>, <u>and let</u> $N \geq 1$.

(1) <u>For any integer</u> $r \leq 0$, $\theta_{Np^\infty}(r)$ <u>belongs to the total quotient ring</u> $Q(\mathbb{Z}_p[[(\mathbb{Z}/Np^\infty)^\times]])$. <u>More precisely we have</u>

$$(1 - \chi_{cyclo}(\sigma)^{1-r}\sigma^{-1})\ \theta_{Np^\infty}(r) \in \mathbb{Z}_p[[(\mathbb{Z}/Np^\infty)^\times]] \quad \underline{\text{in}} \quad Q_p[[(\mathbb{Z}/Np^\infty)^\times]]$$

<u>for all</u> $\sigma \in (\mathbb{Z}/Np^\infty)^\times$.

(Note that $1 - \chi_{cyclo}(\sigma)^{1-r}\sigma^{-1}$ is a non-zero-divisor of $\mathbb{Z}_p[[(\mathbb{Z}/Np^\infty)^\times]]$ if σ is of infinite order. In fact, if σ is of infinite order, $\chi_{cyclo}(\sigma)^{1-r}$ is not a root of 1 but the images of σ

in $\mathbb{Q}_p[(\mathbb{Z}/Np^n)^\times]$ are roots of 1 for all $n \geq 0$, and so the images of $1 - \chi_{cyclo}(\sigma)^{1-r}\sigma^{-1}$ in $\mathbb{Q}_p[(\mathbb{Z}/Np^n)^\times]$ are non-zero-divisors.)

(2) <u>Let</u> r, r' <u>be integers</u> ≤ 0. <u>Then, the ring automorphism</u> $Q(\mathbb{Z}_p[[(\mathbb{Z}/Np^\infty)^\times]]) \xrightarrow{\cong} Q(\mathbb{Z}_p[[(\mathbb{Z}/Np^\infty)^\times]])$ <u>induced by</u>

$$\mathbb{Z}_p[[(\mathbb{Z}/Np^\infty)^\times]] \xrightarrow{\cong} \mathbb{Z}_p[[(\mathbb{Z}/Np^\infty)^\times]] \quad ; \quad \sigma \longmapsto \chi_{cyclo}(\sigma)^{r'-r}\sigma$$

$(\sigma \in (\mathbb{Z}/Np^\infty)^\times)$ <u>sends</u> $\theta_\infty(r)$ to $\theta_\infty(r')$.

The rest of §1.3 is devoted to (classical) proofs of the above theorems.

1.3.8. For $r \geq 1$, define the rational function

$$g_r(t) \in \mathbb{Z}[t, (1-t)^{-1}] \subset \mathbb{Q}(t) \quad \text{by}$$

$$g_r(t) = -(t\frac{d}{dt})^r(\log(1-t)).$$

Here for any $f \in \mathbb{Q}(t)$ and for $r \geq 1$, $(t\frac{d}{dt})^r(\log(f))$ denotes

$$(t\frac{d}{dt})^{r-1}(\frac{f^{-1}df}{t^{-1}dt})$$

where $(t\frac{d}{dt})^{r-1}$ is the $(r-1)$-fold iteration of the operator

$$t\frac{d}{dt} \; : \; \mathbb{Q}(t) \longrightarrow \mathbb{Q}(t) \; ; \; f \longmapsto t\frac{df}{dt}.$$

(The letter log is used in the definition of g_r, but we are working purely algebraically). We have :

(1) $g_1(t) = t(1-t)^{-1}, \quad g_2(t) = t(1-t)^{-2}, \quad \ldots.$

(2) $g_r(t) = \sum_{n\geq 1} n^{r-1}t^n$ <u>in</u> $\mathbb{Q}[[t]]$.

1.3.9. If we neglect the convergence, we would have the following (wrong) proofs of some parts of the theorems 1.3.2, 1.3.4, 1.3.7 in the following way. We have

(*) $\zeta(1-r) = \sum_{n\geq 1} n^{r-1} = g_r(1)$ by 1.3.8 (2).

Thus $\zeta(1-r)$ is the value at $t = 1$ of the rational function $g_r(t)$ over $\mathbb{Q}$, and hence is rational. By $g_r(t) - p^{r-1}g_r(t^p) = \sum_{\substack{n\geq 1 \\ (n,p)=1}} n^{r-1}t^n$ and by $n^{r-1} \equiv n^{r'-1} \mod p^n$ if $(n, p) = 1$ and $r \equiv r' \mod (p-1)p^{n-1}$, $g_r(t) - p^{r-1}g_r(t^p) \mod p^n$ depends only on $r \mod (p-1)p^{n-1}$ in $\mathbb{Z}_p[[t]]$ (this part is a correct argument), and hence (take the value at $t = 1$ neglecting the convergence) $\zeta^{(p)}(r) \mod p^n$ depends only on $r \mod (p-1)p^{n-1}$.

These wrong proofs are in fact "almost correct"; the true proofs of the theorems are given by modifying these wrong proofs, as in the following.

<u>Lemma</u> 1.3.10. <u>For</u> $r, c \in \mathbb{Z}$ <u>such that</u> $r \geq 1$ <u>and</u> $c \neq 0$, <u>let</u>

$$g_{r,c}(t) = g_r(t) - c^r g_r(t^c) \in \mathbb{Q}(t).$$

<u>Let</u> $N \in \mathbb{Z}$, $N \geq 1$, $(c, N) = 1$. <u>Then</u>, $g_{r,c}(t)$ <u>has no poles at</u> N-th <u>roots of</u> 1.

<u>Proof</u>. This follows from

$$g_{r,c}(t) = \left(t\frac{d}{dt}\right)^r \log\{(1-t^c)(1-t)^{-1}\}$$

and the fact $(1-t^c)(1-t)^{-1}$ has not zero or pole at N-th roots of 1.

<u>Lemma</u> 1.3.11. <u>For an integer</u> $c \neq 0$, <u>let</u>

$$R_c = \mathbb{Z}[t^{\pm 1}, \{(1-t^c)(1-t)^{-1}\}^{\pm 1}].$$

(1) $g_{r,c}(t) \in R_c$.

(2) <u>Let</u> $N \in \mathbb{Z}$, $N \geq 1$ <u>and assume</u> $(c, N) = 1$. <u>Then the image of</u> R_c <u>in</u> $\mathbb{Q}[t]/(t^N - 1)$ <u>is contained in</u> $\mathbb{Z}[\frac{1}{c}, t]/(t^N - 1)$. <u>In particular</u> (<u>by</u> (1)), <u>the image of</u> $g_{r,c}(t)$ <u>in</u> $\mathbb{Q}[t]/(t^N - 1)$ <u>is contained in</u> $\mathbb{Z}[\frac{1}{c}, t]/(t^N - 1)$.

<u>Proof</u>. (1) follows from the expression of $g_{r,c}(t)$ given in the proof of 1.3.10.

The statement (2) for c and that for $-c$ are equivalent, so we may assume $c > 0$. It is sufficient to show that $(1-t^c)(1-t)^{-1}$ is invertible in $\mathbb{Z}[\frac{1}{c}, t]/(t^N - 1)$. Let I be the ideal of $\mathbb{Z}[t]$ generated by $(1 - t^c)(1 - t)^{-1}$ and $1 - t^N$. It is sufficient to prove $c \in I$. Since $t^c \equiv t^N \equiv 1 \bmod I$, we have $t \equiv 1 \bmod I$ and hence $1 - t \in I$. Since $(1 - t^c)(1 - t)^{-1} \bmod (1 - t) = c$, we have $c \in I$ as desired.

1.3.12. We start analytic arguments. For $(t, s) \in \mathbb{C} \times \mathbb{C}$ which satisfies either $|t| \leq 1$ and $\mathrm{Re}(s) > 1$, or $|t| < 1$, define

$$\zeta(t, s) = \sum_{n \geq 1} t^n n^{-s}$$

which converges absolutely.

The following lemma will be proved in 1.3.20 later.

Lemma 1.3.13. Let $F = \{t \in \mathbb{R}\ ;\ t \geq 1\}$.

(1) $\zeta(t, s)$ is extended to a holomorphic function on $(\mathbb{C} \setminus F) \times \mathbb{C}$.

(2) For any integer $c \geq 1$, the function

$$\zeta_c(t, s) \underset{\mathrm{def}}{=} \zeta(t, s) - c^{1-s}\zeta(t^c, s)$$

in (t, s) is extended to a holomorphic function on

$$(\mathbb{C} \setminus \{t \in \mathbb{C}\ ;\ t \notin F,\ t^c \in F\}) \times \mathbb{C}.$$

Corollary 1.3.14. (1) If α is a root of 1 in $\mathbb{C}$ and if $\alpha \neq 1$, $s \longmapsto \zeta(\alpha, s)$ is extended to a holomorphic function on $\mathbb{C}$.

(2) If α is a root of 1 in $\mathbb{C}$ and $c \geq 1$ is an integer which is prime to the order N of α, then $s \longmapsto \zeta_c(\alpha, s)$ is extended to a holomorphic function on $\mathbb{C}$.

(3) (A special case of (2).) For any integer $c \geq 1$, the map $s \longmapsto \zeta_c(1, s) = (1 - c^{1-s})\zeta(s)$ is extended to a holomorphic function on $\mathbb{C}$.

From 1.3.14 and from $\zeta(\alpha, s) = \sum_{a \in \mathbb{Z}/N} \zeta_{a(N)}(s)\alpha^a$ (α is a root of 1 of order N), we see easily that $\zeta_{a(N)}(s)$ is extended to a meromorphic function on $\mathbb{C}$ which is holomorphic at $s \neq 1$.

Lemma 1.3.15. (1) $g_r(t) = \zeta(t, 1-r)$ for any integer $r \geq 1$ and for any $t \in \mathbb{C} \setminus F$.

(2) Let $c \in \mathbb{Z}$, $c \geq 1$. Then, $g_{r,c}(t) = \zeta_c(t, 1-r)$ for any integer $r \geq 1$ and for any $t \in \mathbb{C} \setminus \{t \in \mathbb{C} \ ; \ t \notin F, \ t^c \in F\}$.

Proof. These equations hold when $|t| < 1$ and hence hold for t as above.

Corollary 1.3.16. Let $c, N \in \mathbb{Z}$, $c, N \geq 1$ and $(c, N) = 1$. Let

$$\sigma_c : \mathbb{C}[t]/(t^N - 1) \longrightarrow \mathbb{C}[t]/(t^N - 1)$$

be the ring homomorphism $t \longmapsto t^c$ over $\mathbb{C}$. Then for $r \in \mathbb{Z}$, $r \geq 1$, the image of $\sum_{a \in \mathbb{Z}/N} \zeta_{a(N)}(1-r)t^a$ under

$$1 - c^r\sigma_c \ : \ \mathbb{C}[t]/(t^N - 1) \longrightarrow \mathbb{C}[t]/(t^N - 1)$$

coincides with $g_{r,c}(t)$ mof $(t^N - 1)$.

1.3.17. We prove Thm. 1.3.2 (1). Let $r, N \in \mathbb{Z}$, $r, N \geq 1$. Let c be an integer such that $(c, N) = 1$. By 1.3.16, the image of

$$\sum_{a \in \mathbb{Z}/N} \zeta_{a(N)}(1-r)t^a \text{ under } 1 - c^r\sigma_c : \mathbb{C}[t]/(t^N - 1) \longrightarrow \mathbb{C}[t]/(t^N - 1)$$

is containd in $\mathbb{Q}[t]/(t^N - 1)$. Take c such that $c \geq 2$. Then $1 - c^r\sigma_c$ is bijective and induces a bijection $\mathbb{Q}[t]/(t^N - 1) \xrightarrow{\cong} \mathbb{Q}[t]/(t^N - 1)$. This proves $\sum_{a \in \mathbb{Z}/N} \zeta_{a(N)}(1-r)t^a \in \mathbb{Q}[t]/(t^N - 1)$.

1.3.18. We can deduce Thm. 1.3.2 (2) from 1.3.2 (1) by using the following relation between values of partial Riemann zeta functions at positive integers and those at negative integers. (Cf. Chapter III §1.1 for another proof of 1.3.2 (2) not using values at negative integers.)

Let $a, N \in \mathbb{Z}$, $N \geq 1$, and let $\alpha = \exp(2\pi i a/N)$. Let $r \in \mathbb{Z}$. If $r = 0$, assume $\alpha \neq 1$. Then,

$$2^{-1}(\zeta(\alpha,\ 1-r) + (-1)^r\zeta(\alpha^{-1},\ 1-r))$$
$$= \left(\frac{-2\pi i}{N}\right)^{-r} \lim_{s\to r} \Gamma(s)(\zeta_{a(N)}(s) + (-1)^r\zeta_{-a(N)}(s)).$$

<u>The left hand side is equal to</u> $\zeta(\alpha,\ 1-r)$ <u>if</u> $r \geq 2$.

This follows from the functional equations of Dirichlet L-series ([Wa] Chap. 4).

1.3.19. We prove Thm. 1.3.4. We have obtained

(*) $\zeta_c(t,\ 1-r) = g_{r,c}(t)$ in $\mathbb{Q}[t]/(t^N - 1)$.

1.3.4 (1) follows from (*) and 1.3.11. 1.3.4 (2) follows from (*) and $(t\frac{d}{dt})(g_r) = g_{r+1}$. To prove 1.3.4 (3), by replacing N by Np^n, we may assume $p^n | N$. Then 1.3.4 (3) is deduced from 1.3.4 (2), $(t\frac{d}{dt})^r(t^a) = a^r t^a$ for $r \geq 0$ and from the following elementary fact: If $a \in \mathbb{Z}_p^\times$ and $r \equiv 0 \bmod (p-1)p^{n-1}$, then $a^r \equiv 1 \bmod p^n$. Finally 1.3.4 (4) follows from 1.3.4 (3).

1.3.20. We prove 1.3.13. For $(t,\ s) \in (\mathbb{C} \setminus F) \times \mathbb{C}$ with F as in 1.3.13, let

(1) $H(t,\ s) = (2\pi i)^{-1} \int_\gamma (1 - \exp(x)t)^{-1}\exp(x)tx^s x^{-1}dx$,

where γ is the following route in $\mathbb{C}$.

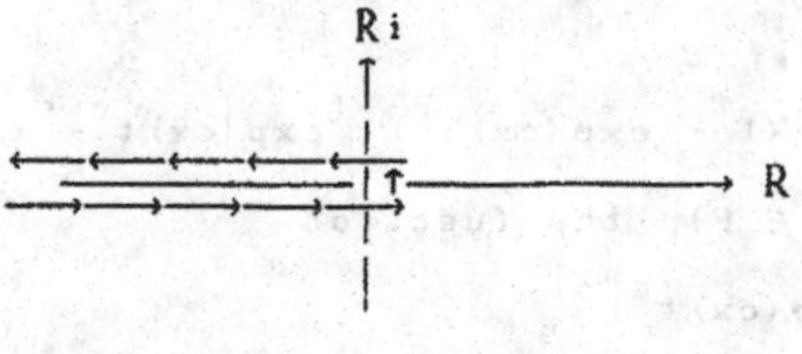

Here, x^s is defined to be $\exp(s\log(x))$ with log the principal branch on $\mathbb{C} \setminus \{x \in \mathbb{R} : x \leq 0\}$. (For each $(t,\ s)$, γ is taken very near to $\{x \in \mathbb{R} : x \leq 0\}$ so that the function $1 - \exp(x)t$ in x does not have zero on γ and on the domain inside γ.)

Then, $H(t,\ s)$ is a holomorphic function on $(\mathbb{C} \setminus F) \times \mathbb{C}$.

<u>Lemma</u> 1.3.21. (1) $\zeta(t,\ s) = \Gamma(1-s)H(t,\ s)$ <u>if</u> $|t| \leq 1$, $t \neq 1$ <u>and</u>

$\mathrm{Re}(s) > 1$.

(2) <u>For any integer</u> $c \geq 1$, $H(t, s) - c^{1-s}H(t^c, s)$ <u>is extended to a holomorphic function on a neighbourhood of</u> $F \times \mathbf{C}$ in $\mathbf{C} \times \mathbf{C}$.

<u>Proof</u> of (1). If $|t| \leq 1$, $t \neq 1$ and $\mathrm{Re}(s) > 1$, $H(t, s)$ is equal to

$$(2\pi i)^{-1} \int_{-\infty}^{0} (1 - \exp(x)t)^{-1}\exp(x)t(-x)^s\exp(-\pi i s)x^{-1}dx$$

$$+ (2\pi i)^{-1} \int_{0}^{-\infty} (1 - \exp(x)t)^{-1}\exp(x)t(-x)^s\exp(\pi i s)x^{-1}dx$$

$$= (2\pi i)^{-1}(\exp(\pi i s) - \exp(-\pi i s)) \int_0^{\infty} (1 - \exp(-y)t)^{-1}\exp(-y)ty^s y^{-1}dy$$

$$= (2\pi i)^{-1}(\exp(\pi i s) - \exp(-\pi i s)) \sum_{n\geq 1} \int_0^{\infty} \exp(-ny)t^n y^s y^{-1}dy$$

$$= (2\pi i)^{-1}(\exp(\pi i s) - \exp(-\pi i s))\Gamma(s) \sum_{n\geq 1} t^n n^{-s}$$

$$= \Gamma(1-s)^{-1}\zeta(t, s).$$

Thus $\zeta(t, s)$ is extended to a meromorphic function on $(\mathbf{C} \setminus F) \times \mathbf{C}$. We show that this function is holomorphic. The pole divisor of this function is contained in the pole divisor

$$(\mathbf{C} \setminus F) \times \{r \in \mathbf{Z} ;\ r \geq 1\}$$

of $\Gamma(1-s)$. However $\zeta(t, s)$ does not have a pole on $\{t \in \mathbf{C} ;\ |t| < 1\} \times \mathbf{C}$ and hence can not have a pole.

<u>Proof</u> of (2). By replacing t by t^c and x by cx in the integration 1.3.20 (1), we have

$$c^{1-s}H(t^c, s) = (2\pi i)^{-1} \int_{\gamma} c(1 - \exp(cx)t^c)^{-1}\exp(cx)t^c x^s x^{-1}dx.$$

For t in $\mathbf{C} \setminus \{t \in \mathbf{C} ;\ t \notin F,\ t^c \in F\}$, the function

$$(1 - \exp(x)t)^{-1} - c(1 - \exp(cx)t^c)^{-1},$$

in x is holomorphic on a neighbourhood of $\{x \in \mathbf{R} ;\ x \leq 0\}$.

§2. <u>Hasse-Weil L-functions</u>.

In this §2, after a preliminary on determinant modules (§2.1), we review conjectures concerning the phase 1 and phase 3 (in the sense of §1.1) for the values of Hasse-Weil zeta functions at integer points in special cases. (If the Hasse-Weil L-function has zero or pole at the integer point in problem, we are interested in the first non-zero coefficient of the Laurent expansion of the Hasse-Weil L-function, and call it also the "zeta value".) The phase 1 is formulated in the famous Beilinson conjectures ([Be_1][Ra]), which saids that zeta values are expressed modulo $\mathbf{Q}^{\times}$ by the regulator maps of K-theory and period integrals. To simplify the review, we consider in this §2 two extreme cases: The case where K-theory appears but period integrals do not appear (§2.2), and the case where period integrals appear but K-theory does not appear which is called the critical case (§2.3). (In the critical case, Beilinson conjectures are reduced to an older conjecture of Deligne [De_3].) In fact to simplify the description more, we restrict ourselves in §2.3 to a special case of the Birch and Swinnerton-Dyer conjecture. In the case considered in §2.2 (resp. §2.3), conjectures in [BK][FP_1][FP_2][Ka_2] on the phase 3 for Hasse-Weil L-functions say that zeta values are related to the etale cohomology by the Chern class maps in K-theory (resp. by the theory of p-adic periods) as follows:

zeta values	—— regulator maps ——	K-theory
| period integrals		| Chern class maps
differential forms	—— p-adic periods ——	etale cohomology

These conjectures will be included in the generalized Iwasawa main conjecture in §3 (cf. 2.3.5).

In this section, we do not discuss the proofs of the results.

§2.1. Determinant modules.

We give here some comments on determinant modules. In this §2.1, Λ denotes a commutative ring.

2.1.1. A perfect complex over Λ is an object of the derived category of Λ-modules which is represented by a bounded complex

$$(\dots \longrightarrow P^q \longrightarrow P^{q+1} \longrightarrow \dots)$$

such that P^q are finitely generated projective Λ-modules for all q.

2.1.2. Let $\mathcal{L}$, $\mathcal{C}$, $\mathcal{E}$ be the following categories. An object of $\mathcal{L}$ is a pair (L, r) where L is an invertible Λ-module and r is a locally constant function $\mathrm{Spec}(\Lambda) \longrightarrow \mathbb{Z}$. A morphism $(P, r) \longrightarrow (P', r')$ is an isomorphism $P \xrightarrow{\cong} P'$ in the case $r = r'$, and does not exist if $r \neq r'$. An object of $\mathcal{C}$ is a perfect complex over Λ (2.1.1) and an object of $\mathcal{E}$ is an exact sequence

$$0 \longrightarrow F'^{\cdot} \longrightarrow F^{\cdot} \longrightarrow F''^{\cdot} \longrightarrow 0$$

of complexes of Λ-modules (they are really comlexes, not regarded as objects of the derived category) which are perfect when regarded as objects of the derived category. Morphisms in $\mathcal{C}$ and $\mathcal{E}$ are isomorphisms (in the evident sense).

Knudsen and Mumford have shown in [KM] that there is a pair $(\det_\Lambda, i)$, which is unique up to canonical isomorphisms, of a functor $\det_\Lambda : \mathcal{C} \longrightarrow \mathcal{L}$ and an isomorphism $i : f \xrightarrow{\cong} g$ where f and g are functors from $\mathcal{E}$ to $\mathcal{L}$ defined by

$$f(0 \longrightarrow F'^{\cdot} \longrightarrow F^{\cdot} \longrightarrow F''^{\cdot} \longrightarrow 0) = \det_\Lambda(\bar{F}^{\cdot})$$

$$g(0 \longrightarrow F'^{\cdot} \longrightarrow F^{\cdot} \longrightarrow F''^{\cdot} \longrightarrow 0) = \det_\Lambda(\bar{F}'^{\cdot}) \otimes_\Lambda \det_\Lambda(\bar{F}''^{\cdot}),$$

satisfying certain conditions. Here $\bar{F}^{\cdot}$ etc. means the object of the derived category associated to $F^{\cdot}$ etc., and $\otimes_\Lambda$ means the functor $\mathcal{L} \times \mathcal{L} \longrightarrow \mathcal{L}$ defined by $(L, r) \otimes_\Lambda (L', r') = (L \otimes_\Lambda L', r + r')$. The "certain conditions" include the following two conditions.

(i) For a finitely generated projective Λ-module P regarded as an object of the derived category concentrated in degree 0, $\det_\Lambda(P)$ is the pair (L, r) defined as follows. If $P^\sim$ denotes the quasi-coherent sheaf on Spec(Λ) corresponding to P, which is locally free of finite rank, r is the rank of $P^\sim$ and L is the Λ-module corresponding to the r-th exterior power of $P^\sim$.

(ii) For an object $E = (0 \longrightarrow P' \longrightarrow P \xrightarrow{\pi} P'' \longrightarrow 0)$ of $\mathcal{S}$ such that P, P', P'' are finitely generated projective Λ-modules regarded as complexes concentrated in degree 0, i_E is the isomorphism which is characterized localy on Spec(Λ) by

$$x_1 \wedge \ldots \wedge x_r \longmapsto (x_1 \wedge \ldots \wedge x_{r'}) \wedge (\pi(x_{r'+1}) \wedge \ldots \wedge \pi(x_r))$$

(r, r', r'' are the ranks of P', P, P'', respectively, $x_1, \ldots, x_r$ are local sections of $P'^\sim$, and $x_{r'+1}, \ldots, x_r$ are local sections of $P^\sim$).

2.1.3. We fix a pair ($\det_\Lambda$, i) in 2.1.2. In this paper, by changing the notation, we denote by $\det_\Lambda$ the functor from $\mathcal{C}$ to the category of invertible Λ-modules which is obtained from the original $\det_\Lambda$ by forgetting the r in (L, r).

For an object C of $\mathcal{C}$ represented by a bounded complex as in 2.1.1, we have a canonical isomorphism

$$\det_\Lambda(C) \cong \bigotimes_{q \in \mathbb{Z}} (\det_\Lambda(P^q))^{(-1)^q}.$$

If the cohomology modules $H^q(C)$ are perfect (when regarded as objects of the derived category concentrated in degree 0), we have a canonical isomorphism

$$\det_\Lambda(C) \cong \bigotimes_{q \in \mathbb{Z}} (\det_\Lambda(H^q(C)))^{(-1)^q}$$

([KM] Rem. b) after Thm. 2).

When one treats these det modules, one should be careful because some canonical isomorphisms are soon multiplied by -1. However, in such delicate situation, we will work modulo sign so this point is not so serious in this paper. (I hope the ambiguity of sign in this paper will be eliminated in the future.)

2.1.4. Let Λ be a regular noetherian ring and let F be a finitely generated Λ-module. Then F has a finite resolution by finitely generated projective Λ-modules. Hence F is a perfect complex when regarded as an object of the derived category concentrated in degree 0, so $\det_\Lambda(F)$ is defined.

Let $Q(\Lambda)$ be the total quotient ring of Λ and assume $F \otimes_\Lambda Q(\Lambda) = 0$ (that is, F is a torsion R-module). Then let $\mathrm{char}_\Lambda(F)$ be the invertible ideal of Λ characterized by the following property: For any prime ideal p of Λ of height one, $\mathrm{char}_\Lambda(F)_p$ coincides with $(pF_p)^n$ where $n = \mathrm{length}_{\Lambda_p}(F_p)$.

2.1.5. ([KM] Thm. 3 (vi).) <u>Let Λ be a regular noethrian ring and let F be a finitely generated torsion Λ-module. Then, with the notation as above, the image of the composite homomorphism</u>

$$\det_\Lambda(F) \subset \det_\Lambda(F) \otimes_\Lambda Q(\Lambda) \cong \det_\Lambda(F \otimes_\Lambda Q(\Lambda)) = \det_{Q(\Lambda)}((0)) = Q(\Lambda)$$

<u>coincides with the fractional ideal</u> $\mathrm{char}(F)^{-1}$.

<u>Proof</u>. Replacing Λ by the local ring Λ_p for p prime ideals of height 1, we may assume that Λ is a discrete valuation ring. Furthermore we may assume $F = \Lambda/a\Lambda$ for a non-zero element a of Λ. Then, $\Lambda/a\Lambda \cong (\underset{\text{deg.}\,-1}{a\Lambda} \longrightarrow \underset{\text{deg.}\,0}{\Lambda})$ in the derived category and hence $\det_\Lambda(\Lambda/a\Lambda) = \Lambda \otimes_\Lambda (a\Lambda)^{-1} = a^{-1}\Lambda$ in its $\otimes_\Lambda Q(\Lambda)$.

§2.2. K-theory and zeta values.

2.2.1. For any scheme X, Quillen's K-group $K_n(X)$ $(n \geq 0)$ are defined ([Qu]). If X is the Spec of the integer ring O_K of a number field K, we have

$$K_0(X) \cong \mathbb{Z} \oplus \mathrm{Pic}(O_K), \qquad K_1(X) \cong (O_K)^{\times}.$$

The ideal class group $\mathrm{Pic}(O_K)$ and the unit group $(O_K)^{\times}$ play central roles in number theory. So it is natural to expect that the higher K-groups K_n are also important, and today we know this is really true. In this §2.2, we review first the classical relationship between the regulator map on $(O_K)^{\times}$ and the property of the Dedekind zeta function at $s = 0$. We then review a special case of known conjectures on the phase 1 (the famous Beilinson conjecture) and on the phase 3 (conjectures in [BK][FP_1][FP_2][Ka_2]; we follow the formulation of [FP_1][FP_2]) for values of Hasse-Weil L-functions. The case considered here is the one where period integrals nor p-adic periods do not appear and K-theory is sufficient to relate (conjecturally) zeta values to etale cohomology.

2.2.2. Recall that in the classical theory of regulators, the unit groups and the ideal class groups are related to Dedekind zeta functions as follows: <u>We have an isomorphism of $\mathbb{R}$-vector spaces</u>

$$(1) \qquad (\log(\| \ \|_v))_v \ : \ (O_K)^{\times} \otimes \mathbb{R} \xrightarrow{\cong} (\oplus_v \mathbb{R})^0$$

<u>where</u> v <u>ranges over all Archimedean place of</u> K,

$$(\oplus_v \ )^0 = \{ (a_v)_v \in \oplus_v \ ; \ \sum_v a_v = 0 \}.$$

<u>and</u> $\| \ \|_v$ <u>is the normalized absolute value at</u> v. A positive real number R_K called the regulator of K is defined to be the volume of the image of $(O_K)^{\times}$ in $(\oplus_v \mathbb{R})^0$ with respect to the lattice $(\oplus_v \mathbb{Z})^0$. We

have

(2) $$\lim_{s \to 0} s^{-e} \zeta_K(s) = -\frac{R_K h_K}{w_K}$$

<u>where</u> $\zeta_K(s)$ <u>is the Dedekind zeta function of</u> K, $e = r_1 + r_2 - 1$ <u>with</u> r_1 (resp. r_2) <u>the number of real</u> (resp. <u>complex</u>) <u>places of</u> K, $h_K = \#(\mathrm{Pic}(O_K))$ <u>is the class number of</u> K, <u>and</u> w_K <u>is the number of roots of</u> 1 <u>in</u> K.

The property $R_K^{-1} \lim_{s \to 0} s^{-e} \zeta_K(s) \in \mathbb{Q}$ is the first phase for the property of $\zeta_K(s)$ at $s = 0$, and the formula (2) is the third phase.

Note that the formula (2) (up to sign) can be translated as follows in terms of determinant modules. <u>There exists an element</u> z <u>of</u> $\det_{\mathbb{Q}}((O_K)^\times \otimes \mathbb{Q})$ <u>which is sent to</u> $\lim_{s \to 0} s^{-e} \zeta_K(s)$ <u>times a</u> $\mathbb{Z}$-<u>basis</u> of $\det_{\mathbb{Z}}((\oplus_v \mathbb{Z})^0)$ by

$$\det_{\mathbb{Q}}((O_K)^\times \otimes \mathbb{Q}) \otimes \mathbb{R} \xrightarrow[\text{by (1)}]{\cong} \det_{\mathbb{Z}}((\oplus_v \mathbb{Z})^0) \otimes \mathbb{R}$$

(this is the first phase part), and

(3) $$[\det_{\mathbb{Z}}((O_K)^\times) : \mathbb{Z}\cdot z] = h_K .$$

Here $\det_{\mathbb{Z}}((O_K)^\times)$ and $\mathbb{Z}z$ are regarded as $\mathbb{Z}$-lattices of the one-dimensional $\mathbb{Q}$-vector space $\det_{\mathbb{Q}}((O_K)^\times \otimes \mathbb{Q})$, and for such two lattices L and L', [L : L'] denotes $\#(L/L'')\#(L'/L'')^{-1}$ where $L'' = L \cap L'$.

2.2.3. Now we consider Hasse-Weil L-functions.

We first recall the L-functions of <u>p-adic Galois representations</u>. Let $\bar{\mathbb{Q}}$ be an algebraic closure of $\mathbb{Q}$. Consider a 4-ple (p, S, V, A) where;

p is a prime number,

S is a finite set of prime numbers containing p,

V is a finite dimensional $\mathbf{Q}_p$-vector space endowed with a continuous action of $\mathrm{Gal}(\bar{\mathbf{Q}}/\mathbf{Q})$ which is unramified outside S, and

A is a commutative ring over Q which is a finite product of finite extensions of Q and which is acting on V commuting with the action of $\mathrm{Gal}(\bar{\mathbf{Q}}/\mathbf{Q})$.

Let $\Lambda = A \otimes \mathbf{Q}_\ell$. For a prime number $\ell \notin S$, let

$$P_{\Lambda,\ell}(V, t) = \det_\Lambda(1 - \sigma_\ell^{-1} t \; ; \; V) \in \Lambda[t]$$

where σ_ℓ is an arithmetic Frobenius of ℓ in $\mathrm{Gal}(\bar{\mathbf{Q}}/\mathbf{Q})$ and $\det_\Lambda$ is the determinant over Λ. (For a homomorphism $h : V \longrightarrow V$ over Λ, $\det_\Lambda(h \; ; V)$ denotes the element a of Λ such that the map $\det_\Lambda(V) \longrightarrow \det_\Lambda(V)$ induced by h coincides with the multiplication by a.)

Assume the following (i)(ii) on (p, S, V, A) are satisfied.

(i) For any prime number $\ell \notin S$, $P_{\Lambda,\ell}(V, t)$ belongs to $A[t]$.

(ii) The product

$$L_{A,S}(V, s) = \prod_\ell P_{\Lambda,\ell}(V, \ell^{-s})^{-1} \qquad (s \in \mathbf{C})$$

where ℓ ranges over all prime numbers not contained in S, converges absolutely in $A \otimes \mathbf{C}$ if the real part Re(s) of s is sufficiently large.

Then, we call the $A \otimes \mathbf{C}$-valued function $L_{A,S}(V, s)$ (for s in the range of the convergence) the L-function of V with respect to A and S. If $A = \mathbf{Q}$, $L_{A,S}(V, s)$ is denoted by $L_S(V, s)$.

Some remarks:

(1) For a 4-ple (p, S, V, A) satisfying (i)(ii) and for $r \in \mathbf{Z}$, the 4-ple (p, S, V(r), S) also satisfies the conditions (i)(ii) (V(r) denotes the r-fold Tate twist of V), and

$$L_{A,S}(V(r), s) = L_{A,S}(V, s+r) .$$

(2) If $0 \longrightarrow V' \longrightarrow V \longrightarrow V'' \longrightarrow 0$ is an exact sequence of

4-ples with common (p, S, A) satisfying (i)(ii), we have

$$L_{A,S}(V, s) = L_{A,S}(V', s)\, L_{A,S}(V'', s) .$$

(3) There is a natural definition of $P_{\Lambda,\ell}(V, t) \in \Lambda[t]$ for $\ell \in S$. Let $\bar{Q}_\ell$ be an algebraic closure of Q_ℓ, take an embedding $\bar{Q} \longrightarrow \bar{Q}_\ell$, and consider the representation of $\mathrm{Gal}(\bar{Q}_\ell/Q_\ell)$ in V via the induced map $\mathrm{Gal}(\bar{Q}_\ell/Q_\ell) \longrightarrow \mathrm{Gal}(\bar{Q}/Q)$. The definition of $P_{\Lambda,\ell}(V, t)$ is;

$$P_{\Lambda,\ell}(V, t) = \det_\Lambda(1 - \sigma_\ell^{-1} t\ ;\ H^0(\mathrm{Gal}(\bar{Q}_\ell/Q_{\ell,ur}), V)) \quad \text{if } \ell \neq p,$$

$$P_{\Lambda,\ell}(V, t) = \det_\Lambda(1 - \varphi_\ell^{-1} t\ ;\ D_{crys}(V)) \quad \text{if } \ell = p .$$

Here $Q_{\ell,ur}$ denotes the maximal unramified extension of Q_ℓ in $\bar{Q}_\ell$, $\sigma_\ell \in \mathrm{Gal}(Q_{\ell,ur}/Q_\ell)$ denotes the arithmetic Frobenius, D_{crys} is a functor of Fontaine which will be reviewed in Chap. II, 1.3.1, and φ_ℓ is the Frobenius operator (loc. cit). Then $P_{\Lambda,\ell}(V, t)$ is independent of the choices of $\bar{Q}_\ell$ and $\bar{Q} \longrightarrow \bar{Q}_\ell$.

If $P_{\Lambda,\ell}(V, t) \in A[t]$ for all $\ell \in S$ and if the conditions (i)(ii) are satisfied, we define $L_A(V, s) = \prod_\ell P_{\Lambda,\ell}(V, \ell^{-s})^{-1}$ where ℓ ranges over all prime numbers (if A = Q, $L_A(V, s)$ is just denoted by L(V, s)).

(4) In the rest of §2, we consider only the case A = Q. We will consider general A in §3.

<u>Standard example</u> (<u>Hasse</u>-<u>Weil</u> L-<u>function</u>). Let X be a proper smooth scheme over Q, let p be a prime number, let

$$V_m = H^m_{et}(X \otimes_Q \bar{Q}, Q_p)$$

and let S be a finite set of prime numbers containing p and all primes at which X has bad reductions. Then, (p, S, V_m, Q) satisfies the conditions (i)(ii). The functions

$$L_S(H^m(X), s) \underset{\mathrm{def}}{=} L_S(V_m, s) \quad \text{for } m \in \mathbb{Z}$$

are called the Hasse-Weil L-functions of X and are independent of p.

It is conjectured that these functions have analytic continuations to whole $\mathbb{C}$ as meromorphic functions.

$L(V_m, s)$ is denoted by $L(H^m(X), s)$ if $P_{Q,\ell}(V_m, t)$ belongs to $\mathbb{Q}[t]$ and is independent of p for any $\ell \in S$ (these conditions are satisfied for example in the case X is an abelian variety over $\mathbb{Q}$, and conjectured to hold always.)

The definitions of L-functions in this 2.2.3 are generalized in a natural way to p-adic representations V of $\mathrm{Gal}(\bar{K}/K)$ and to proper smooth schemes X over K for number fields K (instead of σ_ℓ, we consider the arithmetic Frobenius of a non-zero prime ideal of O_K). However the L-function of such V (resp. X) coincides with the L-function of the p-adic representation of $\mathrm{Gal}(\bar{\mathbb{Q}}/\mathbb{Q})$ induced from V (resp. the L-function of X which we regard as a scheme over $\mathbb{Q}$ via $X \longrightarrow \mathrm{Spec}(K) \longrightarrow \mathrm{Spec}(\mathbb{Q})$). We consider mainly the case $K = \mathbb{Q}$ in this paper, since we do not lose generality for this reason, and since some arguments become simple in the case $K = \mathbb{Q}$.

2.2.4. We fix some notations. Let X be a smooth proper scheme over $\mathbb{Q}$, and fix integers $m, r \geq 0$.

Define the $\mathbb{Q}$-vector spaces Σ and Φ by

$$\Sigma = (H^m(X(\mathbb{C}), \mathbb{Q}) \otimes \mathbb{Q}(2\pi i)^{r-1})^+,$$

$$\Psi = \begin{cases} (K_{2r-m-1}(X) \otimes \mathbb{Q})^{(r)} & \text{if } 2r - m - 1 \neq 0, \\ CH^r(X)_{\hom\sim 0} \otimes \mathbb{Q} & \text{if } 2r - m - 1 = 0. \end{cases}$$

Here: Concerning Σ, $(\ )^+$ means the $\mathrm{Gal}(\mathbb{C}/\mathbb{R})$-fixed part where $\sigma \in \mathrm{Gal}(\mathbb{C}/\mathbb{R})$ acts by $\sigma \otimes \sigma$ on the tensor product. Concerning Ψ, $(\ )^{(r)}$ means the part on which the Adams operators ψ_k act by k^r for any $k \geq 1$ ($K_n(X) \otimes \mathbb{Q} = \bigoplus_{r\geq 0} (K_n(X) \otimes \mathbb{Q})^{(r)}$ holds. Cf. [Wo]). $CH^r(X)$

denotes the Chow group of codimension r cycles, and $(\ )_{hom\sim 0}$ means the subspace of elements which are homologically equivalent to zero.

Fix a prime number p, and let

$$V = H^m_{et}(X \otimes_{\mathbb{Q}} \bar{\mathbb{Q}}, \mathbb{Q}_p) .$$

For a finite set S of prime numbers containing p and all prime numbers at which X has bad reductions, let

$$H^i_S = H^i_{et}(\mathbb{Z}[\tfrac{1}{S}], V)$$

where we regard V as a smooth $\mathbb{Q}_p$-sheaf ([De$_4$] 1.1.1 b) on $\mathrm{Spec}(\mathbb{Z}[\frac{1}{S}])_{et}$ and we denote $H^i_{et}(\mathrm{Spec}(R), \ )$ by $H^i_{et}(R, \ )$ for a commutative ring R. (It is known that H^i_S is a finite dimensional $\mathbb{Q}_p$-vector space.) Define

$$H^i_{lim} = \varinjlim_S H^i_S .$$

2.2.5. Let X, m, r, p be as in 2.2.4. Assume

(i) $r > \inf(m, \dim(X))$.

We have two canonical maps

$$a : \Psi \otimes \mathbb{R} \longrightarrow \Sigma \otimes \mathbb{R}$$

$$b : \Psi \otimes \mathbb{Q}_p \longrightarrow H^1_{lim}$$

where a is the regulator map ([Be$_1$]) and b is the Chern class map ([So$_1$]). (Without the assumption (i), the target group of the regulator map is a certain quotient group of $\Sigma \otimes \mathbb{R}$, cf. [Be$_1$].

For example, if X = Spec(K) for a number field K and (m, r) = (0, 1), then

$$\Sigma = \oplus_v \mathbb{Q}$$

where v ranges over all Archimedean places of K,

$$\Psi = K^\times \otimes \mathbb{Q}, \qquad H^i_S = H^i_{et}(O_K[\tfrac{1}{S}], \mathbb{Q}_p(1)) .$$

$a : \Psi \otimes \mathbb{R} \longrightarrow \Sigma \otimes \mathbb{R}$ is the regulator map $x \longmapsto (\log(\|x\|_v))_v$, and

$b : \Psi \otimes Q_p \longrightarrow H^1_{lim}$ is the map induced by the connecting homomorphisms $(O_K[\frac{1}{S}])^\times \longrightarrow H^1_{et}(O_K[\frac{1}{S}], Z/p^n(1))$ of the Kummer sequences $0 \longrightarrow Z/p^n(1) \longrightarrow G_m \xrightarrow{p^n} G_m \longrightarrow 0$ on $\mathrm{Spec}(O_K[\frac{1}{S}])_{et}$.

2.2.6. Let X, m, r, p be as in 2.2.4 and S be a finite set of prime numbers containing p and all prime numbers at which X has bad reductions. Let V be as in 2.2.4 and let $V^* = \mathrm{Hom}_{Q_p}(V, Q_p)$ on which $\mathrm{Gal}(\bar{Q}/Q)$ acts by $(\sigma, h) \longmapsto h \circ \sigma^{-1}$ $(\sigma \in \mathrm{Gal}(\bar{Q}/Q),\ h \in \mathrm{Hom})$. The following Conj. 2.2.7 (3) (resp. (4)) is the phase 1 (resp. phase 3) for $L_S(V^*(1), s)$ at $s = 0$ (under the assumptions of 2.2.7).

I give some remarks on 2.2.7.

(1) The conjecture 2.2.7 (1)-(3) are parts of Beilinson conjectures ([Ra]) except the part concerning H^i_{lim} which is due to Jannsen [Ja] (cf. also [So$_3$]). The conjecture 2.2.7 (4) is a part of a conjecture in [FP$_1$][FP$_2$].

(2) The assumption (iii) in 2.2.7 is satisfied for example, if (under the other assumptions (i) and (ii)) X has potentially good reductions at any prime numbers.

(3) If X is purely of dimension d, we have by Poincaré duality $V^*(1) \cong H^{2d-m}(X \otimes_Q \bar{Q}, Q_p)(d+1-r)$. If X is projective, we have by hard Lefschetz ([De$_4$] 4.1) $V^*(1) \cong H^m(X \otimes_Q \bar{Q}, Q_p)(m+1-r)$.

(4) Under the assumptions (i)(ii) of 2.2.7, one can prove that

$$H^1_S \xrightarrow{\cong} H^1_{lim}, \qquad H^0_S = 0 .$$

<u>Conjecture</u> 2.2.7. <u>Let</u> X, m, r, p, S <u>be as in</u> 2.2.5 <u>and assume the following</u> (i)-(iii). <u>Let</u> $V = H^m_{et}(X \otimes_Q \bar{Q}, Q_p)(r)$.

(i) $r > \inf(m, \dim(X))$.

(ii) $(m, r) \neq (1, 0), (2\dim(X), \dim(X) + 1)$.

(iii) $P_{A,\ell}(V^*(1), 1) \neq 0$ <u>for all</u> $\ell \in S$. (Cf. 2.2.3 (3), 2.2.6 (2).)

<u>Then</u>:

(1) <u>We have isomorphisms</u>

$$a : \Psi \otimes \mathbb{R} \xrightarrow{\cong} \Sigma \otimes \mathbb{R},$$

$$b : \Psi \otimes \mathbb{Q}_p \xrightarrow{\cong} H^1_{lim},$$

<u>and</u> $H^i = 0$ <u>for</u> $i \neq 1$.

(2) (The analytic continuation of $L_S(V^*(1), s)$ to whole $\mathbb{C}$ is assumed.) <u>Let</u> $e = \mathrm{ord}_{s=0}(L_S(V^*(1), s))$. <u>Then</u> $e = \dim_{\mathbb{Q}}(\Sigma)$.

(3) <u>Under the isomorphism</u>

$$\det_{\mathbb{Q}}(\Psi) \otimes \mathbb{R} \xrightarrow{\cong} \det_{\mathbb{Q}}(\Sigma) \otimes \mathbb{R}$$

<u>induced by</u> a, <u>the image of</u> $\det_{\mathbb{Q}}(\Psi)$ <u>coincides with</u> $\lim_{s\to 0} s^{-e} L_S(V^*(1), s)$ <u>times</u> $\det_{\mathbb{Q}}(\Sigma)$.

(4) <u>Assume</u> $p \neq 2$. <u>Let</u>

$$\Sigma_{\mathbb{Z}} \underset{\mathrm{def}}{=} (H^m(X(\mathbb{C}), \mathbb{Z}) \otimes \mathbb{Z}(2\pi i)^{r-1})^+,$$

<u>and define the finitely generated</u> $\mathbb{Z}_p$<u>-modules</u> $H^i_{S,\mathbb{Z}_p}$ <u>by</u>

$$H^i_{S,\mathbb{Z}_p} \underset{\mathrm{def}}{=} H^i_{et}(\mathbb{Z}[\tfrac{1}{S}], T) \quad \underline{\text{with}}\ T = H^m_{et}(X \otimes_{\mathbb{Q}} \bar{\mathbb{Q}}, \mathbb{Z}_p)(r).$$

<u>Let</u> γ <u>be a</u> $\mathbb{Z}$<u>-basis of</u> $\det_{\mathbb{Z}}(\Sigma_{\mathbb{Z}})$. <u>Let</u> ω_γ <u>be the element of</u> $\det_{\mathbb{Q}}(\Psi)$ <u>which is sent to</u> $(\lim_{s\to 0} s^{-e} L_S(V^*(1), s))\cdot\gamma$ <u>under the isomorphism in</u> (3) (ω_γ <u>exists by</u> (3)), <u>and let</u> z_γ <u>be the image of</u> ω_γ <u>under the isomorphism</u>

$$\det_{\mathbb{Q}}(\Psi) \otimes \mathbb{Q}_p \xrightarrow{\cong} \det_{\mathbb{Q}_p}(H^1_{lim})$$

<u>induced by</u> b. <u>Then</u>

$$[\det_{\mathbb{Z}_p}(H^1_{S,\mathbb{Z}_p}) : \mathbb{Z}_p z_\gamma] = \#(H^0_{S,\mathbb{Z}_p})\cdot\#(H^2_{S,\mathbb{Z}_p}).$$

<u>In other words</u> (2.1.5), z_γ <u>is a</u> $\mathbb{Z}_p$<u>-basis of</u> $(\det_{\mathbb{Z}_p}(R\Gamma(\mathbb{Z}[\tfrac{1}{S}], T)))^{-1}$ <u>via the isomorphism</u> $(\det_{\mathbb{Z}_p}(R\Gamma(\mathbb{Z}[\tfrac{1}{S}], T)))^{-1} \otimes_{\mathbb{Z}_p} \mathbb{Q}_p \cong \det_{\mathbb{Q}_p}(H^1_{lim})$.

2.2.8. For some known results concerning Conj. 2.2.7 (3) (the phase 1), see [Ra].

We review here the case where $X = \mathrm{Spec}(K)$ for a number field K, $m = 0$, and $r \geq 2$. In this case, (i)(ii)(iii) in 2.2.7 are satisfied, and all the statements in 2.2.7 (1)-(3) are known to be true. In fact, concerning (1), the bijectivity of a is by Borel [Bo], and that of b is by Soulé [So_2]. Note $L_S(V^*(1), s) = \zeta_K^{(S)}(s+1-r)$, where $\zeta_K^{(S)}(s)$ is "the Dedekind zeta function of K without the Euler factors at prime ideals lying over S". Hence (2) is classical, and (3) is a result of Borel [Bo]. (It is known [Be_1] that $(K_q(K) \otimes \mathbf{Q})^{(i)}$ for $q, i \in \mathbf{Z}$ is zero if $q \neq 2i-1$ and $(q, i) \neq (0, 0)$.)

2.2.9. We review Conj. 2.2.7 (4) (the phase 3) in the case $X = \mathrm{Spec}(K)$ for a number field K, $m = 0$, and $r \geq 2$.

First assume that K is totally real and r is even. Then Σ, Ψ, H_S^i for all $i \in \mathbf{Z}$ are zero, and by 2.1.5 we see that 2.2.7 (4) coincides with the Lichtenbaum conjecture

$$\zeta_K(1-r)(p) = \#(H^2(O_K[\tfrac{1}{S}], \mathbf{Z}_p(r))) \cdot \#(H^1(O_K[\tfrac{1}{S}], \mathbf{Z}_p(r)))^{-1}$$

proved by Mazur-Wiles [MW] for special K and by Wiles [Wi_2] in general.

For general $r \geq 2$, 2.2.7 (4) is true if $K = \mathbf{Q}$. (This was deduced in [BK] §6 from the solution of Iwasawa main conjecture by Mazur-Wiles, assuming a conjecture [BK] 6.2. This last conjecture was proved by Beilinson [Be] (this is the result of Beilinson introduced at the end of 3.4.9). It seems that the case K is abelian over Q is proved by a similar method.

§2.3. Birch and Swinnerton-Dyer conjecture.

We give a reformulation by Fontaine and Perrin-Riou [FP_1][FP_2] of the Birch and Swinnerton-Dyer conjecture for abelian varieties, assuming that the L-function of the abelian variety does not vanish at s = 1. (We exclude the 2-primary part.)

2.3.1. Generally, let X be a proper smooth scheme over Q, and fix integers m, r ≥ 0 and a prime number p. Let S be a finite set of prime numbers containing p and all prime numbers at which X has bad reductions. Let

$$\Sigma = H^m(X(\mathbb{C}),\ \mathbb{Q}(2\pi i)^{r-1})$$

$$V = H^m_{et}(X \otimes_{\mathbb{Q}} \bar{\mathbb{Q}},\ \mathbb{Q}_p)(r), \qquad H^i_S = H^i(\mathbb{Z}[\tfrac{1}{S}],\ V)$$

be as in 2.2. Let

$$\Omega = \mathrm{fil}^r H^m_{dR}(X/\mathbb{Q})$$

where $\mathrm{fil}^{\cdot}$ is the Hodge filtration defined by

$$\mathrm{fil}^i H^m_{dR}(X/\mathbb{Q}) = \mathrm{Image}(H^m(X,\ \Omega^{\geq i}_{X/\mathbb{Q}}) \longrightarrow H^m_{dR}(X/\mathbb{Q})) \qquad (i \in \mathbb{Z})$$

with $\Omega^{\geq i}_{X/\mathbb{Q}}$ the degree ≥ i part of the de Rham complex $\Omega^{\cdot}_{X/\mathbb{Q}}$.

Then we have canonical homomorphisms

$$a\ :\ \Omega \otimes \mathbb{R} \longrightarrow \Sigma \otimes \mathbb{R},$$

$$b\ :\ H^1_S \longrightarrow \Omega \otimes \mathbb{Q}_p.$$

which are defined by the theory of period integrals (resp. p-adic periods), respectively. (Note that when compared with the maps a and b in §2.2, Ω plays the role of Ψ in §2.2 but the direction of the map b is the converse.) The map a is induced by the period isomorphism

$$H^m_{dR}(X/\mathbb{Q}) \otimes \mathbb{C} \xrightarrow{\ \cong\ } H^m(X(\mathbb{C}),\ \mathbb{Q}) \otimes \mathbb{C}$$

(cf. 1.2.2) as

$$\Omega \otimes \mathbb{R} \xrightarrow{\ \subset\ } H^m_{dR}(X/\mathbb{Q}) \otimes \mathbb{R} \cong (H^m(X(\mathbb{C}),\ \mathbb{Q}) \otimes \mathbb{C})^+$$

$$\longrightarrow (H^m(X(\mathbb{C}),\ \mathbb{Q}) \otimes \mathbb{C})^+ / (H^m(X(\mathbb{C}),\ \mathbb{Q}) \otimes \mathbb{R}(2\pi i)^r)^+ \longleftarrow \Sigma \otimes \mathbb{R}.$$

The definition of the map b is rather difficult. It is the composition

$$H^1(\mathbf{Z}[\tfrac{1}{S}], V) \longrightarrow H^1(\mathbf{Q}_p, V) \xrightarrow{\exp^*} \mathrm{fil}^r H^m_{dR}((X \otimes_{\mathbf{Q}} \mathbf{Q}_p)/\mathbf{Q}_p)$$
$$\cong \mathrm{fil}^r H^m_{dR}(X/\mathbf{Q}) \otimes \mathbf{Q}_p$$

where $\exp^*$ is a homomorphism called the dual exponential map defined in Chap. II §1.2 by using the theory of p-adic periods.

2.3.2. Now assume X is an abelian variety over Q, and let $m = r = 1$. Then, $\Omega = \Gamma(X, \Omega^1_{X/\mathbf{Q}})$. In this case, a is bijective. The above map $\exp^*$ in this special case is defined without using the theory of p-adic periods as is explained later in 2.3.6.

Since $V^*(1) \cong V$ (non-canonically), $L_S(V^*(1), s) = L_S(V, s) = L_S(H^1(X), s+1)$. Assume

(*) $L_S(H^1(X), s)$ has analytic continuation to whole $\mathbf{C}$ as a holomorphic function and $L_S(H^1(X), 1) \neq 0$.

Then the phase 1 (resp. phase 3) about $L_S(H^1(X), 1)$ is 2.3.3 (1) (resp. (2) and (3)) below.

Conjecture 2.3.3 (Birch-Swinnerton-Dyer conjecture in the case $L_S(H^1(X), 1) \neq 0$). Let X be an abelian variety over Q and assume (*).

(1) Under the isomorphism

$$\det_{\mathbf{Q}}(\Omega) \otimes \mathbf{R} \xrightarrow{\cong} \det_{\mathbf{Q}}(\Sigma) \otimes \mathbf{R}$$

induced by a, the image of $\det_{\mathbf{Q}}(\Omega)$ coincides with $L_S(H^1(X), 1)$ times $\det_{\mathbf{Q}}(\Sigma)$.

(2) The map b is an isomorphism and $H^i_S = 0$ for $i \neq 1$.

(3) Assume $p \neq 2$. Let

$$\Sigma_{\mathbf{Z}} = H^1(X(\mathbf{C}), \mathbf{Z})^+.$$

and define finitely generated $\mathbf{Z}_p$-modules $H^i_{S,\mathbf{Z}_p}$ by

$$H^i_{S,\mathbb{Z}_p} = H^i_{et}(\mathbb{Z}[\tfrac{1}{S}], T) \quad \underline{\text{with}} \quad T = H^1_{et}(X \otimes_{\mathbb{Q}} \bar{\mathbb{Q}}, \mathbb{Z}_p)(1)$$

<u>Let</u> γ <u>be a</u> $\mathbb{Z}$-<u>basis of</u> $\det_{\mathbb{Z}}(\Sigma_{\mathbb{Z}})$, <u>let</u> ω_γ <u>be the element of</u> $\det_{\mathbb{Q}}(\Omega)$ <u>which is sent to</u> $L_S(H^1(X), 1)\cdot\gamma$ <u>under the isomorphism in</u> (1) (ω_γ <u>exists by</u> (1)), <u>and let</u> z_γ <u>be the element of</u> $\det_{\mathbb{Q}_p}(H^1_S)$ <u>which is sent to</u> ω_γ <u>under the isomorphism</u>

$$\det_{\mathbb{Q}_p}(H^1_S) \xrightarrow{\cong} \det_{\mathbb{Q}}(\Omega) \otimes \mathbb{Q}_p$$

<u>induced by</u> b. <u>Then</u>

$$[\det_{\mathbb{Z}_p}(H^1_{S,\mathbb{Z}_p}) : \mathbb{Z}_p z_\gamma] = \#(H^2_{S,\mathbb{Z}_p})$$

<u>In other words,</u> z_γ <u>is a</u> $\mathbb{Z}_p$-<u>basis of</u> $\{\det_{\mathbb{Z}_p}(R\Gamma(\mathbb{Z}[\tfrac{1}{S}], T))\}^{-1}$.

<u>Remark</u> 2.3.4. (1) The Birch and Swinnerton-Dyer conjecture is usually stated in a different form ([Ta$_1$]).

(2) For results on the Birch and Swinnerton-Dyer conjectures of elliptic curves, cf. Kolyvagin [Ko] and Rubin [Ru].

2.3.5. In Conj. 2.2.7 and 2.3.3, we obtain a canonical conjectural Λ-basis $z_\gamma \otimes \gamma^{-1}$ of the invertible $\mathbb{Z}_p$-module

$$\{\det_{\mathbb{Z}_p}(R\Gamma(\mathbb{Z}[\tfrac{1}{S}], T))\}^{-1} \otimes_{\mathbb{Z}} \{\det_{\mathbb{Z}}(\Sigma_{\mathbb{Z}})\}^{-1}.$$

This basis coincides up to sign with the <u>"zeta element" of the</u> $\mathbb{Z}_p$-<u>sheaf</u> T which appears (conjecturally) in our Iwasawa main conjecture 3.2.2.

2.3.6. The definition of the map $\exp^* : H^1(\mathbb{Q}_p, V) \longrightarrow \Gamma(X, \Omega^1_{X/\mathbb{Q}_p})$ for an abelian variety X over $\mathbb{Q}_p$ and for $V = H^1(X \otimes_{\mathbb{Q}_p} \bar{\mathbb{Q}}_p, \mathbb{Q}_p)(1)$ is as follows. Note that $\Gamma(X, \Omega^1_{X/\mathbb{Q}_p})$ is identified with the $\mathbb{Q}_p$-dual of the tangent space Lie(X) of X at the origin. We have the exponential map of the p-adic Lie group $X(\mathbb{Q}_p)$

$$\exp \ : \ \mathrm{Lie}(X) \xrightarrow{\simeq} X(\mathbb{Q}_p) \otimes \mathbb{Q} .$$

On the other hand, the exact sequence (defining ${}_{p^n}X$)

$$0 \longrightarrow {}_{p^n}X \longrightarrow X \xrightarrow{p^n} X \longrightarrow 0$$

induces the connecting map of the Galois cohomology

$$X(\mathbb{Q}_p) \longrightarrow H^1(\mathbb{Q}_p, {}_{p^n}X)$$

and by taking $\mathbb{Q} \otimes \varprojlim$ we obtain

$$X(\mathbb{Q}_p) \longrightarrow H^1(\mathbb{Q}_p, V_pX) \quad \text{where } V_pX = T_pX \otimes_{\mathbb{Z}_p} \mathbb{Q}_p$$

with T_pX the p-adic Tate module of X. Denote the composition of this map and the above exponential map still as

$$\exp \ : \ \mathrm{Lie}(X) \longrightarrow H^1(\mathbb{Q}_p, V_pX) .$$

By the local Tate duality (cf. Chap. II 1.4) and by the fact there is a canonical isomorphism $V_pX \simeq V^*(1)$ with V as above, $H^1(\mathbb{Q}_p, V)$ is canonically isomorphic to the $\mathbb{Q}_p$-dual of $H^1(\mathbb{Q}_p, V_pX)$. Hence by taking the $\mathbb{Q}_p$-dual of the last map exp, we have a map

$$\exp^* \ : \ H^1(\mathbb{Q}_p, V) \longrightarrow \Gamma(X, \Omega^1_{X/\mathbb{Q}_p}) .$$

§3. Generalized Iwasawa main conjecture.

We formulate our generalized Iwasawa main conjecture 3.2.2. We discuss in §3.3 the relationship with the classical Iwasawa main conjecture, and in §3.4 important problems arising from our conjecture. In §3.5, we discuss the case of our conjecture for varieties over finite fields.

In this §3, p denotes a fixed prime number.

§3.1. Preliminaries.

3.1.1. We will consider a commutative topological ring Λ which is either

(1) a pro-p-ring (that is, a compact ring which is an inverse limit of finite rings whose orders are powers of p),

or

(2) a $\mathbf{Q}_p$-algebra which is a finite product of finite extensions of $\mathbf{Q}_p$.

3.1.2. Let Λ be as in 3.1.1. We define the category $D_{ctf}(\Lambda)$, and the category $D_{ctf}(X, \Lambda)$ for any noetherian scheme X, as follows.

(1) First, assume that Λ is a finite ring. Let $D_{ctf}(\Lambda)$ be the full subcategory of the derived category $D(\Lambda)$ of Λ-modules, consisting of perfect complexes (2.1.1). For a noetherian scheme X, let $D(X, \Lambda)$ be the derived category of sheaves of Λ-modules on X_{et}, and let $D_{ctf}(X, \Lambda)$ be the full subcategory of $D(X, \Lambda)$ consisting of objects $\mathcal{F}$ satisfying the following two conditions (i)(ii).

(i) The cohomology sheaves $\mathcal{H}^q(\mathcal{F})$ are constructible ([SGA 4]) for all $q \in \mathbf{Z}$ and are zero for almost all $q \in \mathbf{Z}$.

(ii) For any $x \in X$, the stalk $\mathcal{F}_{\bar{x}}$ is a perfect complex over Λ.

(2) If Λ satisfies the condition (1) in 3.1.1, we define $D_{ctf}(\Lambda)$ (resp. $D_{ctf}(X, \Lambda)$ for a noetherian scheme X) as follows. An object of $D_{ctf}(\Lambda)$ (resp. $D_{ctf}(X, \Lambda)$) is a system $(\mathcal{F}_I, \rho_{I,J})$ where I ranges over the set of open ideals of Λ, $\mathcal{F}_I$ is an object of $D_{ctf}(\Lambda/I)$ (resp. $D_{ctf}(X, \Lambda/I)$), and $\rho_{I,J}$ is an isomorphism $\mathcal{F}_J \otimes^L_{\Lambda/J} \Lambda/I \xrightarrow{\cong} \mathcal{F}_I$ given for each pairs of open ideals I, J such that $I \supset J$, satisfying the conditions $\rho_{I,I}$ = identity, and $\rho_{I,I'} \circ \rho_{I',I''} = \rho_{I,I''}$ if $I \supset I' \supset I''$. ($\otimes^L$ denotes the tensor product in the derived category.) Morphisms in

$D_{ctf}(\Lambda)$ (resp. $D_{ctf}(X, \Lambda)$) are defined in a natural way.

(3) Finally assume Λ satisfies the condition (2) in 3.1.1. Write $F = \prod_i F_i$ with F_i finite extensions of $\mathbf{Q}_p$, let O_i be the integer ring of F_i for each i, and let $O_\Lambda = \prod_i O_i$. We define

$$D_{ctf}(\Lambda) = D_{ctf}(O_\Lambda) \otimes \mathbf{Q} \quad (\text{resp. } D_{ctf}(X, \Lambda) = D_{ctf}(X, O_\Lambda) \otimes \mathbf{Q}).$$

Here for an additive category $\mathcal{C}$, $\mathcal{C} \otimes \mathbf{Q}$ is the category whose objects are the same as $\mathcal{C}$ but the Hom in $\mathcal{C} \otimes \mathbf{Q}$ is $\otimes \mathbf{Q}$ of the Hom in $\mathcal{C}$.

The following is known.

3.1.3. <u>Let Λ be a finite ring of order a power of</u> p.

(1) <u>Let Z be a regular noetherian scheme of dimension ≤ 1 on which</u> p <u>is invertible, and let $f : X \longrightarrow Y$ be a Z-morphism of schemes of finite type over</u> Z. <u>Then Rf_* sends $D_{ctf}(X, \Lambda)$ into $D_{ctf}(Y, \Lambda)$.</u>

(2) <u>Consider the following cases.</u>

(i) X <u>is a scheme over $\mathbf{Z}[\frac{1}{p}]$ of finite type and</u> $p \neq 2$.

(ii) X <u>is a scheme over $\mathbf{R}$ of finite type and</u> $p \neq 2$.

(iii) X <u>is a scheme over $\mathbf{Q}_\ell$ of finite type for a prime number ℓ.</u>

<u>Let $\mathcal{F}$ be an object of $D_{cft}(X, \Lambda)$. Then in these cases (i)-(iii), we have</u> $R\Gamma(X, \mathcal{F}) \in D_{cft}(\Lambda)$, <u>and</u>

$$R\Gamma(X, \mathcal{F}) \otimes^L_\Lambda \Lambda' \xrightarrow{\cong} R\Gamma(X, \mathcal{F}') \quad \text{where } \mathcal{F}' = \mathcal{F} \otimes^L_\Lambda \Lambda'$$

<u>for any finite ring Λ' of order a power of</u> p <u>and for any homomorphism</u> $\Lambda \longrightarrow \Lambda'$. <u>In the case</u> (i), <u>we have</u>

(*) $\quad \operatorname{rank}_\Lambda(R\Gamma(X, \mathcal{F})) = -\operatorname{rank}(R\Gamma(X \otimes \mathbf{R}, \mathcal{F}(-1))).$

<u>In the case</u> (iii) <u>with</u> $\ell = p$ (resp. $\ell \neq p$), <u>we have</u>

(**) $\quad \operatorname{rank}_\Lambda(R\Gamma(X, \mathcal{F})) = -\operatorname{rank}_\Lambda(R\Gamma(X \otimes_{\mathbf{Q}_p} \bar{\mathbf{Q}}_p, \mathcal{F}))$

(resp. $\operatorname{rank}_\Lambda(R\Gamma(X, \mathcal{F})) = 0$).

Proof. (1) is by Deligne [De_2] and [SGA 4] Chap. 17. (2) except the part concerning ranks is proved by the same method by using the fact that when $X = \mathrm{Spec}(\mathbb{Z}[\frac{1}{p}])$, $\mathrm{Spec}(\mathbb{R})$, or $\mathrm{Spec}(\mathbb{Q}_\ell)$, $H^i(X, \mathcal{F})$ are finite for all i and zero for almost all i ([Ma], [Se]). The problem on ranks is reduced to the case where Λ is a field and then follows from Tate [Ta_1] Thm. 2.2 and Serre [Se] Chap. II, §5.7.

3.1.4. In 3.1.3, replace the assumption Λ is finite of order a power of p, by the assumption Λ is as in 3.1.1 (1). Then, in 3.1.3 (1) (resp. (2)), we define the object $Rf_*(\mathcal{F})$ (resp. $R\Gamma(X, \mathcal{F})$) of $D_{ctf}(Y, \Lambda)$ (resp. $D_{ctf}(\Lambda)$) for an object $\mathcal{F} = (\mathcal{F}_I)_I$ of $D_{ctf}(X, \Lambda)$, to be $(Rf_*(\mathcal{F}_I))_I$ (resp. $(R\Gamma(X, \mathcal{F}_I))_I$.)

If the assumption Λ is finite of order p is replaced by the assumption Λ is as in 3.1.1 (2), then in 3.1.3 (1) (resp. (2)), we define $Rf_*(\mathcal{F} \otimes \mathbb{Q})$ to be $Rf_*(\mathcal{F}) \otimes \mathbb{Q}$ (resp. $R\Gamma(X, \mathcal{F} \otimes \mathbb{Q})$ to be $R\Gamma(X, \mathcal{F}) \otimes \mathbb{Q}$) for an object $\mathcal{F}$ of $D_{ctf}(X, O_\Lambda)$.

§3.2. The generalized Iwasawa main conjecture.

Let p be a fixed prime number.

3.2.1. We will consider triples $(X, \Lambda, \mathcal{F})$, such that

(i) X is a scheme of finite type over $\mathbb{Z}[\frac{1}{p}]$,

(ii) Λ is as in 3.1.1,

(iii) $\mathcal{F}$ is an object of $D_{ctf}(X, \Lambda)$.

For such $(X, \Lambda, \mathcal{F})$, we define the invertible Λ-module $\Delta_\Lambda(X, \mathcal{F})$ by

$$\Delta_\Lambda(X, \mathcal{F}) = \{\det_\Lambda(R\Gamma(X, \mathcal{F}))\}^{-1} \otimes_\Lambda \{\det_\Lambda(R\Gamma(X \otimes_{\mathbb{Z}} \mathbb{R}, \mathcal{F}(-1)))\}^{-1}.$$

Conjecture 3.2.2. Let p be an odd prime number. Then, for any triple $(X, \Lambda, \mathcal{F})$ as in 3.2.1, we can define a Λ-basis $z_\Lambda(X, \mathcal{F})$ of $\Delta_\Lambda(X, \mathcal{F})$ (called the zeta element of $(X, \Lambda, \mathcal{F})$) satisfying the following conditions (i) - (v). (The condition (v), which is concerned with values of zeta functions and which involves the word "motive", will be presented only partially here, and will be presented in detail in Chapter III §2 after a preliminary on motives).

(i) Let $(X, \Lambda, \mathcal{F})$ and $(X, \Lambda', \mathcal{F}')$ be triples as in 3.2.1 with common X, let $\Lambda \longrightarrow \Lambda'$ be a homomorphism (it is allowed that Λ is as in 3.1.1 (1) and Λ' is as in 3.1.1 (2)), and let $h : \mathcal{F} \otimes^L_\Lambda \Lambda' \xrightarrow{\cong} \mathcal{F}'$ be a Λ'-isomorphism. Then the isomorphism

$$\Delta_\Lambda(X, \mathcal{F}) \otimes^L_\Lambda \Lambda' \xrightarrow{\cong} \Delta_{\Lambda'}(X, \mathcal{F}')$$

induced by h (3.1.3 (2)) sends $z_\Lambda(X, T) \otimes 1$ to $z_{\Lambda'}(X, T')$.

(ii) Let $(X, \Lambda, \mathcal{F})$ be a triple as in 3.2.1, let Y be a scheme of finite type over $\mathbb{Z}[\frac{1}{p}]$, and let $f : X \longrightarrow Y$ be a morphism. Then

$$z_\Lambda(X, \mathcal{F}) = z_\Lambda(Y, Rf_*\mathcal{F}) \quad \text{in} \quad \Delta_\Lambda(X, \mathcal{F}) = \Delta_\Lambda(Y, Rf_*\mathcal{F}).$$

(iii) Assume Λ is a finite ring. Let $(X, \Lambda, \mathcal{F})$, $(X, \Lambda, \mathcal{F}')$, $(X, \Lambda, \mathcal{F}'')$ be triples as in 3.2.1 with common X and Λ, and assume we are given an exact sequence of complexes of injective Λ-sheaves

(*) $\quad 0 \longrightarrow I' \longrightarrow I \longrightarrow I'' \longrightarrow 0$

such that I (resp. I', resp. I'') is a representative of $\mathcal{F}$ (resp. $\mathcal{F}'$, resp. $\mathcal{F}''$). Then,

$$z_\Lambda(X, \mathcal{F}) = z_\Lambda(X, \mathcal{F}') \otimes z_\Lambda(X, \mathcal{F}'')$$

via the isomorphism $\Delta_\Lambda(X, \mathcal{F}) \cong \Delta_\Lambda(X, \mathcal{F}') \otimes^L_\Lambda \Delta_\Lambda(X, \mathcal{F}'')$ induced by the exact sequences of complexes

$$0 \longrightarrow \Gamma(X, I') \longrightarrow \Gamma(X, I) \longrightarrow \Gamma(X, I'') \longrightarrow 0$$

$$0 \longrightarrow \Gamma(X \otimes \mathbb{R}, I'(-1)) \longrightarrow \Gamma(X \otimes \mathbb{R}, I(-1)) \longrightarrow \Gamma(X \otimes \mathbb{R}, I''(-1)) \longrightarrow 0.$$

(iv) <u>If</u> Λ <u>is a finite ring and</u> X <u>is a scheme over</u> F_ℓ <u>for some prime number</u> $\ell \neq p$, $z_\Lambda(X, \mathcal{F})$ <u>is defined as follows. Let</u> $\bar{X} = X \otimes_{F_\ell} \bar{F}_\ell$ <u>with</u> $\bar{F}_\ell$ <u>an algebraic closure of</u> F_ℓ, <u>take a complex of injective</u> Λ-<u>sheaves</u> $I^\cdot$ <u>which is a representative of</u> $\mathcal{F}$, <u>and consider the exact sequence of complexes</u>

$$(*) \qquad 0 \longrightarrow \Gamma(X, I^\cdot) \longrightarrow \Gamma(\bar{X}, I^\cdot) \xrightarrow{1-\sigma_\ell} \Gamma(\bar{X}, I^\cdot) \longrightarrow 0 .$$

<u>Here</u> σ_ℓ <u>is the arithmetic Frobenius in</u> $\mathrm{Gal}(\bar{F}_\ell/F_\ell)$, <u>that is,</u> $\sigma_\ell(x) = x^\ell$ <u>for</u> $x \in \bar{F}_\ell$. <u>Consider the isomorphisms</u>

$$(**) \qquad \Delta_\Lambda(X, \mathcal{F}) \cong \det_\Lambda(R\Gamma(\bar{X}, \mathcal{F})) \otimes^L (\det_\Lambda(R\Gamma(\bar{X}, \mathcal{F})))^{-1} \cong \Lambda$$

<u>where the first isomorphism is by</u> (*) <u>and the second is induced by the identity map between the two</u> $R\Gamma(\bar{X}, \mathcal{F})$. <u>The element</u> $z_\Lambda(X, \mathcal{F})$ <u>is defined to be the image of</u> $1 \in \Lambda$ <u>under the composite isomorphism</u> (**).

(v) (This is a condition on the <u>relationship with zeta values of motives over</u> $\mathbb{Q}$. The full description of this condition is given in Chap. III §2. Descriptions of certain special cases are given in 3.2.6 below.)

<u>Remark</u> 3.2.3. (1) We can formulate a conjecture including the case $p = 2$ (cf. 3.4.14).

(2) To find zeta elements in 3.2.2, by the condition (ii), it is in fact sufficient to find them in the case $X = \mathrm{Spec}(\mathbb{Z}[\frac{1}{p}])$.

(3) Roughly speaking, the condition 3.2.2 (v) is that if $X = \mathrm{Spec}(\mathbb{Z}[\frac{1}{S}])$ for a finite set S of prime numbers including p and $\mathcal{F}$ is a smooth sheaf on X which comes from a motive over $\mathbb{Q}$, then the zeta element "expresses" a zeta value up to sign. This sign ± 1 should be determined, but I have not yet well understood this sign.

(4) Roughly speaking, the condition 3.2.2 (v) is given by

generalizing motives in the theory of Fontaine and Perrin-Riou [FP_1] [FP_2] to "motives with coefficients".

3.2.4. To give a partial description of (v), we discuss a little about motives over $\mathbf{Q}$. For the idea "motif" of Grothendieck, whose definition is not yet fixed in the present mathematics, see for example [Ra][FP_1], etc. In fact, we do not discuss what motives are. In any case, a motif (a mixed motif in a more precise terminology) M over $\mathbf{Q}$ is "something" which "comes from the algebraic geometry over $\mathbf{Q}$" and which gives rise to a triple (M_B, M_{dR}, M_p) having the following structures (fix a prime number p):

(1) M_B (called the Betti realization of M) is a finite dimensional $\mathbf{Q}$-vector space endowed with an action of $\mathrm{Gal}(\mathbf{C}/\mathbf{R})$.

(2) M_{dR} (called the de Rham realization of M) is a finite dimensional $\mathbf{Q}$-vector space endowed with a decreasing filtration $(M^i_{dR})_{i \in \mathbf{Z}}$ such that $M^i_{dR} = M_{dR}$ for $i << 0$ and $M^i_{dR} = 0$ for $i >> 0$.

(3) M_p (called the p-adic etale realization of M) is a smooth $\mathbf{Q}_p$-sheaf on $\mathrm{Spec}(\mathbf{Q}_p)_{et}$ which comes from a smooth $\mathbf{Q}_p$-sheaf on $\mathrm{Spec}(\mathbf{Z}[\frac{1}{S}])_{et}$ for some finite set S of prime numbers. If we fix an algebraic closure $\bar{\mathbf{Q}}$ of $\mathbf{Q}$, M_p is simply a finite dimensional $\mathbf{Q}_p$-vector space endowed with a continuous action of $\mathrm{Gal}(\bar{\mathbf{Q}}/\mathbf{Q})$ which is unramified at almost all prime numbers (this interpretation is obtained by taking the stalk at $\mathrm{Spec}(\bar{\mathbf{Q}})$).

(4) (Relationship between M_B and M_{dR}.) We are given an isomorphism of $\mathbf{C}$-vector spaces $M_B \otimes \mathbf{C} \cong M_{dR} \otimes \mathbf{C}$ compatible with the action of $\mathrm{Gal}(\mathbf{C}/\mathbf{R})$. (Here $\sigma \in \mathrm{Gal}(\mathbf{C}/\mathbf{R})$ acts on $M_B \otimes \mathbf{C}$ as $\sigma \otimes \sigma$ and on $M_{dR} \otimes \mathbf{C}$ as $1 \otimes \sigma$.)

(5) (Relationship between M_B and M_p.) We are given an isomorphism of

$\mathbf{Q}_p$-vector spaces $M_B \otimes \mathbf{Q}_p \cong \Gamma(\mathrm{Spec}(\mathbf{C}), M_p)$ which is compatible with the action of $\mathrm{Gal}(\mathbf{C}/\mathbf{R})$.

(6) (Relationship between M_{dR} and M_p.) We are given an isomorphism of filtered $\mathbf{Q}_p$-vector spaces $M_{dR} \otimes \mathbf{Q}_p \cong D_{dR}(M_p)$. Here we used a functor D_{dR} whose definition is reviewed in Chap. II §1.1 later.

For example, if X is a smooth proper scheme over $\mathbf{Q}$, it is believed that there is a motif $M = H^m(X)(r)$ for each $m, r \in \mathbf{Z}$, whose realizations are

$M_B = H^m(X(\mathbf{C}), \mathbf{Q}) \otimes \mathbf{Q}(2\pi i)^r$ with the natural action of $\mathrm{Gal}(\mathbf{C}/\mathbf{R})$,

$M_{dR} = H^m_{dR}(X/\mathbf{Q})$ with the filtration $M^i_{dR} = \mathrm{fil}^{i+r} H^m_{dR}(X/\mathbf{Q})$

(fil denotes the Hodge filtration (2.3.1)),

$M_p = H^m_{et}(X \otimes_{\mathbf{Q}} \bar{\mathbf{Q}}, \mathbf{Q}_p)(r)$ with the natural action of $\mathrm{Gal}(\bar{\mathbf{Q}}/\mathbf{Q})$.

3.2.5. It is believed that motives over $\mathbf{Q}$ form an abelian $\mathbf{Q}$-category ($\mathbf{Q}$-category means that Hom are $\mathbf{Q}$-vector spaces and compositions of morphisms are $\mathbf{Q}$-bilinear), and the above realization functors are exact and faithful.

Let A be a commutative ring over $\mathbf{Q}$ which is a finite product of finite extensions of $\mathbf{Q}$. By "an A-motif over $\mathbf{Q}$", we mean a motif over $\mathbf{Q}$ endowed with an action of A.

It is believed that if M is an A-motif and S is a finite set of primes containing p such that the action of $\mathrm{Gal}(\bar{\mathbf{Q}}/\mathbf{Q})$ on M_p is unramified, then $(A \otimes \mathbf{Q}_p, M_p, S, A)$ satisfies the conditions (i)(ii) in 2.2.3 and $L_{A,S}(M_p, s)$ is extended to a meromorphic function on whole $\mathbf{C}$.

It is believed that the category of A-motives over $\mathbf{Q}$ have functors tensor product over A, the A-dual $\underline{\mathrm{Hom}}_A(\ ,\ A)$, the Tate twist, which commute with realization functors.

In what follows, we assume all these believes are the truths.

3.2.6. The condition 3.2.2 (v) has the following form. Assume we are given an A-motif M over Q where A is a finite product of finite extensions of Q. Let $\Lambda = A \otimes \mathbf{Q}_p$ and let $M_p^*(1) = \mathrm{Hom}_\Lambda(M_p, \Lambda)(1)$ where $\mathrm{Gal}(\bar{\mathbf{Q}}/\mathbf{Q})$ acts on Hom by $(\sigma, h) \longmapsto h\circ\sigma^{-1}$ ($h \in \mathrm{Hom}$, $\sigma \in \mathrm{Gal}$). (Note $M_p^*(1)$ is the p-adic etale realization of the A-motif $\underline{\mathrm{Hom}}_A(M, A)(1)$ by the believes in 3.2.5.) Let S be a finite set of primes containing p such that M_p is unramified outside S. Then the condition 3.2.2 (v) is concerned with the relationship between the zeta element $z_\Lambda(\mathrm{Spec}(\mathbf{Z}[\frac{1}{S}]), M_p)$ and the value

$$\lim_{s\to 0} s^{-e} L_{A,S}(M_p^*(1), s)$$

where $e : \mathrm{Spec}(A \otimes \mathbf{C}) \longrightarrow \mathbf{Z}$ is the order of $L_S(M_p^*(1), s)$ at $s = 0$. The full description of 3.2.2 (v) is given in Chap. III §2. We describe it here in two special cases, the strictly K-theoretic case and the strictly critical case, which include (conjecturally) the cases treated in §2.2 and §2.3, respectively.

Let

$$\Sigma = M_B(-1)^+$$

where (-1) means the inverse Tate twist $\otimes_\mathbf{Q} \mathbf{Q}(2\pi i)^{-1}$ and $(\ )^+$ is the $\mathrm{Gal}(\mathbf{C}/\mathbf{R})$-fixed part,

$$H_S^i = H^i(\mathbf{Z}[\tfrac{1}{S}], M_p).$$

In both the strictly K-theoretic case and the strictly critical case, we will have (by our definition)

$$H_S^i = 0 \quad \text{for all } i \neq 1$$

and hence

$$(1) \qquad \Delta_\Lambda(\mathrm{Spec}(\mathbf{Z}[\tfrac{1}{S}]), M_p) \cong \det{}_\Lambda(H_S^1) \otimes_\Lambda \{\det{}_\Lambda(\Sigma)\}^{-1}.$$

(In fact, $\Sigma \otimes Q_p = H^0(R, M_p(-1)) = R\Gamma(R, M_p(-1))$ by 3.2.3 (2).)

For example, if $A = Q$, $M = H^m(X)(r)$ for a smooth proper scheme X over Q and if X, m, r, S satisfy the assumption of Conj. 2.2.7 (resp. 2.3.3), then the triple $(\mathrm{Spec}(Z[\frac{1}{S}]), Q_p, M_p)$ is conjectured to be strictly K-theoretic (resp. strictly critical), and the condition 3.2.2 (v) for this triple is

$$z_{Q_p}(\mathrm{Spec}(Z[\tfrac{1}{S}]), M_p) = \pm z_\gamma \otimes \gamma^{-1}$$

with γ and z_γ as in 2.2.7 (resp. 2.3.3).

We first talk about the strictly critical case. Let M be a motif over Q, and let $\Omega = M^0_{dR}$. We define canonical homomorphisms

$$a : \Omega \otimes R \longrightarrow \Sigma \otimes R$$

$$b : H^1_S \longrightarrow \Omega \otimes Q_p$$

as follows. We define a as the composite map

$$\Omega \otimes R \subset M_{dR} \otimes R \cong (M_B \otimes C)^+ \longrightarrow (M_B \otimes C)^+/(M_B \otimes R)^+ \xleftarrow{\cong} (M_B(-1))^+ \otimes R$$

in which the first isomorphism comes from 3.2.3 (1). We define b as the composite map

$$H^1_S \longrightarrow H^1(Q_p, M_p) \longrightarrow D^0_{dR}(M_p) \cong M^0_{dR} \otimes Q_p$$

in which the second arrow is the "dual exponential map" defined in Chap. II §1.2 and the isomorphism comes from 3.2.3 (3).

We say M is strictly critical if the homomorphisms a and b are bijective and $H^i_S = 0$ for all $i \neq 1$. (We conjecture the last condition is automatically satisfied if a and b are bijective).

In the strictly critical case, the condition 3.2.2 (v) becomes the following (2) (3).

(2) <u>The order of</u> $L_{A,S}(M_p^*(1), s)$ <u>at</u> $s = 0$ <u>(which is a function</u> $\mathrm{Spec}(A \otimes C) \longrightarrow Z$) <u>is zero</u>.

(3) <u>Consider the isomorphisms</u>

$$\det_A(\Omega) \otimes_A \det_A(\Sigma)^{-1} \otimes \mathbb{R} \cong A \otimes \mathbb{R}$$

$$\det_A(\Omega) \otimes_A \det_A(\Sigma)^{-1} \otimes \mathbb{Q}_p \cong \Delta_\Lambda(\mathrm{Spec}(\mathbb{Z}[\tfrac{1}{S}]), M_p)$$

<u>induced by</u> a <u>and</u> b, <u>respectively</u>. <u>Then</u>, $z_\Lambda(\mathrm{Spec}(\mathbb{Z}[\frac{1}{S}]), M_p)$ <u>comes from an element of</u> $\det_A(\Omega) \otimes_A \det_A(\Sigma)^{-1}$ <u>whose image in</u> $A \otimes \mathbb{R}$ <u>is</u> $\pm L_{A,S}(M^*(1), 0)$. <u>Here</u> $\pm$ <u>means a function</u> $\mathrm{Spec}(A) \longrightarrow \{\pm 1\}$.

We next talk about the strictly K-theoretic case. For this we have to introduce a K-group Ψ associated to M. In the case $M = H^m(X)(r)$ for a smooth proper scheme X over $\mathbb{Q}$ and for m, $r \in \mathbb{Z}$, Ψ is defined as in 2.2.4. A conjectural construction of Ψ in the general case is $\Psi = \mathrm{Ext}^1(\mathbb{Q}, M)$ where Ext^1 is the ext group of the conjectural abelian category of motives over $\mathbb{Q}$. In our situation where M is an A-motif, Ψ should be endowed with an A-module structure. Forthermore, we should have a regulator map (conjecturely), which is, in the case $\Omega = 0$ an $A \otimes \mathbb{R}$-homomorphism

$$a : \Psi \otimes \mathbb{R} \longrightarrow \Sigma \otimes \mathbb{R}.$$

We should have also a Chern class map, which should be a Λ-homomorphism

$$b : \Psi \otimes \mathbb{Q}_p \longrightarrow H^1_S.$$

(If $M = H^m(X)(r)$, these a and b are those in §2.1. Cf. [FP_1] or Chap. III §2 for the general case.) We say M is strictly K-theoretic if $\Omega = 0$, a and b are isomorphisms, and $H^i_S = 0$ for $i \neq 1$. (We conjecture that the last condition is satisfied if $\Omega = 0$ and a and b are bijective).

In the strictly K-theoretic case, the condition 3.2.2 (v) becomes the following (4) and (5).

(4) The order e of $L_{A,S}(M_p^*(1), s)$ at $s = 0$ regarded as a function $\mathrm{Spec}(A \otimes \mathbb{C}) \longrightarrow \mathbb{Z}$ coincides with the composite map $\mathrm{Spec}(A \otimes \mathbb{C}) \longrightarrow \mathrm{Spec}(A) \longrightarrow \mathbb{Z}$ in which the second arrow is $\mathrm{rank}_A(\Sigma)$.

(5) Consider the isomorphisms

$$\det_A(\Psi) \otimes_A \det_A(\Sigma)^{-1} \otimes \mathbb{R} \cong A \otimes \mathbb{R}$$

$$\det_A(\Psi) \otimes_A \det_A(\Sigma)^{-1} \otimes \mathbb{Q}_p \cong \Delta_\Lambda(\mathrm{Spec}(\mathbb{Z}[\tfrac{1}{S}]), M_p)$$

induced by a and b, respectively. Then, $z_\Lambda(\mathrm{Spec}(\mathbb{Z}[\frac{1}{S}]), M_p)$ comes from an element of $\det_A(\Psi) \otimes_A \det_A(\Sigma)^{-1}$ whose image in $A \otimes \mathbb{R}$ is $\pm \lim_{s \to 0} s^{-e} L_{A,S}(M^*(1), s)$. Here $\pm$ means a function $\mathrm{Spec}(A) \longrightarrow \{\pm 1\}$.

§3. 3. Zeta elements of Galois representations in invertible modules.

Let p be a fixed odd prime number.

3. 3. 1. In this §3. 3, we consider triples $(X, \Lambda, \mathcal{F})$ (3. 2. 1) such that $X = \mathrm{Spec}(\mathbb{Z}[\frac{1}{S}])$ for a finite set S of prime numbers containing p and $\mathcal{F}$ is an invertible smooth Λ-sheaf (which is put in degree 0 as an object of $D_{ctf}(X, \Lambda)$).

We define a canonical Λ-basis $z_\Lambda(X, \mathcal{F})$ of $\Delta_\Lambda(X, \mathcal{F})$ in the case $S = \{p\}$ and give a plan of the definition for S general.

Classical Iwasawa theory and cyclotomic units play essential roles in this §3. 3.

In this §3. 3, we do not ask if $z_\Lambda(X, \mathcal{F})$ discussed here has the properties required in 3. 2. 2 (v). This question will be discussed later (see 3. 4. 4) and we will obtain a partial affirmative answer.

This §3. 3 is written under the influence of the definitions of "cyclotomic elements in p-adic Galois cohomology" of Soulé [So_2] and Deligne [De_5].

3.3.2. Our plan to define $z_\Lambda(X, \mathcal{F})$ is to deduce it from the following "universal case".

Fix an algebraic closure $\bar{Q}$ of Q, let $Q^{ab,S}$ be the union of all finite abelian extensions of Q in $\bar{Q}$ which are unramified outside S, and let

$$G = \mathrm{Gal}(Q^{ab,S}/Q) .$$

By "the universal case", we mean the case of the triple $(X, \Lambda_0, \mathcal{F}_0)$ with $X = \mathrm{Spec}(Z[\frac{1}{S}])$ and

$$(1) \quad \Lambda_0 = Z_p[[G]] ,$$

$$\mathcal{F}_0 = \varprojlim_L (f_L)_* f_L^*(Z_p(1)_X) ,$$

where L ranges over all finite extensions of Q contained in $Q^{ab,S}$, f_L is the canonical morphism $\mathrm{Spec}(O_L[\frac{1}{S}]) \longrightarrow \mathrm{Spec}(Z[\frac{1}{S}])$, and the limit is taken with respect to trace maps. If we have $z_{\Lambda_0}(X, \mathcal{F}_0)$, we will obtain $z_\Lambda(X, \mathcal{F})$ as follows, for any Λ as in 3.1.1 and for any invertible smooth Λ-sheaf $\mathcal{F}$ on X.

Let $\chi_{\mathcal{F}} : G \longrightarrow \Lambda^\times$ be the action of G on $H^0(\bar{Q}, \mathcal{F}) = H^0(Q^{ab,S}, \mathcal{F})$. Consider the homomorphism

$$\Lambda_0 \longrightarrow \Lambda \ : \ \sigma \longmapsto \chi_{cyclo}(\sigma)\chi_{\mathcal{F}}(\sigma)^{-1} \quad (\sigma \in G) .$$

Fix a Λ-basis ν of $\Gamma(\bar{Q}, \mathcal{F}(-1))$. Then we have an isomorphism

$$h_\nu : \mathcal{F}_0 \otimes_{\Lambda_0} \Lambda \xrightarrow{\ \cong\ } \mathcal{F}$$

in $D_{ctf}(X, \Lambda)$ defined as follows. We assume Λ is a finite ring. (The definition of h_ν for general Λ follows from this case by taking $\varprojlim$ and $\otimes\, Q$.) Then there exists L such that $f_L^*(\mathcal{F})$ is a constant sheaf. We have a homomorphism

$$f_L^*(Z_p(1)_X) \longrightarrow f_L^*(\mathcal{F}) \quad ; \quad x \longmapsto x\nu .$$

This homomorphism induces $(f_L)_* f_L^*(Z_p(1)_X) \longrightarrow (f_L)_* f_L^*(\mathcal{F}) \longrightarrow \mathcal{F}$

where the second arrow is the trace map, and this composite map induces an isomorphism $h_\nu : \mathcal{F}_0 \otimes_{\Lambda_0} \Lambda \xrightarrow{\cong} \mathcal{F}$.

Assume we have Λ-basis $z_{\Lambda_0}(X, \mathcal{F}_0)$ of $\Delta_{\Lambda_0}(X, \mathcal{F}_0)$. Then we define a Λ-basis $z_\Lambda(X, \mathcal{F})$ as the image of $z_{\Lambda_0}(X, \mathcal{F}_0)$ under the isomorphism $\Delta_{\Lambda_0}(X, \mathcal{F}_0) \otimes_{\Lambda_0} \Lambda \xrightarrow{\cong} \Delta_\Lambda(X, \mathcal{F})$ induced by h_ν. This element $z_\Lambda(X, \mathcal{F})$ will be independent of the choice of ν, by 3.1.3 (2) (if we replace ν by $a\nu$ with $a \in \Lambda^\times$, then $z_\Lambda(X, \mathcal{F})$ is multiplied by a^e with

$$e = -\operatorname{rank}_\Lambda(R\Gamma(X, \mathcal{F})) - \operatorname{rank}_\Lambda((R\Gamma(R, \mathcal{F}(-1))) = 0) .$$

3.3.3. In the rest of §3.3, we denote Λ_0 and $\mathcal{F}_0$ of 3.3.2 (1) by Λ and $\mathcal{F}$, respectively. We define a canonical Λ-basis $z_\Lambda(X, \mathcal{F})$ of $\Delta_\Lambda(X, \mathcal{F})$ in the case $S = \{p\}$, and give a plan of the definition for S general. Note

$$H^i(X, \mathcal{F}) = \varprojlim_N H^i(Z[\alpha_N, \tfrac{1}{S}], Z_p(1))$$

where N ranges over integers ≥ 1 such that S coincides with the set of prime divisors of N, α_N denotes a primitive N-th root of 1 in $\bar{Q}$, and the limit is taken with respect to the trace maps. This group is zero for $i \neq 1, 2$. Note also that

$$H^i(R, \mathcal{F}(-1)) = \varprojlim_N H^i(R \otimes Q(\alpha_N), Z_p)$$

with N and α_N as above, this group is zero if $i \neq 0$, and $H^0(R, \mathcal{F}(-1))$ is a free $\Lambda/(1-\tau)$-module of rank 1 where $\tau \in G$ denotes the complex conjugation.

3.3.4. Let $\tau \in G$ be the complex conjugation, and let

$$\Lambda_+ = \Lambda/(1-\tau)\Lambda, \quad \Lambda_- = \Lambda/(1+\tau)\Lambda$$

$$\mathcal{F}_+ = \mathcal{F}/(1-\tau)\mathcal{F}, \quad \mathcal{F}_- = \mathcal{F}/(1+\tau)\mathcal{F}.$$

Then

$$\Lambda \xrightarrow{\cong} \Lambda_+ \times \Lambda_-, \qquad \mathcal{F} \xrightarrow{\cong} \mathcal{F}_+ \times \mathcal{F}_-$$

$$\Delta_\Lambda(X, \mathcal{F}) \xrightarrow{\cong} \Delta_{\Lambda_+}(X, \mathcal{F}_+) \times \Delta_{\Lambda_-}(X, \mathcal{F}_-) .$$

So we should consider $z_{\Lambda_\varepsilon}(X, T_\varepsilon)$ for $\varepsilon = \pm$, and should define $z_\Lambda(X, \mathcal{F})$ as the inverse image of $(z_{\Lambda_+}(X, \mathcal{F}_+),\ z_{\Lambda_-}(X, \mathcal{F}_-))$ under the last isomorphism.

3.3.5. As a preliminary step, we define elements

$$z_{\Lambda_+}(X, \mathcal{F}_+) \in H^1(\mathbb{Z}[\tfrac{1}{S}], \mathcal{F}_+) \otimes_{\Lambda_+} H^0(\mathbb{R}, \mathcal{F}_+(-1))^{-1}$$

$$z_{\Lambda_-}(X, \mathcal{F}_-) \in Q(\Lambda_-)$$

as follows (without assuming any conjecture). Here $H^0(\mathbb{R}, \mathcal{F}_+(-1))^{-1}$ denotes the inverse of the invertible Λ_+-module $H^0(\mathbb{R}, \mathcal{F}_+(-1))$, and $Q(\Lambda_-)$ denotes the total quotient ring of Λ_-.

First we consider $z_{\Lambda_+}(X, \mathcal{F}_+)$. For an integer $N \geq 2$, let c_N be the "cyclotomic unit" defined by

$$c_N = (1-\alpha)(1-\alpha^{-1}) \in \mathbb{R}^\times \quad \text{with} \quad \alpha = \exp(2\pi i N^{-1}) .$$

Then c_N is a unit of the integer ring $\mathbb{Z}[\alpha]^+$ of $\mathbb{Q}(\alpha)^+$ (so, a "unit" in the usual sense in algebraic number theory) if and only if N is a composite number, and is a unit of the ring $\mathbb{Z}[\alpha, \frac{1}{N}]^+$ if N is a power of a prime. (Here $(\ )^+$ denotes the $\mathrm{Gal}(\mathbb{C}/\mathbb{R})$-fixed part). We define $z_{\Lambda_+}(X, \mathcal{F}_+)$ to be the element corresponding to the homomorphism of Λ_+-modules

$$H^0(\mathbb{R}, \mathcal{F}_+(-1)) \longrightarrow H^1(\mathbb{Z}[\tfrac{1}{S}], \mathcal{F}_+)$$

which is defined as the inverse limit of

$$H^0(\mathbb{Q}(\alpha_N)^+ \otimes \mathbb{R}, \mathbb{Z}_p) = \oplus_\iota H^0(\mathbb{R}, \mathbb{Z}_p) \longrightarrow H^1(\mathbb{Z}[\alpha_N, \tfrac{1}{S}]^+, \mathbb{Z}_p(1))$$

$$(a_\iota)_\iota \longmapsto \sum_\iota 2^{-1} a_\iota (\iota^{-1}(c_N))$$

$(a_\iota \in H^0(R, \mathbb{Z}_p))$ where N ranges over integers ≥ 1 such that S coincides with the set of prime divisors of N, α_N denotes a primitive N-th root of 1, ι ranges (for each N) over all homomorphisms $\mathbb{Q}(\alpha_N)^+ \longrightarrow \mathbb{R}$, and for $x \in (\mathbb{Z}[\alpha_N, \frac{1}{S}]^+)^\times$, (x) denotes the image of x in $H^1(\mathbb{Z}[\alpha_N, \frac{1}{S}]^+, \mathbb{Z}_p(1))$ under the connecting maps of the Kummer sequences $0 \longrightarrow \mathbb{Z}/p^i(1) \longrightarrow \mathbb{G}_m \xrightarrow{p^i} \mathbb{G}_m \longrightarrow 0$.

Next we consider $z_-(X, \mathcal{F}_-) \in Q(\Lambda_-)$. Let

$$z = \varprojlim_N \sum_{a \in (\mathbb{Z}/N)^\times} \zeta_{a(N)}(0) \cdot \sigma_a^{-1} \in \varprojlim_N \mathbb{Q}[[\mathrm{Gal}(\mathbb{Q}(\alpha_N)/\mathbb{Q})]].$$

where N ranges over integers ≥ 1 such that S coincides with the set of prime divisors of N, and σ_a is the element of $\mathrm{Gal}(\mathbb{Q}(\alpha_N)/\mathbb{Q})$ such that $\sigma_a(\alpha_N) = \alpha_N^a$. This element z is regarded as an element of $Q(\Lambda)$, and has the property

$$(1 - \chi_{\mathrm{cyclo}}(\sigma)\sigma^{-1})z \in \Lambda \quad \text{for any} \quad \sigma \in \mathrm{Gal}(\mathbb{Q}^{ab, S}/\mathbb{Q})$$

(1.3.7 (1)). We define $z_-(X, \mathcal{F}_-) \in Q(\Lambda_-)$ as the image of $z \in Q(\Lambda)$ in $Q(\Lambda_-)$.

3.3.6. I remark here that cyclotomic units are "great elements" which are related to zeta values as in the following (1), and there is a parallelism between the above definitions of $z_+(X, \mathcal{F}_+)$ and $z_-(X, \mathcal{F}_-)$: For $\alpha = \exp(2\pi i N^{-1})$ $(N \geq 2)$ and for a non-zero element a of $\mathbb{Z}/N$, we have

$$(1) \qquad -2^{-1}\log((1 - \alpha^a)(1 - \alpha^{-a})) = \lim_{s \to 0} s^{-1}(\zeta_{a(N)}(s) + \zeta_{-a(N)}(s)).$$

3.3.7. Now we try to regard the element $z_{\Lambda_\varepsilon}(X, \mathcal{F}_\varepsilon)$ $(\varepsilon = \pm)$ as a Λ_ε-basis of $\Delta_{\Lambda_\varepsilon}(X, \mathcal{F}_\varepsilon)$. Since Λ_ε is in general not a noetherian ring and some argument becomes simple with the noetherian assumption, we consider the noetherian rings $\Lambda_{Np^\infty} = \mathbb{Z}_p[[G_{Np^\infty}]]$ with

$G_{Np^\infty} = \varprojlim_n \mathrm{Gal}(\mathbf{Q}(\alpha_{Np^n})/\mathbf{Q})$ for integers $N \geq 1$ such that S coincides with the set of prime divisors of Np. Then $Q(\Lambda_{Np^\infty})$ is a finite product of fields. Let

$$\Lambda_{Np^\infty,+} = \Lambda_{Np^\infty}/(1-\tau)\Lambda_{Np^\infty}, \qquad \Lambda_{Np^\infty,-} = \Lambda_{Np^\infty}/(1+\tau)\Lambda_{Np^\infty},$$

$$\mathcal{F}_{Np^\infty} = \mathcal{F} \otimes_\Lambda \Lambda_{Np^\infty}, \qquad \mathcal{F}_{Np^\infty,\varepsilon} = \mathcal{F} \otimes_\Lambda \Lambda_{Np^\infty,\varepsilon} \qquad (\varepsilon = \pm).$$

Then it is known that

(1) $H^i(X, \mathcal{F}_{Np^\infty,+}) \otimes_{\Lambda_{Np^\infty,+}} Q(\Lambda_{Np^\infty,+})$ is an invertible $Q(\Lambda_{Np^\infty,+})$-module if $i = 1$, and is zero if $i \neq 1$.

(2) $H^i(X, \mathcal{F}_{Np^\infty,-}) \otimes_{\Lambda_{Np^\infty,-}} Q(\Lambda_{Np^\infty,-}) = 0$ for all i.

(Indeed, $H^i = 0$ for $i \geq 3$ is automatic ([Ma]), $H^0 = 0$ is seen easily, $H^2 \otimes Q = 0$ is reduced to the finiteness of ideal class groups, and the statement for $H^1 \otimes Q$ follows from 3.1.3 (2).)

From (1) and (2), we obtain isomorphisms

$$(3) \qquad \Delta_{\Lambda_{Np^\infty,+}}(X, \mathcal{F}_{Np^\infty,+}) \otimes_{\Lambda_{Np^\infty,+}} Q(\Lambda_{Np^\infty,+})$$
$$\cong H^1(X, \mathcal{F}_{Np^\infty,+}) \otimes_{\Lambda_{Np^\infty,+}} H^0(\mathbf{R}, \mathcal{F}_{Np^\infty,+}(-1))^{-1} \otimes_{\Lambda_{Np^\infty,+}} Q(\Lambda_{Np^\infty,+}),$$

$$(4) \qquad \Delta_{\Lambda_{Np^\infty,-}}(X, \mathcal{F}_{Np^\infty,-}) \otimes_{\Lambda_{Np^\infty,-}} Q(\Lambda_{Np^\infty,-}) \cong Q(\Lambda_{Np^\infty,-}).$$

The case $S = \{p\}$ of the following 3.3.8 is the classical Iwasawa main conjecture proved by Mazur-Wiles [MW].

<u>Conjecture</u> 3.3.8. <u>Let</u> $\varepsilon = +$ (resp. $-$). <u>Then the image of</u> $z_\varepsilon(X, \mathcal{F}_\varepsilon)$ (3.3.5) <u>in</u> $\Delta_{\Lambda_{Np^\infty,\varepsilon}}(X, \mathcal{F}_{Np^\infty,\varepsilon}) \otimes_{\Lambda_{Np^\infty,\varepsilon}} Q(\Lambda_{Np^\infty,\varepsilon})$ <u>under</u> 3.3.7 (3) (resp. (4)) <u>is a</u> $\Lambda_{Np^\infty,\varepsilon}$<u>-basis of</u> $\Delta_{\Lambda_{Np^\infty,\varepsilon}}(X, \mathcal{F}_{Np^\infty,\varepsilon})$.

3.3.9. In fact, in the case $S = \{p\}$ (in this case, $\Lambda_\varepsilon = \Lambda_{Np^\infty,\varepsilon}$ holds), Λ_ε is a regular ring, and the case $\varepsilon = +$ (resp. $-$) of 3.3.8 is equivalent by 2.1.5 to the familiar form of the classical Iwasawa main conjecture

$$\mathrm{char}_{\Lambda_+}(A/B) = \mathrm{char}_{\Lambda_+}(C) \quad \text{with}$$

$$A = H^1(X, \mathcal{F}_+) \cong \varprojlim_n ((\mathbb{Z}[\alpha_{p^n}, \tfrac{1}{p}]^+)^\times \otimes \mathbb{Z}_p)$$

$$B = \text{Image of } H^0(\mathbb{R}, \mathcal{F}_+(-1)) \longrightarrow H^1(X, \mathcal{F}_+) \qquad (3.3.5)$$

$$C = H^2(X, \mathcal{F}_+) \cong \varprojlim_n \mathrm{Pic}(\mathbb{Z}[\alpha_{p^n}]^+)(p)$$

where the two $\cong$ comes from the Kummer sequence

(resp.

$$\Lambda_- \cdot z_{\Lambda_-}(X, \mathcal{F}_-) = \mathrm{char}_{\Lambda_-}(A)\,\mathrm{char}_{\Lambda_-}(B)^{-1} \quad \text{with}$$

$$A = H^2(X, \mathcal{F}_-) \cong \varprojlim_n \mathrm{Pic}(\mathbb{Z}[\alpha_{p^n}])(p)^-$$

$$B = \mathrm{char}_{\Lambda_-}(\mathbb{Z}_p(1)).$$

where the isomorphism comes from the Kummer sequence and (-) denotes the part on which the complex conjugation acts by -1).

So we have defined a Λ-basis $z_\Lambda(X, \mathcal{F})$ of $\Delta_\Lambda(X, \mathcal{F})$ in the case $S = \{p\}$. For S general, the result of Mazur-Wiles [MW] and Wiles [Wi$_2$] prove 3.3.8 in the case $\Lambda_{Np^\infty, \varepsilon}$ is a regular ring (in this case, by 2.1.5, 3.3.8 is rewritten as an equation concerning "Char" as above which was proved in [MW][Wi$_2$]). However if $\Lambda_{Np^\infty, \varepsilon}$ is not regular, I can not deduce 3.3.8 from results in [MW][Wi$_2$].

§3.4. Problems.

Let p be an odd prime number.

3.4.1. We discuss some problems which naturally arise from our Iwasawa main conjecture. We discuss the following questions.

(1) Is the definition of zeta elements in a special case given in §3.3 compatible with the conditions (v) in Conj. 3.2.2? (See 3.4.4.)

(2) How to construct zeta elements? Here I confess that I do not know a good method. (See 3.4.11.)

(3) How useful are zeta elements for arithmetic algebraic geometry? (See 3.4.13.)

3.4.2. In the following 3.4.3-3.4.9, let $(X, \Lambda, \mathcal{F})$ be as in 3.3.1 and let A be a finite product of finite extensions of $\mathbf{Q}$ such that $\Lambda = A \otimes \mathbf{Q}_p$. Fix an algebraic closure $\bar{\mathbf{Q}}$ of $\mathbf{Q}$ and let $\mathbf{Q}^{ab, S}$ be as in 3.3.2. Let

$$V = \Gamma(\mathrm{Spec}(\bar{\mathbf{Q}}), \mathcal{F}) .$$

3.4.3. Some philosophy of motif tells us that if $\mathcal{F}$ comes from an A-motif M, then there exist a finite subextension L of $\mathbf{Q}^{ab, S}/\mathbf{Q}$, a homomorphism $\eta : A' \underset{\mathrm{def}}{=} \mathbf{Q}[\mathrm{Gal}(L/\mathbf{Q})] \longrightarrow A$, and $r \in \mathbf{Z}$ such that $M \cong H^0(\mathrm{Spec}(L))(r) \otimes_{A'} A$ as an A-motif. In this case,

$$\chi_{\mathcal{F}}(\sigma) = \chi_{\mathrm{cyclo}}(\sigma)^r \eta(\sigma)^{-1} \quad \text{for all } \sigma \in \mathrm{Gal}(\mathbf{Q}^{ab, S}/\mathbf{Q}) .$$

where we denote the composite homomorphism $\mathrm{Gal}(\mathbf{Q}^{ab, S}/\mathbf{Q}) \longrightarrow \mathrm{Gal}(L/\mathbf{Q}) \xrightarrow{\text{by } \eta} A^{\times}$ by the same letter η.

3.4.4. In 3.4.5-3.4.9, let $r \in \mathbf{Z}$ and assume

$$\chi_{\mathcal{F}}(\sigma) = \chi_{\mathrm{cyclo}}(\sigma)^r \lambda(\sigma)$$

for a homomorphism $\lambda : \mathrm{Gal}(\mathbf{Q}^{ab, S}/\mathbf{Q}) \longrightarrow A^{\times}$ which is continuous with respect to the discrete topology of $A^{\times}$.

We will discuss the problem 3.4.1 (1) as follows. Here $\tau \in \mathrm{Gal}(\mathbf{Q}^{ab, S}/\mathbf{Q})$ is the complex conjugation.

The case $r \geq 1$ and $\chi_{\mathcal{F}}(\tau) = 1$ ...... in 3.4.8 .

The case $r \geq 1$ and $\chi_{\mathcal{F}}(\tau) = -1$ ..... in 3.4.9 .

The case $r \leq 0$ and $\chi_{\mathcal{F}}(\tau) = 1$ ..... in Chap. III, 1.2.6.

The case $r \leq 0$ and $\chi_{\mathcal{F}}(\tau) = -1$ .... in Chap. III, §3.

We consider the L-function (2.2.3). We see easily that the conditions (i)(ii) of 2.2.3 are satisfied and:

Lemma 3.4.5. Let N be an integer ≥ 1 such that S coincides with the set of prime divisors of N and such that λ factors through $\mathrm{Gal}(\mathbf{Q}(\alpha_N)/\mathbf{Q})$ where α_N is a primitive N-th root of 1 in $\bar{\mathbf{Q}}$. Then we have

$$L_{A,S}(V^*(1), s) = \sum_{a \in (\mathbf{Z}/N)^\times} \zeta_{a(N)}(s+1-r)\lambda(\sigma_a)$$

where σ_a is the element of $\mathrm{Gal}(\mathbf{Q}^{ab,S}/\mathbf{Q})$ such that $\sigma_a(\alpha_N) = \alpha_N^a$.

We consider first the case where $r \geq 1$ and $\chi_{\mathcal{F}}(\tau) = 1$. By Thm. 1.3.7 and 3.4.5, we have

Proposition 3.4.6. Assume $r \geq 1$ and $\chi_{\mathcal{F}}(\tau) = 1$. Let z_- be the element of $\mathbf{Z}_p[[\mathrm{Gal}(\mathbf{Q}^{ab,S}/\mathbf{Q})]][(1 - \chi_{cyclo}(\mu)\mu^{-1})^{-1}]/(\tau+1)$ denoted by $z_-(X, \mathcal{F}_-)$ in 3.3.5, where μ is any element of $\mathrm{Gal}(\mathbf{Q}^{ab,S}/\mathbf{Q})$ such that $\chi_{cyclo}(\mu)$ is of infinite order. Then the image of z_- under

$$\mathbf{Z}_p[[\mathrm{Gal}(\mathbf{Q}^{ab,S}/\mathbf{Q})]][(1 - \chi_{cyclo}(\mu)\mu^{-1})^{-1}]/(\tau+1) \longrightarrow \Lambda$$

$$\sigma \longmapsto \chi_{cyclo}(\sigma)\chi_{\mathcal{F}}(\sigma)^{-1} \qquad (\sigma \in \mathrm{Gal}(\mathbf{Q}^{ab,S}/\mathbf{Q}))$$

coincides with $L_{A,S}(V^*(1), 0)$.

By Soulé [So₂], we have

Proposition 3.4.7. Assume $r \geq 2$ and $\chi_{\mathcal{F}}(\tau) = 1$. Then, $H^i(\mathbf{Z}[\frac{1}{S}], \mathcal{F}) = 0$ for all $i \in \mathbf{Z}$.

Indeed, for N as in 3.4.5, there is a surjection $H^i(\mathbf{Z}[\alpha_N, \frac{1}{S}], \mathbf{Q}_p(r)) \cong H^i(\mathbf{Z}[\alpha_N, \frac{1}{S}], \mathcal{F}) \longrightarrow H^i(\mathbf{Z}[\frac{1}{S}], \mathcal{F})$ (the trace map), and $H^2(\mathbf{Z}[\alpha_N, \frac{1}{S}], \mathbf{Q}_p(r)) = 0$ by Soulé [So₁]. Hence $H^i(\mathbf{Z}[\frac{1}{S}], \mathcal{F}) = 0$ for all $i \neq 1$, and we have $H^1(\mathbf{Z}[\frac{1}{S}], \mathcal{F}) = 0$ by the equation (•) in 3.1.3 (2).

3.4.8. Assume $r \geq 2$, $\chi_{\mathcal{F}}(\tau) = 1$, and assume the philosophy of motif described in 3.4.3 is correct. Then $\Sigma = 0$, $\Omega = 0$ for the A-motif which yields $\mathcal{F}$. By this and by 3.4.7, we see that we are in the strictly critical situation (3.2.6).

Hence 3.4.6 saids that if Conj. 3.3.8 is true, the canonical isomorphism $\Delta_\Lambda(X, \mathcal{F}) \cong \Lambda$ (which comes from $\det_\Lambda(0) \cong \Lambda$) sends $z_\Lambda(X, \mathcal{F})$ to $L_{A,S}(V^*(1), 0)$.

This shows the compatibility of the construction of §3.3 with 3.2.2 (v) in the case $r \geq 2$ and $\chi_{\mathcal{F}}(\tau) = 1$.

We can obtain a similar conclusion in the case $r = 1$ and $\chi_{\mathcal{F}}(\tau) = 1$, by a further careful study, but we do not discuss this here.

3.4.9. We next consider the case $r \geq 1$ and $\chi_{\mathcal{F}}(\tau) = -1$.

Assume the philosophy of motif described in 3.4.3 is correct, and consider the A-motif M which yields $\mathcal{F}$. So, $M = M' \otimes_{A'} A$ where $M' = H^0(\mathrm{Spec}(L))(r)$ and $A' = \mathbf{Q}[\mathrm{Gal}(L/\mathbf{Q})]$ for a finite subextension L of $\mathbf{Q}^{ab,S}/\mathbf{Q}$ and for a homomorphism $A' \longrightarrow A$. We have $M'_B = \bigoplus_\iota \mathbf{Q}(2\pi i)^r$ where ι ranges over all embeddings $L \longrightarrow \mathbf{C}$, and $M_B = M'_B \otimes_{A'} A$. We should have also

$$\Psi = \Psi' \otimes_{A'} A \quad \text{where} \quad \Psi' = K_{2r-1}(L) \otimes \mathbf{Q}.$$

Let

$$a : \Psi \otimes \mathbf{R} \longrightarrow \Sigma \otimes \mathbf{R} \quad \text{with} \quad \Sigma = M_B(-1)^+$$

$$b : \Psi \otimes \mathbf{Q}_p \longrightarrow H^1_{\lim}$$

be the homomorphisms induced from the regulator map $K_{2r-1}(L) \longrightarrow M'_B(-1)^+ \otimes \mathbf{R}$ and the Chern class map $K_{2r-1}(L) \longrightarrow \varprojlim_S H^1(O_L[\frac{1}{S}], \mathbf{Q}_p(r))$.

Now assume $r \geq 2$. Then by Borel [Bo] and Soulé [So_2], the maps a and b are bijective, and $H^2(\mathbf{Z}[\frac{1}{S}], \mathcal{F}) = 0$. Thus we are in the strictly K-theoretic case. Furthermore, Ψ is an invertible Λ-module by Borel [Bo].

By Beilinson ($[Be_2]$), there exists an A-basis of $\Psi \otimes_A \Sigma^{-1}$ whose image in $A \otimes \mathbb{R}$ with respect to the map a coincides with $\pm \lim_{s \to 0} s^{-1} L_{A,S}(V^*(1), 0)$, and whose image in $H^1_{lim} \otimes_A \Sigma^{-1}$ under the map b coincides with the following element. This element is the image of $z_+(X, \mathcal{F}_+)$ in 3.3.5 under the map

$$H^1(X, \mathcal{F}_+) \otimes_\Lambda H^0(R, \mathcal{F}_+(-1))^{-1} \longrightarrow H^1(X, \mathcal{F}) \otimes_\Lambda H^0(R, \mathcal{F}(-1))^{-1}$$
$$\cong H^1(X, \mathcal{F}) \otimes_A \Sigma^{-1}$$

where Λ_+ and $\mathcal{F}_+$ are as in 3.3.5 and where the first map is induced by $\mathcal{F}_+ \otimes_{\Lambda_+} \Lambda \cong \mathcal{F}$ with

$$\Lambda_+ \longrightarrow \Lambda \;;\quad \sigma \longmapsto \chi_{cyclo}(\sigma)\chi_{\mathcal{F}}(\sigma)^{-1} \qquad (\sigma \in Gal(Q^{ab,S}/Q)\,.$$

This shows the compatibility of the construction in §3 with 3.2.2 (v) in the case $r \geq 2$ and $\chi_{\mathcal{F}}(\tau) = -1$.

We can obtain a similar conclusion in the case $r = 1$ and $\chi_{\mathcal{F}}(\tau) = -1$, by a further careful study basing on 3.3.6, but we do not discuss this here.

<u>Remark</u> 3.4.10. As we have seen in 3.4.6-3.4.8, <u>the phase 2 (the p-adic property) of Riemann zeta function considered in §1.3 is included in our Iwasawa main conjecture</u>.

Roughly speaking, how our approach includes the phase 2 is the following: If r and r′ are integers satisfying $r \equiv r' \bmod (p-1)p^{n-1}$, then $\mathbb{Z}/p^n(r) \cong \mathbb{Z}/p^n(r')$ as a sheaf on $Spec(\mathbb{Z}[\frac{1}{p}])$. Hence the zeta element of $\mathbb{Z}_p(r)$ that of $\mathbb{Z}_p(r')$ coincide mod p^n. From this, the congruence between zeta values (1.3.4 (3)) arises. The correspondence $r \longmapsto \mathbb{Z}_p(r)$ for $r \in \mathbb{Z}$ is extended to $r \in \varprojlim \mathbb{Z}/(p-1)p^n\mathbb{Z}$. This explains the existence of the p-adic interporation of special values of Riemann zeta function (1.3.4 (4)).

In Chap. III §3, we will discuss the general relationship between our Iwasawa main conjecture and p-adic L-functions.

3.4.11. How to construct p-adic zeta elements is clearly an important problem, but I do not have any good idea. Is it possible to construct them not considering zeta values (just considering p-adic etale cohomology)?

Here is one comment about the construction: it seems possible to reduce the construction of the p-adic zeta element of M_p for an A-motif M to the strictly K-theoretic case. The idea is as follows. Let S be a finite set of primes containing p, let A be a finite product of finite extensions of Q, and let M be an A-motif such that M_p is unramified outside S. We want to define $z_\Lambda(X, M_p)$ where $\Lambda = A \otimes Q_p$, $X = \mathrm{Spec}(\mathbb{Z}[\frac{1}{S}])$. Then, it is conjectured that for a sufficiently large $r \geq 0$, all the motives $M(r) \otimes_Q H^0(\mathrm{Spec}(Q(\beta_n)))$ are strictly K-theoretic for any $n \geq 0$ where β_n is a primitive p^n-th root of 1. Let $O_\Lambda \subset \Lambda$ be as in 3.1.2, and take a smooth O_Λ-sheaf $\mathcal{F}$ on X such that $\mathcal{F} \otimes_{O_\Lambda} \Lambda = M_p$. Let $G = \mathrm{Gal}((\bigcup_n Q(\beta_n))/Q)$ and let

$$\Lambda' = O_\Lambda[[G]], \qquad \mathcal{F}' = \varprojlim_n f_{n*} f_n^*(\mathcal{F}(r)) \in D_{ctf}(X, \Lambda')$$

where f_n denotes the canonical morphism $\mathrm{Spec}(\mathbb{Z}[\beta_n, \frac{1}{S}]) \longrightarrow \mathrm{Spec}(\mathbb{Z}[\frac{1}{S}])$. Then,

(1) $\quad M_p \cong \mathcal{F}' \otimes_{\Lambda'} \Lambda \quad$ with respect to the O_Λ-homomorphism

$$\Lambda' \longrightarrow \Lambda \; ; \; \sigma \longmapsto \chi_{cyclo}(\sigma)^r \quad (\sigma \in G).$$

Assume we have already defined the zeta element of the p-adic etale realization $f_{n*} f_n^*(\mathcal{F}(r)) \otimes_{O_\Lambda} \Lambda$ of the $A[\mathrm{Gal}(Q(\beta_n)/Q)]$-motif $M(r) \otimes_Q H^0(\mathrm{Spec}(Q(\beta_n)))$ for any $n \geq 0$. Then, $z_{\Lambda'}(X, \mathcal{F}')$ should be obtained as the inverse limit of them. And $z_\Lambda(X, \mathcal{F})$ should be obtained as the image of $z_{\Lambda'}(X, \mathcal{F}')$ under (1).

There are two questions:

(i) How to construct zeta elements in the strictly K-theoretic case.

(ii) Does the zeta element $z_{\Lambda}(X, \mathcal{F})$ obtained in the above way satisfy 3.2.2 (v)?

The question (ii) and the similar question 3.4.1 (1) are illustrated as "Are Tate twists of great p-adic elements still great?". (The word "great" expresses that they are related to zeta values.). This question (ii) (whose answer is "yes" if our Iwasawa main conjecture is true) is one of the deepest problems arising from our Iwasawa main conjecture.

The question (i) is clearly also important. My only suggestion is that we may learn how to construct $\omega_{\gamma} \in \det_{A}(\Psi)$ (then z_{γ} will be obtained as the image of ω_{γ} under the Chern class map) from S. Bloch, A. Beilinson, and other people who have discovered important elements in K-theory related to zeta values ([Bl] lect. 8, 9, [Be_1], etc.). It seems to me that the only known general method to discover such important elements is to open our mouths and wait for such elements to drop from the sky. However I do not know why these people with small mouths can catch such elements so often.

3.4.12. There are many Λ-basis of $\Delta_{\Lambda}(X, \mathcal{F})$, but our conjecture says that there is a special important basis.

Such special basis should have a beautiful definition (as in §3.3, and also as in Chap. III §1.2 where some zeta elements arise from elliptic units having beautiful definitions). The definitions of the known important elements in K-theory related to zeta values are also beautiful (cf. [Ra]). How to understand this?

The existence of such special element is similar to the existence of a special woman (resp. a man) for a man (resp. a woman). Is this analogy helpful for us in finding the zeta elements?

3.4.13. I explain that <u>zeta elements have big powers and useful</u> (once they are found to exist) <u>in the study of arithmetic geometry</u>.

For example, let X be an abelian variety over $\mathbf{Q}$ of dimension g, let S be a finite set of primes containing p such that X has good reductions outside S. and assume $L_S(H^1(X), s)$ has a meromorphic continuation to whole $\mathbf{C}$ and has order 0 at $s = 1$. Let $V = H^1(X \otimes_{\mathbf{Q}} \bar{\mathbf{Q}}, \mathbf{Q}_p)(1)$. Then the conjecture in §2.3 (or the conjectural property 3.2.2 (v) of the zeta element for the triple $(\mathrm{Spec}(\mathbf{Z}[\frac{1}{S}]), \mathbf{Q}_p, V))$ saids the following: There exists an element z of $\overset{g}{\Lambda} H^1(\mathbf{Z}[\frac{1}{S}], V)$ whose image in the one dimensional $\mathbf{Q}_p$-vector space $\overset{g}{\Lambda}\,\Gamma(X, \Omega^1_{X/\mathbf{Q}}) \otimes \mathbf{Q}_p$ under $\overset{g}{\Lambda} \exp^*$ is non-zero. Assume we have discovered such element z. (In fact in some very special cases, we will construct z in a forthcoming paper [Ka_3] using the idea in 3.4.11.) Then we obtain a strong consequence, the finiteness of $X(\mathbf{Q})$, as is shown below. So, I think that the philosophy of zeta elements is useful as a working hypothesis for the study of arithmetic geometry.

Now we deduce the finiteness of $X(\mathbf{Q})$ from the existence of z. We have that the map $H^1(\mathbf{Z}[\frac{1}{S}], V) \longrightarrow \Gamma(X, \Omega^1_{X/\mathbf{Q}}) \otimes \mathbf{Q}_p$ is surjective. Consider the commutative diagram

$$\begin{array}{ccccc} X(\mathbf{Q}) \otimes \mathbf{Q} & \xrightarrow{\subset} & X(\mathbf{Q}_p) \otimes \mathbf{Q} & \xleftarrow[\cong]{\exp} & \mathrm{Lie}(X) \otimes \mathbf{Q}_p \\ \downarrow & & \downarrow & & \downarrow \\ H^1(\mathbf{Q}, V_pX) & \longrightarrow & H^1(\mathbf{Q}_p, V_pX) & \xrightarrow{(*)} & \mathrm{Hom}_{\mathbf{Q}_p}(H^1(\mathbf{Z}[\frac{1}{S}], V), \mathbf{Q}_p) \end{array}$$

Here the left and middle vertical arrows are defined as in 2.3.6, and

(∗) is defined by Tate duality. The right vertical arrow is the $\mathbb{Q}_p$-dual of the surjection in problem, and hence it is injective. On the other hand, the composite of the lower rows is zero by the global reciprocity law of the Tate duality. Hence $X(\mathbb{Q}) \otimes \mathbb{Q} \longrightarrow X(\mathbb{Q}_p) \otimes \mathbb{Q}$ is zero, and this implies the finiteness of $X(\mathbb{Q})$.

3.4.14. In this section, we excluded the case $p = 2$ for the simplicity of the story. We can formulate our Iwasawa main conjecture for $p = 2$ as follows. Consider a triple $(X, \Lambda, \mathcal{F})$ as in 3.2.1. We consider the case $X = \mathrm{Spec}(\mathbb{Z}[\frac{1}{p}])$. (Recall that in 3.2.2, zeta elements of general case is reduced to this case by 3.2.2 (ii).) We define an object $R\Gamma_c(X, \mathcal{F})$ of $D_{ctf}(\Lambda)$ as follows. We describe the definition assuming Λ is a finite ring (the general case is reduced to this case in the evident way). Take an injective resolution $\mathcal{F} \longrightarrow I^{\cdot}$ of $\mathcal{F}$, and let $R\Gamma_c(X, \mathcal{F})$ be mapping fiber of

(1) $\quad \Gamma(X, I^{\cdot}) \longrightarrow \Gamma(\mathbb{R}, I^{\cdot}) \oplus \Gamma(\mathbb{Q}_p, I^{\cdot}),$

that is, the simple complex associated to the double complex (1). We define

$$\Delta_{c,\Lambda}(X, \mathcal{F}) = (\det_\Lambda R\Gamma_c(X, \mathcal{F}))^{-1} .$$

Let $\mathcal{F}^* = R\underline{\mathrm{Hom}}_\Lambda(\mathcal{F}, \Lambda)$. Then, if $p \neq 2$, the duality theory of etale cohomology in algebraic number theory ([Ma]) says that there is a canonical Λ-isomorphism

$$\Delta_{c,\Lambda}(X, \mathcal{F}) \cong \Delta_\Lambda(X, \mathcal{F}^*(1)) .$$

Hence in the case $p \neq 2$, we can rewrite 3.2.2 for $(X, \Lambda, \mathcal{F})$ by using $\Delta_{c,\Lambda}(X, \mathcal{F}^*(1))$. And we can formulate the extension of 3.2.2 including the case $p = 2$ in the evident way by using $\Delta_{c,\Lambda}(X, \mathcal{F})$.

I prefered Δ to Δ_c in the text of §3 because some great elements live in H^1 of $R\Gamma$ (as we have seen) but not in the cohomology of $R\Gamma_c$.

§3.5. Varieties over finite fields.

3.5.1. Recall that the point of the departure of the classical Iwasawa theory was the following belief of Iwasawa concerning the deep analogy between curves over finite fields and Spec of rings of integers of number fields; the expression of the zeta functions of the former in terms of the Tate modules of Jacobian varieties (or in terms of the etale cohomology theory) should have counter parts for zeta functions of the latter.

In this §3.5, let $(X, \Lambda, \mathcal{F})$ be a triple as in 3.2.2 (iv) (= the geometric case) with Λ as in 3.1.1 (2). The purpose of §3.5 is to show (Prop. 3.5.7) that the zeta element $z_\Lambda(X, \mathcal{F})$ defined in 3.2.2 (iv) is related to zeta values really (not conjecturally) under a mild condition given in 3.5.3. For example, if $H^m(X \otimes_{F_\ell} \bar{F}_\ell, \mathcal{F}) = 0$ for all $m \in \mathbb{Z}$, Prop. 3.5.7 says that the canonical identifications

$$\Delta_\Lambda(X, \mathcal{F}) = (\det_\Lambda(\{0\}))^{-1} = \Lambda$$

send $z_\Lambda(X, \mathcal{F})$ to the value at $t = 1$ of the zeta function $Z_\Lambda(X, D_\Lambda(\mathcal{F}), t)$, up to sign. Here

$$D_\Lambda(\mathcal{F}) = R\underline{Hom}_\Lambda(\mathcal{F}, f^!\Lambda)$$

where f is the structural morphism $X \longrightarrow \mathrm{Spec}(F_\ell)$, and the zeta function $z_\Lambda(X, ?, t)$ is defined by

$$Z_\Lambda(X, \mathcal{F}, t) = \prod_{q,x} \det_\Lambda(1 - \sigma_x^{-1} t^{\deg(x)} ; \mathcal{H}^q(\mathcal{F})_{\bar{x}})^{(-1)^q} \in \Lambda[[t]]$$

$$= \det_\Lambda(1 - \sigma_\ell^{-1} t ; R\Gamma_c(X \otimes_{F_\ell} \bar{F}_\ell, \mathcal{F}))^{-1}$$

where q ranges over integers, x ranges over closed points of X, $\mathcal{H}^q(\mathcal{F})$ is the q-th cohomology sheaf of $\mathcal{F}$, $(\ )_{\bar{x}}$ is the stalk at a geometric point over x, $\sigma_\ell \in \mathrm{Gal}(\bar{F}_\ell/F_\ell)$ is the arithmetic Frobenius, and the last equation is by Grothendieck [Gr].

3.5.2. Let $\mathscr{C}(\mathcal{F})$ be the complex

$$H^0(X, \mathcal{F}) \xrightarrow{\cup\theta} H^1(X, \mathcal{F}) \xrightarrow{\cup\theta} H^2(X, \mathcal{F}) \xrightarrow{\cup\theta} \cdots\cdots$$

where $\cup\theta$ means the cup product with the generator θ of $H^1(F_\ell, Z_p) = \mathrm{Hom}_{cont}(\mathrm{Gal}(\bar{F}_\ell/F_\ell), Z_p)$ which sends $\sigma_k \in \mathrm{Gal}(\bar{F}_\ell/F_\ell)$ to $1 \in Z_p$.

For $i \in Z$, let P^i (resp. Q^i) be the kernel (resp. cokernel) of

$$1 - \sigma_\ell \ : \ H^i(\bar{X}, \mathcal{F}) \longrightarrow H^i(\bar{X}, \mathcal{F}) .$$

In the spectral sequence

$$E_2^{i,j} = H^i(F_\ell, H^j(\bar{X}, \mathcal{F})) \Rightarrow E_\infty^{i+j} = H^{i+j}(X, \mathcal{F}) ,$$

$E^{i,j}$ is P^j if $i = 0$, Q^j if $i = 1$, and is zero if $i \geq 2$. Hence we obtain short exact sequences

$$(1) \qquad 0 \longrightarrow Q^{i-1} \longrightarrow H^i(X, \mathcal{F}) \longrightarrow P^i \longrightarrow 0 \qquad (i \in Z),$$

and $\cup\theta$ coincides with the composite map

$$H^i(X, \mathcal{F}) \longrightarrow P^i \xrightarrow{\alpha_i} Q^i \longrightarrow H^{i+1}(X, \mathcal{F})$$

where the first and the last arrows are those of (1), and α_i is the composite map

$$P^i \xrightarrow{incl} H^i(\bar{X}, \mathcal{F}) \xrightarrow{proj} Q^i$$

with incl the inclusion map and proj the projection to the quotient. Hence we have

<u>Lemma</u> 3.5.3. <u>The following two conditions</u> (1)(2) <u>are equivalent</u>.

(1) <u>The complex</u> $\mathscr{C}(\mathcal{F})$ <u>is exact</u>.

(2) <u>The maps</u> $\alpha_i : P^i \longrightarrow Q^i$ <u>are bijective for all</u> i.

<u>Remark</u> 3.5.4. The above condition (2) (and hence (1) also) is satisfied if the actions of the Frobenius σ_ℓ on $H^i(\bar{X}, \mathcal{F})$ are semi-simple for all i. It is conjectured that this semi-simplicity holds if $\mathcal{F}$ comes from "a motif over X".

The following lemma follows by Poincaré duality between $R\Gamma$ for $\mathcal{F}$

and $R\Gamma_c$ for $D_X(\mathcal{F})$.

<u>Lemma</u> 3.5.5. <u>Assume the equivalent conditions in</u> 3.5.3 <u>are satisfied</u>. <u>Then the order</u> e <u>of</u> $Z_\Lambda(X, D_\Lambda(\mathcal{F}), t)$ <u>at</u> $t = 1$ <u>coincides with</u>

$$\sum_{i\in\mathbb{Z}} (-1)^i \mathrm{rank}_\Lambda(P^i) = \sum_{i\in\mathbb{Z}} (-1)^i \mathrm{rank}_\Lambda(Q^i)$$

<u>as a function</u> $\mathrm{Spec}(\Lambda) \longrightarrow \mathbb{Z}$.

3.5.6. Assuming the equivalent conditions in 3.5.3 are satisfied, we define an isomorphism

$$\lambda \; : \; (\det_\Lambda(R\Gamma(X, \mathcal{F})))^{-1} \longrightarrow \Lambda$$

as the composite

$$(\det_\Lambda(R\Gamma(X, \mathcal{F})))^{-1} \cong \bigotimes_{i\in\mathbb{Z}} (\det_\Lambda(H^i(X, \mathcal{F})))^{(-1)^{i-1}}$$

$$\cong (\det_\Lambda(\mathcal{C}(\mathcal{F})))^{(-1)} \cong \Lambda$$

where the last isomorphism is by the acyclicity of $\mathcal{C}(\mathcal{F})$ (that is, $\mathcal{C}(\mathcal{F}) = 0$ in the derived category). This isomorphism λ is in general different from the isomorphism discussed in 3.2.2 (vi) (the difference of them is the theme of Prop. 3.5.7). If $R\Gamma(X, \mathcal{F}) = 0$, the very simple isomorphism coming from $\det_\Lambda(0) \cong \Lambda$ is λ.

The map λ coincides up to sign with the composite isomorphism

$$(\det_\Lambda(R\Gamma(X, \mathcal{F})))^{-1} \cong \bigotimes_{i\in\mathbb{Z}} (\det_\Lambda(H^i(X, \mathcal{F})))^{(-1)^{i-1}}$$

$$\cong \bigotimes_{i\in\mathbb{Z}} (\det_\Lambda(P^i) \otimes (\det_\Lambda(Q^i))^{-1})^{(-1)^{i-1}} \cong \Lambda$$

where the second isomorphism is induced by the short exact sequences 3.5.2 (1) and the last isomorphism is induced by $\alpha_i : P^i \xrightarrow{\cong} Q^i$.

<u>Proposition</u> 3.5.7. <u>Assume the equivalent conditions in</u> 3.5.3 <u>are satisfied</u>. <u>Then the isomorphism</u> λ <u>sends</u> $z_\Lambda(X, \mathcal{F})$ <u>to</u>

$$\pm ((1-t)^{-e} Z_\Lambda(X, D_\Lambda(\mathcal{F}), t))_{t=1} \in \Lambda^\times$$

where $e : \mathrm{Spec}(\Lambda) \longrightarrow \mathbb{Z}$ is the order of $Z_\Lambda(X, D_\Lambda(\mathcal{F}), t)$ at $t = 1$.

This proposition follows from

Lemma 3.5.8. Let R be a finite product of fields, let C be a perfect complex of R-modules and let $h : C \longrightarrow C$ be a homomorphism. For $i \in \mathbb{Z}$, let P^i (resp. Q^i) be the kernel of $H^i(1-h)$: $H^i(C) \longrightarrow H^i(C)$ and assume that for any i, the composite map $\alpha_i : P^i \xrightarrow{\text{incl}} H^i(C) \xrightarrow{\text{proj}} Q^i$ is bijective. Let $e : \operatorname{Spec}(R) \longrightarrow \mathbb{Z}$ be the order of $f(t) \underset{\text{def}}{=} \prod_{i\in\mathbb{Z}} \det_R(1 - ht \; ; \; H^q(C) \longrightarrow H^q(C))^{(-1)^q}$ at $t = 1$. Then e is equal to $\sum_{i\in\mathbb{Z}} (-1)^i \operatorname{rank}_R(P^i) = \sum_{i\in\mathbb{Z}} (-1)^i \operatorname{rank}_R(Q^i)$, and the composite map

$$(*) \qquad R \cong \det_R(C) \otimes_R \det_R(C)^{-1}$$
$$\cong \bigotimes_{i\in\mathbb{Z}} \{\det_R(H^i(C))^{(-1)^i} \otimes_R \det_R(H^i(C))^{(-1)^{i-1}}\}$$
$$\cong \bigotimes_{i\in\mathbb{Z}} \{\det_R(P^i)^{(-1)^i} \otimes_R \det_R(Q^i)^{\otimes(-1)^{i-1}}\} \cong R$$

sends 1 to $\pm \lim_{t\to 0} (1-t)^{-e} f(t)$. Here in $(*)$, the first isomorphism comes from the identity map between the two C, the second comes from $\det_R(C) \cong \bigotimes_{i\in\mathbb{Z}} \det_R(H^i(C))^{\otimes(-1)^i}$, the third comes from the exact sequences $0 \longrightarrow P^i \longrightarrow H^i(C) \xrightarrow{1-h} H^i(C) \longrightarrow Q^i \longrightarrow 0$, and the last comes from the isomorphisms α_i.

Proof. This is reduced to the case where C is concentrated in degree 0, and furthermore $1 - h$ is either an isomorphism or the zero map.

Remark 3.5.9. Consider a triple $(X, \mathbb{Z}_p, \mathfrak{F})$ with X as in 3.2.2 (iv). Assume $\mathcal{F} \otimes_{\mathbb{Z}_p} \mathbb{Q}_p$ satisfies the equivalent conditions in 3.5.3. Then, 3.5.7 for $\mathcal{F} \otimes_{\mathbb{Z}_p} \mathbb{Q}_p$ shows that $[(1-t)^{-e} Z_{\mathbb{Q}_p}(X, D_\Lambda(\mathcal{F} \otimes_{\mathbb{Z}_p} \mathbb{Q}_p), t)]_{t=1}$ is a $\mathbb{Z}_p^\times$-multiple of $\prod_{q\in\mathbb{Z}} \#(H^q(\mathcal{C}(\mathcal{F})))^{(-1)^q}$.

Chapter II. Local theory.

Recall our philosophy is that, in the Iwasawa theory for Hasse-Weil L-functions, the theory of p-adic periods related to B_{dR} is the most important local theory.

In this chapter, K denotes a complete discrete valuation field of characteristic 0 with perfect residue field of characteristic $p > 0$. We denote by O_K the valuation ring of K, and by m_K the maximal ideal of K.

We denote by K_0 the field of fractions of the ring W(k) of p-Witt vectors, and regard K_0 as a subfield of K in the canonical way.

We fix an algebraic closure $\bar{K}$ of K. We denote by K_{ur} the maximal unramified extension of K in $\bar{K}$.

$H^q(K, \ )$ denotes the continuous Galois cohomology $H^q_{cont}(Gal(\bar{K}/K), \ )$ [Ta$_2$]. Let $\mathbb{Z}_p(1) = H^0(K, \mathbb{Z}_p(1))$.

§1. p-adic Galois representations of p-adic fields.

This §1 is a review of the theory of p-adic periods related to B_{dR}, except that §1.2 and §1.4 contain a new material "the dual exponential map". Cf. Fontaine [Fo$_1$], Fontaine and Messing [FM], Faltings [Fa], Bloch-Kato [BK], Illusie [Il].

§1.1. B_{dR} and de Rham representations.

We review the field B_{dR} and the theory of Fontaine on de Rham representations. The idea to use crystalline cohomology theory for the definition of B_{dR} (different from the original method in [Fo$_1$])

is due to [FM] (this comment applies also to the definition of B_{crys} introduced in §1.3).

§1.1.1. The definition of B_{dR}. For $n \geq 1$, let

$$B_{n,\bar{K}/K} = H^0(\mathrm{Spec}(O_{\bar{K}}/p^n)/\mathrm{Spec}(O_K/p^n)_{crys}, \mathcal{O}_{crys})$$

where $\mathrm{Spec}(O_{\bar{K}}/p^n)/\mathrm{Spec}(O_K/p^n)_{crys}$ is the crystalline site of the scheme $\mathrm{Spec}(O_{\bar{K}}/p^n)$ over the base scheme $\mathrm{Spec}(O_K/p^n)$ with respect to the canonical divided power structure on $p(O_K/p^n)$, and $\mathcal{O}_{crys}$ denotes the structural sheaf of the crystalline site. Then, the canonical map $B_{n,\bar{K}/K} \longrightarrow O_{\bar{K}}/p^n$ is surjective. Let $J_{n,\bar{K}/K}$ be the kernel of this surjection, and define the ideal $J^{[r]}_{n,\bar{K}/K}$ of $B_{n,\bar{K}/K}$ for $r \geq 0$ as the r-th divided power of $J_{n,\bar{K}/K}$. Then for $r \geq 0$, $J^{[r]}_{n,\bar{K}/K}/J^{[r+1]}_{n,\bar{K}/K}$ is a free $O_{\bar{K}}/p^n$-module of rank 1 generated by the class of $(x^{p^n} - \pi)^{[r]}$ for any prime element π of O_K and any element x of $B_{n,\bar{K}/K}$ such that the image of x^{p^n} in $O_{\bar{K}}/p^n$ coincides with the class of π. Let

$$B_{\infty,\bar{K}/K} = \varprojlim_n B_{n,\bar{K}/K}, \qquad J^{[r]}_{\infty,\bar{K}/K} = \varprojlim_n J^{[r]}_{n,\bar{K}/K}.$$

Define

$$B^+_{dR} = \varprojlim_r \left((B_{\infty,\bar{K}/K}/J^{[r]}_{\infty,\bar{K}/K}) \otimes \mathbf{Q}\right).$$

Then B^+_{dR} is a complete discrete valuation ring, and for $r \geq 0$, the kernel of the canonical surjection $B^+_{dR} \longrightarrow (B_\infty/J^{[r]}_\infty) \otimes \mathbf{Q}$ coincides with the r-th power of the maximal ideal of B^+_{dR}.

B_{dR} is defined to be the field of fractions of B^+_{dR}. It is hence a complete discrete valuation field, and $\mathrm{Gal}(\bar{K}/K)$ acts on B_{dR} preserving the valuation.

Note that B^+_{dR} has a K-algebra structure.

It is known that for any finite extension L of K, the canonical map from the B_{dR} of K to the B_{dR} of L is an isomorphism. (We identify them.)

1.1.2. <u>The functor D_{dR}</u>. D_{dR} is a functor from the category of finite dimensional $\mathbb{Q}_p$-vector spaces endowed with a continuous action of $Gal(\bar{K}/K)$, to the category of finite dimensional K-vector spaces endowed with a descending filtration with index set $\mathbb{Z}$. It is defined by

$$D_{dR}(V) = H^0(K, V \otimes_{\mathbb{Q}_p} B_{dR})$$

where $Gal(\bar{K}/K)$ acts on the tensor product diagonally, with the filtration

$$D^i_{dR}(V) = H^0(K, V \otimes_{\mathbb{Q}_p} B^i_{dR}) \qquad (i \in \mathbb{Z})$$

where B^i_{dR} is the part of B_{dR} consisting of elements with (normalized additive) valuation $\geq i$.

1.1.3. <u>De Rham representations</u>. Let V be a finite dimensional $\mathbb{Q}_p$-vector space endowed with a continuous action of $Gal(\bar{K}/K)$. Then it is known that

$$\dim_K(D_{dR}(V)) \leq \dim_{\mathbb{Q}_p}(V) .$$

V is called a de Rham representation of $Gal(\bar{K}/K)$ if the equality holds here. If V is a de Rham representation, the canonical map

$$D_{dR}(V) \otimes_K B_{dR} \longrightarrow V \otimes_{\mathbb{Q}_p} B_{dR}$$

is bijective and preserves the filtrations (fil^i on the left is $\sum_{r+s=i} fil^r \otimes fil^s$ and fil^i on the right is $V \otimes fil^i$). de Rham representations are stable under tensor products, exterior powers, duals, and direct sums, and D_{dR} commutes with these operations for de Rham representations. Subrepresentations and quotient representations of a de Rham representation are de Rham.

If V is a de Rham representation and L is a finite extension of K in $\bar{K}$, then V is also de Rham as a representation of $Gal(\bar{K}/L)$ and

$$D_{dR}(V) \otimes_K L \longrightarrow D_{dR}(L, V), \quad D^i_{dR}(V) \otimes_K L \longrightarrow D^i_{dR}(L, V)$$

are isomorphisms, where $D_{dR}(L, \ )$ denotes the functor D_{dR} for representaions of $Gal(\bar{K}/L)$.

The following 1.1.4 was conjectured by Fontaine [Fo_1], and proved in special cases by Fontaine [Fo_1] and Fontine-Messing [FM].

Theorem 1.1.4 (Faltings [Fa]). *Let X be a smooth variety over K and let $m \in \mathbb{Z}$. Then the p-adic etale cohomology $H^m_{et}(X \otimes_K \bar{K}, \mathbb{Q}_p)$ is a de Rham representation of $Gal(\bar{K}/K)$, and $D_{dR}(H^m(X \otimes_K \bar{K}, \mathbb{Q}_p))$ is canonically isomorphic, as a filtered K-vector space, to the de Rham cohomology $H^m_{dR}(X/K) = H^m(X, \Omega^{\cdot}_{X/K})$.*

Here the filtration on $H^m_{dR}(X/K)$ is defined as in [De_1]. (If X is proper over K, $fil^r H^m_{dR}(X/K) = Image(H^m(X, \Omega^{\geq r}_{X/K}) \longrightarrow H^m_{dR}(X/K))$.)

1.1.5. Finally we review the relationship between D_{dR} and Tate twists. For a finite dimensional $\mathbb{Q}_p$-vector space V endowed with a continuous action of $Gal(\bar{K}/K)$ and for $r \in \mathbb{Z}$, $D_{dR}(V(r))$ is identified as a K-vector space with $D_{dR}(V)$ but the filtration is given by $D^i_{dR}(V(r)) = D^{i+r}_{dR}(V)$. This identification is given as follows. There is a canonical $\mathbb{Q}_p[Gal(\bar{K}/K)]$-homomorphism $\varepsilon : \mathbb{Q}_p(1) \longrightarrow B_{dR}$ (see below) such that the image of any non-zero element of $\mathbb{Q}_p(1)$ under ε is a prime element of B_{dR}, and the bijection

$$\mathbb{Q}_p(r) \otimes_{\mathbb{Q}_p} B_{dR} \xrightarrow{\cong} B_{dR} \quad ; \quad x^{\otimes r} \otimes y \longmapsto \varepsilon(x)^r y$$

($x \in \mathbb{Q}_p(1)$, $x \neq 0$, $y \in B_{dR}$) induces a bijection $D_{dR}(V(r)) \longrightarrow D_{dR}(V)$ which we take as an identification.

We review the definition of ε. This map is induced by the homomorphism $\varepsilon : \mathbb{Z}_p(1) \longrightarrow B_{\infty,\bar{K}/K}$ defined in the following way. Let $\beta = (\beta_n)_{n\geq 1}$ be an element of $\mathbb{Z}_p(1)$, where β_n are p^n-th roots of 1 in $\bar{K}$ with the property $(\beta_{n+1})^p = \beta_n$ for $n \geq 1$. For each $n \geq 1$, take an element x_n of $B_{n,\bar{K}/K}$ whose image in $O_{\bar{K}}/p^n$ coincides with the class of β_n. Then, $y_n = (x_n)^{p^n}$ depends only on β_n and is independent of the choice of x_n, and $y \underset{\text{def}}{=} (y_n)_n$ belongs to $\varprojlim_n B_{n,\bar{K}/K} = B_{\infty,\bar{K}/K}$. Furthermore, $y \equiv 1 \bmod J^{[1]}_{\infty,\bar{K}/K}$. We define $\varepsilon(\beta)$ to be the image of y under the logarithm $1 + J^{[1]}_{\infty,\bar{K}/K} \longrightarrow J^{[1]}_{\infty,\bar{K}/K}$.

§1.2. Dual exponential maps.

1.2.1. For a de Rham representation V of $\mathrm{Gal}(\bar{K}/K)$, canonical maps

$$\exp : D_{dR}(V)/D^0_{dR}(V) \longrightarrow H^1(K, V), \qquad \exp^* : H^1(K, V) \longrightarrow D^0_{dR}(V)$$

called the exponential map and the dual exponential map, respectively, will play important roles relating Galois cohomology to "de Rham objects". The purpose of §1.2 is to define the dual exponential map. The exponential map was defined in [BK] and will be reviewed in §1.3, and a duality between the exponential map and the dual exponential map will be shown in 1.4.1 (4).

1.2.2. Let $\chi_{cyclo} : \mathrm{Gal}(\bar{K}/K) \longrightarrow \mathbb{Z}_p^\times$ be the action on p^n-th roots of 1 for $n \geq 1$, and let

$$\log(\chi_{cyclo}) \in H^1(K, \mathbb{Z}_p) = \mathrm{Hom}_{cont}(\mathrm{Gal}(\bar{K}/K), \mathbb{Z}_p)$$

be the homomorphism $\sigma \longmapsto \log(\chi_{cyclo}(\sigma))$.

<u>Proposition</u> 1.2.3. <u>Let</u> V <u>be a de Rham representation of</u> $\mathrm{Gal}(\bar{K}/K)$. <u>Then the cup product with</u> $\log(\chi_{cyclo})$ <u>gives isomorphisms</u>

$$D^i_{dR}(V) = H^0(K, V \otimes_{\mathbb{Q}_p} B^i_{dR}) \xrightarrow{\ \cong\ } H^1(K, V \otimes_{\mathbb{Q}_p} B^i_{dR})$$

<u>for all</u> $i \in \mathbb{Z}$.

Here $V \otimes B^i_{dR}$ is endowed with the topology explained in 1.2.5 below.

1.2.4. We define the dual exponential map $\exp^*$ of a de Rham representation V of $\mathrm{Gal}(\bar{K}/K)$ as the composite map

$$H^1(K, V) \longrightarrow H^1(K, V \otimes B^+_{dR}) \xrightarrow{\cong} D^0_{dR}(V),$$

where the second arrow is the inverse of the cup product with $\log(\chi_{cyclo})$ (1.2.3).

1.2.5. The topology of $V \otimes B^i_{dR}$ is defined as follows.

First, we define the topology of B^+_{dR} by regarding it as the inverse limit of the topological rings $(B_{\infty,\bar{K}/K}/J^{[r]}_{\infty,\bar{K}/K}) \otimes \mathbf{Q}$, where the topology of $(B_{\infty,\bar{K}/K}/J^{[r]}_{\infty,\bar{K}/K}) \otimes \mathbf{Q}$ is defined by taking $(p^n(B_{\infty,\bar{K}/K}/J^{[r]}_{\infty,\bar{K}/K}))_{n\geq 0}$ as a fundamental system of neighbourhoods of 0. For any finite extension L of K, the bijection from the B^+_{dR} of K to the B^+_{dR} of L is a homeomorphism.

For $i \in \mathbf{Z}$, we endow B^i_{dR} with the topology for which $x \longmapsto ax$; $B^+_{dR} \longrightarrow B^i_{dR}$ is a homeomorphism for any generator a of the B^+_{dR}-module B^i_{dR}. (The existence of such topology is deduced from the following fact; $J^{[r]}_{\infty,\bar{K}/K}/J^{[r+1]}_{\infty,\bar{K}/K}$ is an invertible module over $O_{\mathbf{C}_p} = \varprojlim_n O_{\bar{K}}/p^n O_{\bar{K}}$ for any $r \geq 0$ (1.1.1), and hence for any r, $i \geq 0$ and for any $a \in J^{[i]}_{\infty,\bar{K}/K}$ which generates the B^+_{dR}-module B^i_{dR}, the map $x \longmapsto ax$; $J^{[r]}_{\infty,\bar{K}/K}/J^{[r+1]}_{\infty,\bar{K}/K} \longrightarrow J^{[r+i]}_{\infty,\bar{K}/K}/J^{[r+i+1]}_{\infty,\bar{K}/K}$ is injective and the cokernel is killed by some power of p. If $i \geq 0$, this topology of B^i_{dR} coincides with the topology as a subspace of B^+_{dR}.)

Finally we endow $V \otimes B^i_{dR}$ with the topology such that for any $\mathbf{Q}_p$-basis $(e_j)_j$ of V, the map $\prod_j B^i_{dR} \longrightarrow V \otimes B^i_{dR}$; $(x_j)_j \longmapsto \sum_j e_j \otimes x_j$ is a homeomorphism.

1.2.6. To prove Prop. 1.2.3, we review some facts on B_{dR} and on de

Rham representations.

Let $\mathbb{C}_p$ be the completion of $\bar{K}$. Then, a canonical $\mathbb{C}_p$-linear isomorphism of $\mathrm{Gal}(\bar{K}/K)$-modules

$$\mathbb{C}_p(i) \cong B^i_{dR}/B^{i+1}_{dR} \qquad \text{for } i \in \mathbb{Z}$$

is defined as follows. For $i = 0$, it is induced by the isomorphisms $B_{n,\bar{K}/K}/J_{n,\bar{K}/K} \xrightarrow{\cong} O_{\bar{K}}/p^n$. For $i = 1$, it is induced by the map $\varepsilon : \mathbb{Q}_p(1) \longrightarrow B^1_{dR}/B^2_{dR}$. For i general, it is induced from the cases $i = 0, 1$, by the multiplication in B_{dR}. For a de Rham representation V, by taking the graded quotients of the isomorphism preserving filtrations $V \otimes_{\mathbb{Q}_p} B_{dR} \cong D_{dR}(V) \otimes_K B_{dR}$ (1.1.3), we have the "Hodge-Tate decomposition"

$$(1) \qquad V \otimes_{\mathbb{Q}_p} B^i_{dR}/B^{i+1}_{dR} \cong \bigoplus_{r+s=i} gr^r(D_{dR}(V)) \otimes_K \mathbb{C}_p(s).$$

1.2.7. We prove Prop. 1.2.3. The key is the following result of Tate [Ta_1]: <u>The continuous cohomology</u> $H^m(K, \mathbb{C}_p(r))$ <u>is zero if</u> $(m, r) \neq (0, 0), (1, 0)$, <u>and</u> $H^0(K, \mathbb{C}_p)$ (resp. $H^1(K, \mathbb{C}_p)$) <u>is a one-dimensional</u> K-<u>vector space generated by</u> 1 (resp. $\log(\chi_{cyclo})$).

From this and 1.2.6 (1), we obtain an exact sequence

$$0 \longrightarrow H^1(K, V \otimes B^{i+1}_{dR}) \longrightarrow H^1(K, V \otimes B^i_{dR}) \longrightarrow D^i_{dR}(V)/D^{i+1}_{dR}(V) \longrightarrow 0.$$

From this, we have for sufficiently small i

$$H^1(K, V \otimes B^i_{dR}) \cong \varprojlim_n H^1(K, V \otimes B^n_{dR}) = 0,$$

and then by this exact sequence we obtain 1.2.3.

§1.3. <u>Exponential maps.</u>

We review the ring B_{crys} and the functor D_{crys} of Fontaine ([Fo_1] [FM]), and review the exponential maps of de Rham representations ([BK]).

1.3.1. The subring B_{crys} of B_{dR} is defined as follows. Consider the ring $B_{\infty,\bar{K}/K_0}$ (note that $\bar{K}$ is also the algebraic closure of K_0). Then $B_{\infty,\bar{K}/K_0} \longrightarrow B_{\infty,\bar{K}/K}$ is injective and the image of $\varepsilon : \mathbb{Z}_p(1) \longrightarrow B_{\infty,\bar{K}/K}$ is contained in $B_{\infty,\bar{K}/K_0}$. The definition of B_{crys} is:

$$B_{crys} = B_{\infty,\bar{K}/K_0}[p^{-1}, t^{-1}] \subset B_{dR},$$

where t is the image of any basis of $\mathbb{Z}_p(1)$ under ε (then B_{crys} is independent of the choice of t).

The Frobenius operator $\varphi : B_{crys} \longrightarrow B_{crys}$ is induced from the Frobenius $O_{\bar{K}}/p \longrightarrow O_{\bar{K}}/p$; $x \longmapsto x^p$, the Frobenius $W_n(k) \longrightarrow W_n(k)$, and the isomorphisms $H^0(\mathrm{Spec}(O_{\bar{K}}/p^n)/\mathrm{Spec}(W_n(k))_{crys}, \mathcal{O}_{crys}) \xrightarrow{\cong} H^0(\mathrm{Spec}(O_{\bar{K}}/p)/\mathrm{Spec}(W_n(k))_{crys}, \mathcal{O}_{crys})$.

1.3.2. For a finite dimensional $\mathbb{Q}_p$-vector space V endowed with a continuous action of $\mathrm{Gal}(\bar{K}/K)$, let

$$D_{crys}(V) = H^0(K, V \otimes_{\mathbb{Q}_p} B_{crys}).$$

Then $D_{crys}(V)$ is a finite dimensional K_0-vector space endowed with a Frobenius linear operator $1 \otimes \varphi$, which is denoted by φ. ("Frobenius linear" means that $\varphi(ax) = \varphi(a)\varphi(x)$ for any $a \in K_0$ and $x \in D_{crys}(V)$.)

1.3.3. The sequence

$$(1) \qquad 0 \longrightarrow \mathbb{Q}_p \xrightarrow{\alpha} B_{crys} \oplus B_{dR}^{+} \xrightarrow{\beta} B_{crys} \oplus B_{dR} \longrightarrow 0$$

is exact where $\alpha(x) = (x, x)$ and $\beta(x, y) = ((1 - \varphi)(x), x - y)$.

The proof of the exactness of (1) is given in [BK] by using the crucial exact sequences of Fontaine and Messing [FM]

$$(2) \qquad 0 \longrightarrow \mathbb{Q}_p(r) \longrightarrow J^{[r]}_{\infty,\bar{K}/K_0} \otimes \mathbb{Q} \xrightarrow{1 - p^{-r}\varphi} B_{\infty,\bar{K}/K_0} \otimes \mathbb{Q} \longrightarrow 0.$$

$(r \geq 0)$. The proof of the exactness of (2) is not written in [FM], but recently de Shalit wrote the proof in [dS_3].

For a finite dimensional Q_p-vector space V endowed with a continuous action of $Gal(\bar{K}/K)$, the exact sequence $V \otimes_{Q_p}$ (the sequence (1)) yields an exact sequence

(3) $$0 \longrightarrow H^0(K, V) \xrightarrow{\alpha} D_{crys}(V) \oplus D^0_{dR}(V) \xrightarrow{\beta} D_{crys}(V) \oplus D_{dR}(V) \xrightarrow{\delta} H^1(K, V).$$

1.3.4. For V as in 1.3.3, we define the exponential map

$$\exp : D_{dR}(V)/D^0_{dR}(V) \longrightarrow H^1(K, V) \quad \text{by}$$

$$\exp(x \bmod D^0_{dR}(V)) = \delta(0, x) \quad \text{for} \quad x \in D_{dR}(V).$$

<u>Example</u> 1.3.5. Let A be an abelian variety over K, and let $V_pA = T_pA \otimes Q$ where T_pA is the p-adic Tate module of A. Then V_pA is a de Rham representation of $Gal(\bar{K}/K)$, Lie(A) is identified with $D_{dR}(V_pA)/D^0_{dR}(V_pA)$ ([Fo$_1$]) and $\exp : D_{dR}(V_pA)/D^0_{dR}(V_pA) \longrightarrow H^1(K, V_pA)$ is identified with the composite map $\mathrm{Lie}(A) \longrightarrow A(K) \otimes Q \longrightarrow H^1(K, V_pA)$ where the first arrow is the exponential map of the p-adic Lie group A(K) and the second comes from the Kummer sequence for A. (Cf. Chap. I, §2.3).

<u>Example</u> 1.3.6. Let G be a p-divisible group over O_K, let T_pG be the Tate module of G, and let $V_pG = T_pG \otimes Q$. Then V_pG is a de Rham representation of $Gal(\bar{K}/K)$, Lie(G) is identified with $D_{dR}(V_pG)/D^0_{dR}(V_pG)$ ([Fo$_1$]), and $\exp : D_{dR}(V_pG)/D^0_{dR}(V_pG) \longrightarrow H^1(K, V_pG)$ is identified with the composite map $\mathrm{Lie}(G) \longrightarrow (\varprojlim_n G(O_K/p^n)) \otimes Q \longrightarrow H^1(K, V_pG)$, where the first arrow is the exponential map of the formal group over O_K associated to the connected part of G ([Ta$_2$] §2.2) and the second arrow comes from the Kummer sequence $0 \longrightarrow G_n \longrightarrow G \xrightarrow{p^n} G \longrightarrow 0$ of G.

For the exponential maps in these examples, cf. [BK] §3.

§1.4. Tate duality.

In this §1.4, assume k is a finite field.

The part (1)-(3) of the following 1.4.1 is a duality theorem due to Tate (see [Se] Chap. 2), and (4) is proved in this §1.4.

Theorem 1.4.1. Let ℓ be a prime number.

(1) There exists a canonical isomorphism

$$H^2(K, \mathbf{Q}_\ell(1)) \cong \mathbf{Q}_\ell$$

with a characterization given in 1.4.2 below.

(2) Let V be a finite dimensional $\mathbf{Q}_\ell$-vector space endowed with a continuous action of $\mathrm{Gal}(\bar{K}/K)$. Then, $H^q(K, V)$ are finite dimensional $\mathbf{Q}_\ell$-vector spaces for all q, and are zero if $q \neq 0, 1, 2$.

(3) Let V be as in (2), and let V^* be the dual representation of $\mathrm{Gal}(\bar{K}/K)$, that is, $V^* = \mathrm{Hom}_{\mathbf{Q}_\ell}(V, \mathbf{Q}_\ell)$ on which $\sigma \in \mathrm{Gal}(\bar{K}/K)$ acts by $h \longmapsto h\circ\sigma^{-1}$. Then, the cup product

$$H^q(K, V) \times H^{2-q}(K, V^*(1)) \longrightarrow H^2(K, \mathbf{Q}_\ell(1)) \cong \mathbf{Q}_\ell$$

is a perfect pairing.

(4) Assume $\ell = p$. Let V be as in (2) and assume V is a de Rham representation. Then $\exp^* : H^1(K, V) \longrightarrow D_{dR}(V)$ coincides with the composite map

$$H^1(K, V) \xrightarrow{\cong} \mathrm{Hom}_{\mathbf{Q}_p}(H^1(K, V^*(1)), \mathbf{Q}_p)$$
$$\longrightarrow \mathrm{Hom}_{\mathbf{Q}_p}(D_{dR}(V^*(1)), \mathbf{Q}_p) \xrightarrow{\cong} D_{dR}(V).$$

Here the first arrow is by the duality in (3), the second arrow is the $\mathbf{Q}_p$-dual of exponential map of $V^*(1)$, and the third arrow is induced by the canonical pairing

$$D_{dR}(V) \times D_{dR}(V^*(1)) \longrightarrow D_{dR}(\mathbf{Q}_p(1)) \cong K \xrightarrow{\text{trace}} \mathbf{Q}_p$$

1.4.2. The isomorphism in 1.4.1 (1) is characterized by the following property: Let π be a prime element of K, let $\{\pi\} \in H^1(K, \mathbf{Q}_\ell(1))$ be the image of π under the map $K^\times \longrightarrow H^1(K, \mathbf{Q}_\ell(1))$ induced by the connecting maps of the Kummer sequences $0 \longrightarrow \mathbf{Z}/\ell^n(1) \longrightarrow \mathbf{G}_m \xrightarrow{\ell^n} \mathbf{G}_m \longrightarrow 0$, and let

$$\chi \in H^1(K, \mathbf{Q}_\ell) = \mathrm{Hom}_{\mathrm{cont}}(\mathrm{Gal}(\bar{K}/K), \mathbf{Q}_\ell)$$

be a homomorphism which factors through the canonical surjection $\mathrm{Gal}(\bar{K}/K) \longrightarrow \mathrm{Gal}(K_{ur}/K)$. Then the isomorphism in 1.4.1 (1) sends the cup product $\{\pi\} \cup \chi$ to $\chi(\mathrm{Frob}_K) \in \mathbf{Q}_\ell$ where $\mathrm{Frob}_K \in \mathrm{Gal}(K_{ur}/K)$ is the arithmetic Frobenius (i.e. the unique element which induces $x \longmapsto x^{\#(k)}$ on the residue field $\bar{k}$ of K_{ur}).

Now we prove 1.4.1 (4).

<u>Lemma</u> 1.4.3. <u>Let</u> $\delta : H^1(K, B^+_{dR}/B^1_{dR}) \longrightarrow H^2(K, \mathbf{Q}_p(1))$ <u>be the connecting homomorphism of the exact sequence</u>

$$0 \longrightarrow \mathbf{Q}_p(1) \longrightarrow B^{\varphi=p}_{crys} \cap B^+_{dR} \longrightarrow B^+_{dR}/B^1_{dR} \longrightarrow 0.$$

<u>Then for any de Rham representation</u> V <u>of</u> $\mathrm{Gal}(\bar{K}/K)$, <u>we have</u>

$$\exp(a) \cup b = \delta(\langle a, \exp^*(b)\rangle \log(\chi_{cyclo})) \quad \underline{\text{in}} \quad H^2(K, \mathbf{Q}_p(1))$$

<u>for any</u> $a \in D_{dR}(V)$ <u>and</u> $b \in H^1(K, V^*(1))$.

Here $\langle\ ,\ \rangle$ is the pairing $D_{dR}(V) \times D_{dR}(V^*(1)) \longrightarrow D_{dR}(\mathbf{Q}_p(1)) = K$.

<u>Proof</u>. Take $i < 0$ such that $a \in D^i_{dR}(V)$. By a formal argument on cohomology, we see that $\exp(a) \cup b$ is equal to the image of b under

$$H^1(K, V^*(1)) \longrightarrow H^1(K, V^*(1) \otimes B^+_{dR}) \xrightarrow{a\cup} H^1(K, \mathbf{Q}_p(1) \otimes B^i_{dR})$$
$$= H^1(K, B^{i+1}_{dR}) \longrightarrow H^1(K, B^{i+1}_{dR}/B^1_{dR}) \xleftarrow{\cong} H^1(K, B^+_{dR}/B^1_{dR})$$
$$\xrightarrow{\delta} H^2(K, \mathbf{Q}_p(1))$$

where $a\cup$ denotes the cup product with a. Since the image of b in $H^1(K, V^*(1) \otimes B^+_{dR})$ is $\exp^*(b)\log(\chi_{cyclo})$, we obtain 1.4.3.

1.4.4. By 1.4.3, for the proof of 1.4.1 (4), it is sufficient to prove that the composite

$$H^1(K, B^+_{dR}/B^1_{dR}) \xrightarrow{\delta} H^2(K, Q_p(1)) \cong Q_p$$

sends $a \cdot \log(\chi_{cyclo})$ to $-\mathrm{Tr}_{K/Q_p}(a)$ for any $a \in K$ (the minus sign arises from the fact $x \cup y = -y \cup x$ in $H^2(K, Q_p(1))$ for $x \in H^1(K, V)$, $y \in H^1(K, V^*(1))$). By the case $V = Q_p(1)$ and $b = \log(\chi_{cyclo})$ of 1.4.3, we have

$$\delta(a \cdot \log(\chi_{cyclo})) = \exp(a) \cup \log(\chi_{cyclo}) \quad \text{in} \quad H^2(K, Q_p(1))$$

for $a \in K$ where exp is the exponential map $K \longrightarrow H^1(K, Q_p(1))$. Hence we are reduced to

Lemma 1.4.5. The pairing 1.4.1 (3) for $V = Q_p(1)$

$$H^1(K, Q_p(1)) \times H^1(K, Q_p) \longrightarrow H^2(K, Q_p(1)) \cong Q_p$$

sends $(\exp(a), \log(\chi_{cyclo}))$ to $-\mathrm{Tr}_{K/Q_p}(a)$ for any $a \in K$.

Proof. By using the trace maps for K/Q_p, we are reduced to the case $K = Q_p$. In this case, 1.4.5 is well known in the local class field theory of Q_p.

§2. Generalized explicit reciprocity law for Lubin-Tate groups.

Explicit reciprocity law is an important theme on the arithmetic of local fields, and has been studied by many people. Classically it is a formula which describes the Hilbert symbols in an explicit way by using differential forms. Our viewpoint is that this mysterious relationship between Hilbert symbols and differential forms is related to the mysterious relationship between p-adic etale cohomology and differential forms in the theory of p-adic periods.

(This viewpoint was stressed in [Ka_1]. Note Hilbert symbols are closely related to Galois cohomology theory__a branch of etale cohomology theory.)

I believe that there exist explicit reciprocity laws for all p-adic representations of Gal($\bar{K}/K$), though I can not formulate them. For a de Rham representation V, this law should be some explicit description of the relationship between $D_{dR}(V)$ and the Galois cohomology of V, or more precisely, some explicit descriptions of the maps exp and exp^* of V. In this §2, we give an explicit description of exp^* for $T^{\otimes(-r)}(1) \otimes Q$ $(r \geq 1)$, where T is the Tate module of a Lubin-Tate group (Thm. 2.1.7). This 2.1.7 contains the explicit reciprocity law of Wiles [Wi_1] as the case r = 1. It also contains the explicit reciprocity law in [BK] §2, in which the Lubin-Tate group was the formal multiplicative and we considered the Galois cohomology of unramified extensions of Q_p.

The chapter 1 of de Shalit [dS_2] is an excellent introduction to Lubin-Tate groups and the explicit reciprocity law of Wiles.

There have been so many works on explicit reciprocity laws. I am sorry that I do not give a list of the works. Some works are listed in [dS_1][dS_2][Ka_1].

§2.1. The statement of the result.

We do not review the the definitions of Lubin-Tate groups. See [dS_2] for fundamental facts on Lubin-Tate groups.

2.1.1. Let F be a subfield of K such that $[F : Q_p] < \infty$ and assume that a prime element of F is a prime element in K.

Fix a prime element π of F. Let Γ be the Lubin-Tate group over O_F

corresponding to the pair (F, π), and let G be the connected p-divisible group over O_K obtained from Γ by the extension of scalars $O_F \longrightarrow O_K$. We denote the action of $a \in O_F$ on G as $[a] : G \longrightarrow G$.

2.1.2. Let T be the Tate module of G, and let $V = T \otimes \mathbb{Q}$. For $m \in \mathbb{Z}$, let $T^{\otimes m}$ (resp. $V^{\otimes m}$, resp. $\mathrm{coLie}(G)^{\otimes m}$) be the m-th tensor powers of T (resp. V, resp. $\mathrm{coLie}(G)$) as an invertible O_F (resp. F, resp. O_K)-module. For $n \geq 0$, let K_n be the extension of K corresponding to $\mathrm{Ker}(\mathrm{Gal}(\bar{K}/K) \longrightarrow \mathrm{Aut}(T/\pi^n T))$.

In the following, ξ denotes an O_F-basis of T. For $n \geq 0$, let

$$\xi_n = \pi^{-n}\xi \bmod T \in V/T = G(\bar{K}).$$

2.1.3. For $r \geq 1$, as is explained in 2.2.3 later, we have a canonical isomorphism

$$(1) \qquad D^0_{dR}(V^{\otimes(-r)}(1)) \cong \mathrm{coLie}(G)^{\otimes r} \otimes_{O_K} K$$

which induces

$$(2) \qquad D^0_{dR}(K_n, V^{\otimes(-r)}(1)) \cong \mathrm{coLie}(G)^{\otimes r} \otimes_{O_K} K_n \quad \text{for any } n \geq 0.$$

($D_{dR}(K_n, \ )$ denotes D_{dR} for representations of $\mathrm{Gal}(\bar{K}/K_n)$; cf. 1.1.3).

2.1.4. We denote by $\mathcal{O}(G)$ the ring of functions on G (that is, the inverse limit of the affine rings of the finite group schemes $\mathrm{Ker}(p^n \ ; G \longrightarrow G)$, so $\mathcal{O}(G) \cong O_K[[t]]$). We denote by $\phi : \mathcal{O}(G) \longrightarrow \mathcal{O}(G)$ the unique ring homomorphism which is the identity on $\mathcal{O}(\Gamma)$ and which satisfies $\phi(f) = f^q \bmod \pi\mathcal{O}(G)$ for any $f \in \mathcal{O}(G)$ where q is the order of the residue field of F.

2.1.5. For a norm compatible system $u = (u_n)_n \in \varprojlim_n K_n^{\times}$, we define

$$s_{r,n}(u, \xi) \in H^1(K_n, T^{\otimes(-r)}(1)) \qquad (n \geq 0, \ r \in \mathbb{Z})$$

as the image of u under

$$\varprojlim_m K_m^{\times} \longrightarrow \varprojlim_m H^1(K_m, \mathbb{Z}_p(1)) \xrightarrow{(*)} \varprojlim_m H^1(K_m, T^{\otimes(-r)}(1)/\pi^m)$$

$$\xrightarrow{\text{trace}} \varprojlim_m H^1(K_n, T^{\otimes(-r)}(1)/\pi^m) \cong H^1(K_n, T^{\otimes(-r)}(1))$$

where the first three $\varprojlim$ are taken with respect to norms or traces, and $(*)$ is defined by the cup products with $(\xi_m)^{\otimes(-r)}$.

2.1.6. For u as above, we denote by $g_{u,\xi}$ the Coleman power series of u with respect to ξ, which is an invertible element of $\mathcal{O}(G)[\frac{1}{h}]$ where h is a generator of the augumentation ideal $\mathrm{Ker}(\mathcal{O}(G) \longrightarrow O_K)$, characterized by the following property: $(\phi^{-n}(g_{u,\xi}))(\xi_n) = u_n$ for all $n \geq 1$. ([Co])

Now the main theorem of §2 is the following

<u>Theorem</u> 2.1.7. <u>(The explicit reciprocity law)</u>. <u>For</u> $n,\ r \geq 1$,

$$\exp^* : H^1(K_n, T^{\otimes(-r)}(1)) \longrightarrow D^0_{dR}(K_n, V^{\otimes(-r)}(1))$$
$$\cong \mathrm{coLie}(G)^{\otimes r} \otimes_{O_K} K_n$$

<u>sends</u> $\dot{s}_{r,n}(u, \xi)$ <u>to</u>

$$\frac{1}{(r-1)!}\pi^{-nr}\,\omega^{\otimes r} \otimes \{(\tfrac{d}{\omega})^r \log(\phi^{-n}(g_{u,\xi}))\}(\xi_n).$$

Here ω is any O_K-basis of $\mathrm{coLie}(G) = \mathrm{Hom}_{O_K}(\mathrm{Lie}(G), O_K)$,

$$(\tfrac{d}{\omega})^r \log(\ ) = (\tfrac{d}{\omega})^{r-1}(\frac{d\log(\ )}{\omega})$$

where $(\frac{d}{\omega})^{r-1}$ is the $r-1$ fold iteration of $\frac{d}{\omega} : f \longmapsto \frac{df}{\omega}$ (this ratio is taken by regarding ω as an invariant differential form on G).

The case $r = 1$ of Thm. 2.1.7 coincides with the explicit reciprocity law of Wiles ([Wi_1]).

§2.2. <u>Lubin-Tate groups and crystalline cohomology</u>.

We give the definition the isomorphism 2.1.3 (1) by using the theories in [Ta_2][Fo_1][Fo_2]. We then state a theorem 2.2.7, and show that Thm. 2.1.7 is deduced from 2.2.7.

2.2.1. Let $\Omega^1(G)$ be the space of differential forms on G. Note

$$\Omega^1(G) \cong \mathcal{O}(G) \otimes_{O_K} \mathrm{coLie}(G).$$

For $n \geq 0$, let S_n be topological ring which is p-adically complete and separated characterized by the following property: For any $i \geq 1$, $\mathrm{Spec}(S_n/p^i S_n)$ is the PD envelope of $\mathrm{Spec}(O_{K_n}/p^i O_{K_n})$ in $\mathrm{Spec}(\mathcal{O}(G)/p^i\mathcal{O}(G))$ with respect to the embedding induced by ξ_n : $\mathrm{Spec}(O_{K_n}) \longrightarrow G$.

For $r \geq 1$, let $J_{S_n}^{[r]} \subset S_n$ be the inverse limit (for varying i) of the r-th divided powers of $\mathrm{Ker}(S_n/p^i S_n \longrightarrow O_{K_n}/p^i O_{K_n})$.

Since G is formally smooth over O_K, the fundamental theorem in crystalline cohomology theory shows

$$R\Gamma(\mathrm{Spec}(O_{K_n}/\pi^i)/\mathrm{Spec}(O_K/\pi^i)_{crys}, J^{[r]})$$
$$\cong [J_{S_n}^{[r]}/\pi^i \xrightarrow{d} J_{S_n}^{[r-1]}/\pi^i \otimes_{\mathcal{O}(G)} \Omega^1(G)] \qquad (i \geq 1)$$

where $J = \mathrm{Ker}(\mathcal{O}_{crys} \longrightarrow \mathcal{O}_{zar})$.

2.2.2. For $\omega \in \mathrm{coLie}(G)$, let

$$\ell_\omega \in J_{S_0}$$

be the logarithm of G associated to ω. Then $d\ell_\omega = \omega$. Let

$$\ell_{\omega,\xi,n} = [\pi^n]^*\ell_\omega \in J_{S_n} \quad \text{for } n \geq 0.$$

Then $d\ell_{\omega,\xi,n} = \pi^n\omega$. In particular, $d\ell_{\omega,\xi,n} \equiv 0 \bmod \pi^n$. So, by 2.2.1, $\ell_{\omega,\xi,n} \bmod \pi^n$ is regarded as an element of $H^0(\mathrm{Spec}(O_{K_n}/\pi^n)/\mathrm{Spec}(O_K/\pi^n))_{crys}, J)$. We denote the image of this element in $H^0(\mathrm{Spec}(O_{\bar{K}}/\pi^n)/\mathrm{Spec}(O_K/\pi^n)_{crys}, J) = J_{n,\bar{K}/K} = J_{\infty,\bar{K}/K}/\pi^n$ by the same letter $\ell_{\omega,\xi,n}$.

2.2.3. We have a pairing

(1) $\mathrm{coLie}(G) \otimes_{O_F} T \longrightarrow J_{\infty,\bar{K}/K}$; $\omega \otimes \xi \longrightarrow (\ell_{\omega,\xi,n})_n$,

which induces

(2) $\mathrm{coLie}(G) \longrightarrow T^{\otimes(-1)} \otimes_{O_F} J_{\infty,\bar{K}/K}$.

By [Ta_2][Fo_1][Fo_2], we have the following (3) and (4):

(3) The map (2) induces an isomorphism of invertible B^+_{dR}-modules

$$\mathrm{coLie}(G) \otimes_{O_K} B^+_{dR} \xrightarrow{\cong} V^{\otimes(-1)} \otimes_F B^1_{dR} .$$

(4) The kernel of $V^{\otimes(-1)} \otimes_{Q_p} B^+_{dR} \longrightarrow V^{\otimes(-1)} \otimes_F B^+_{dR}$ is isomorphic to a direct sum of finite copies of B^+_{dR} as a B^+_{dR}-module with an action of $\mathrm{Gal}(\bar{K}/K)$.

From (3) (rewp. (4)), by considering the r-th tensor power over B^+_{dR} (resp. $F \otimes_{Q_p} B^+_{dR}$), we have the following (5) (resp. (6)).

(5) $\mathrm{coLie}(G)^{\otimes r} \otimes_{O_K} B^+_{dR} \xrightarrow{\cong} V^{\otimes(-r)} \otimes_F B^r_{dR}$

(6) The kernel of $V^{\otimes(-r)} \otimes_{Q_p} B^+_{dR} \longrightarrow V^{\otimes(-r)} \otimes_F B^+_{dR}$ is isomorphic to a direct sum of finite copies of B^+_{dR} as a B^+_{dR}-module with an action of $\mathrm{Gal}(\bar{K}/K)$.

By (5) and (6), taking $H^0(K, \ )$, we have

$$\begin{aligned} \mathrm{coLie}(G)^{\otimes r} \otimes_{O_K} K &\xrightarrow{\cong} H^0(K, V^{\otimes(-r)} \otimes_F B^r_{dR}) \\ &= H^0(K, V^{\otimes(-r)} \otimes_F B^1_{dR}) \\ &= H^0(K, V^{\otimes(-r)}(1) \otimes_F B^+_{dR}) . \\ &\xleftarrow{\cong} H^0(K, V^{\otimes(-r)}(1) \otimes_{Q_p} B^+_{dR}) = D^0_{dR}(V^{\otimes(-r)}(1)) \end{aligned}$$

and this gives the isomorphism in 2.1.3 (1).

2.2.4. We will need integral structures of some vector spaces which appeared in 2.2.3. The arguments in [Ta_2] in fact prove the following integral versions of his results introduced in 1.2.7.

Let L be a complete discrete valuation field of characteristic 0 with perfect residue field of characteristic $p > 0$. Let $q \geq 0$ and $r \in \mathbb{Z}$. Then:

(1) If $q \geq 2$ or if $r \neq 0$, $H^q(L, \hat{O}_{\bar{L}}(r))$ is killed by some power of p.

(2) If $q \geq 2$ or if $r \neq 0$, there exists an integer $m \geq 0$ such that p^m kills $H^q(L, (O_{\bar{L}}/p^n)(r))$ for all $n \geq 0$.

(3) $O_L \xrightarrow{\cong} H^0(L, \hat{O}_{\bar{L}})$.

(4) The map $O_L \longrightarrow H^1(L, \hat{O}_{\bar{L}})$: $a \longrightarrow a \cdot \log(\chi_{cyclo})$ is injective and its cokernel is killed by some power of p.

From these facts, we have that for $q = 0, 1$, for $r \geq 1$, and for any finite extension L of K, the kernels and the cokernels of the arrows in the following diagram are killed by some power of p.

$$\mathrm{coLie}(G)^{\otimes r} \otimes_{O_K} O_L \longrightarrow H^q(L, T^{\otimes(-r)} \otimes_{O_F} J^{[r]}_{\infty,\bar{K}/K}/J^{[r+1]}_{\infty,\bar{K}/K})$$
$$\longrightarrow H^q(L, T^{\otimes(-r)} \otimes_{O_F} J_{\infty,\bar{K}/K}/J^{[r+1]}_{\infty,\bar{K}/K}).$$

Here for $q = 0$, the first arrow is the injection induced by the map 2.2.3 (2). In the case $q = 1$, the first arrow is the composite of the map for $q = 0$ and the cup product with $\log(\chi_{cyclo})$.

2.2.5. Let

$$\Theta = \varprojlim (\ldots \longrightarrow \Omega^1(G) \longrightarrow \Omega^1(G) \longrightarrow \Omega^1(G))$$

where the arrows are the trace map $\mathrm{Tr}_\pi : \Omega^1(G) \longrightarrow \Omega^1(G)$ associated to the multiplication $[\pi] : G \longrightarrow G$. (So $\mathrm{Tr}_\pi \circ [\pi]^* = \pi$ on $\Omega^1(G)$.)

2.2.6. We define the canonical maps

$$\mu_{\xi,n} : S_n \otimes_{\mathcal{O}(G)} \Omega^1(G) \longrightarrow H^1(K_n, J_{\infty,\bar{K}/K})$$
$$\mu_{r,\xi,n} : S_n \otimes_{\mathcal{O}(G)} \Omega^1(G) \longrightarrow H^1(K_n, (T^{\otimes(-r)} \otimes_{O_F} J_{\infty,\bar{K}/K}/J^{[r+1]}_{\infty,\bar{K}/K})/\pi^n)$$
$$\theta_{r,\xi,n} : \Theta \longrightarrow H^1(K_n, T^{\otimes(-r)} \otimes_{O_F} J_{\infty,\bar{K}/K}/J^{[r+1]}_{\infty,\bar{K}/K})$$

$(n \geq 0,\ r \in \mathbb{Z})$ as follows. Let

$$S_n \otimes_{\mathcal{O}(G)} \Omega^1(G) \longrightarrow H^1(\mathrm{Spec}(O_{K_n}/\pi^i)/\mathrm{Spec}(O_K/\pi^i)_{\mathrm{crys}},\ J) \qquad (i \geq 1)$$

be the map defined by the isomorphism in 2.2.1. We have also a map

$$R\Gamma(\mathrm{Spec}(O_{K_n}/\pi^i)/\mathrm{Spec}(O_K/\pi^i)_{\mathrm{crys}},\ J)$$
$$\longrightarrow R\Gamma(\mathrm{Gal}(\bar{K}/K_n),\ R\Gamma(\mathrm{Spec}(O_{\bar{K}}/\pi^i)/\mathrm{Spec}(O_K/\pi^i)_{\mathrm{crys}},\ J))$$
$$= R\Gamma(K_n,\ J_{\infty,\bar{K}/K}/\pi^i).$$

Here the last equation comes from

$$H^q(\mathrm{Spec}(O_{\bar{K}}/\pi^i)/\mathrm{Spec}(O_K/\pi^i)_{\mathrm{crys}},\ J^{[r]}) = 0 \quad \text{for all } q \geq 1 \text{ and all } r.$$

([Fo$_3$] §3 Thm. 1 (i) in which the case $K = K_0$ is considered; the same proof works in the general case.) By composing these maps, we get the map $\mu_{\xi,n}$.

We define $\mu_{r,\xi,n}$ to be the composite

$$S_n \otimes_{\mathcal{O}(G)} \Omega^1(G) \xrightarrow{\text{by } \mu_{\xi,n}} H^1(K_n,\ J_{\infty,\bar{K}/K}/J^{[r+1]}_{\infty,\bar{K}/K})$$
$$\xrightarrow{\cup\, \xi_n^{\otimes(-r)}} H^1(K_n,\ (T^{\otimes(-r)} \otimes_{O_F} J_{\infty,\bar{K}/K}/J^{[r+1]}_{\infty,\bar{K}/K})/\pi^n).$$

We define $\theta_{r,\xi,n}$ to be the composite

$$\Theta = \varprojlim_m \Omega^1(G) \xrightarrow{(\mu_{r,\xi,m})_m} \varprojlim_m H^1(K_m,\ (T^{\otimes(-r)} \otimes_{O_F} J_{\infty,\bar{K}/K}/J^{[r+1]}_{\infty,\bar{K}/K})/\pi^m)$$
$$\xrightarrow{\text{trace}} \varprojlim_m H^1(K_n,\ (T^{\otimes(-r)} \otimes_{O_F} J_{\infty,\bar{K}/K}/J^{[r+1]}_{\infty,\bar{K}/K})/\pi^m)$$
$$\xleftarrow{\cong} H^1(K_n,\ T^{\otimes(-r)} \otimes_{O_F} J_{\infty,\bar{K}/K}/J^{[r+1]}_{\infty,\bar{K}/K}).$$

<u>Theorem</u> 2.2.7. <u>For</u> $n,\ r \geq 1$, <u>the composite map</u>

$$\Theta \xrightarrow{\theta_{r,\xi,n}} H^1(K_n,\ T^{\otimes(-r)} \otimes_{O_F} J_{\infty,\bar{K}/K}/J^{[r+1]}_{\infty,\bar{K}/K})$$
$$\longrightarrow H^1(K_n,\ V^{\otimes(-r)} \otimes_F B^1_{dR}/B^{r+1}_{dR}) \cong \mathrm{coLie}(G)^{\otimes r} \otimes_{O_K} K_n$$

<u>sends</u> $(\eta_m)_m \in \Theta$ $(\eta_m \in \Omega^1(G)$ <u>for each</u> $m)$ <u>to</u>

$$-\frac{1}{(r-1)!}\,\pi^{-nr}\,\omega^{\otimes r} \otimes \left(\left(\frac{d}{\omega}\right)^{r-1}\left(\frac{\eta_n}{\omega}\right)\right)(\xi_n).$$

Thm. 2.1.7 is reduced to Thm. 2.2.7 by

Lemma 2.2.8. The diagram

$$\begin{array}{ccccccc}
\mathcal{O}(G)^{\times} & \xrightarrow{\xi_n} O_{K_n}^{\times} & \longrightarrow & H^1(K_n, \mathbb{Z}_p(1)) & \longrightarrow & H^1(K_n, T^{\otimes(-r)}(1)/\pi^n) \\
\big\downarrow \mathrm{dlog} & & & \big\downarrow & & \big\downarrow \\
\Omega^1(G) & \xrightarrow{-\mu_{\xi,n}} & & H^1(K_n, J_{\infty,\bar{K}/K}) & \longrightarrow & H^1(K_n, (T^{\otimes(-r)} \otimes_{O_F} J_{\infty,\bar{K}/K})/\pi^n)
\end{array}$$

is commutative for any $n \geq 0$.

2.2.9. Before we prove 2.2.8, we give an explicit description of the map $\mu_{\xi,n}$. Consider the map

$$\mathcal{O}(G) \otimes_{O_K} B_{\infty,\bar{K}/K} \longrightarrow \hat{O}_{\bar{K}}$$

obtained from $\mathcal{O}(G) \xrightarrow{\xi_n} O_{K_n} \longrightarrow \hat{O}_{\bar{K}}$. Let $\bar{S}_n$ be the topological ring which is p-adically complete and separated characterized by the following property: For any $i \geq 1$, $\mathrm{Spec}(\bar{S}_n/p^i\bar{S}_n)$ is the PD envelope of $\mathrm{Spec}(O_{\bar{K}}/p^iO_{\bar{K}})$ in $\mathrm{Spec}((\mathcal{O}(G) \otimes_{O_K} B_{\infty,\bar{K}/K})/p^i)$. Then, $\bar{S}_n$ is endowed with a natural action of $\mathrm{Gal}(\bar{K}/K_n)$ and has the following structure: There is an isomorphism over $B_{\infty,\bar{K}/K}$ preserving the divided power structures

$$\varprojlim_m (B_m\langle t\rangle) \xrightarrow{\cong} \bar{S}_n \; ; \; t \longmapsto h \otimes 1 - 1 \otimes x$$

where $\langle\ \rangle$ means the PD-polynomial ring, t is an indeterminate, and we fixed a generator h of the augumentation ideal $\mathrm{Ker}(\mathcal{O}(G) \longrightarrow O_K)$ and an element x of $B_{\infty,\bar{K}/K}$ whose image in $\hat{O}_{\bar{K}}$ coincides with $h(\xi_n)$. Let $J_{\bar{S}_n} = \mathrm{Ker}(\bar{S}_n \longrightarrow \hat{O}_{\bar{K}})$. Then

$$0 \longrightarrow J_{\infty,\bar{K}/K} \longrightarrow J_{\bar{S}_n} \xrightarrow{d} \bar{S}_n \otimes_{\mathcal{O}(G)} \Omega^1(G) \longrightarrow 0$$

is exact (this follows from $d(h \otimes 1 - 1 \otimes x) = 1 \otimes dh$ and from the fact dh is a basis of the invertible $\mathcal{O}(G)$-module $\Omega^1(G)$).

The map $\mu_{\xi,n}$ is obtained as the connecting map on the cohomology for this sequence.

2.2.10. We prove 2.2.8. The commutativity of the right square is clear, so we consider the left square.

Let $h \in \mathcal{O}(G)^{\times}$. Take any $m \geq 0$. Take $v \in O_{\hat{K}}^{\times}$ such that $v^{p^m} = h(\xi_n)$. Then the image of h in $H^1(K_n, \mathbb{Z}/p^m(1))$ is represented the cocycle $\mathrm{Gal}(\bar{K}/K) \longrightarrow \mathbb{Z}/p^m(1)$; $\sigma \longmapsto \frac{\sigma(v)}{v}$. Take $\tilde{v} \in (B_{\infty,\bar{K}/K})^{\times}$ whose image in $O_{\hat{K}}^{\times}$ coincides with v. Recalling the definition of

$$\mathbb{Z}/p^m(1) \longrightarrow J_{\infty,\bar{K}/K}/p^m \quad ; \quad \alpha \longmapsto \log(\tilde{\alpha}^{p^m})$$

($\tilde{\alpha}$ is a lifting of the p^m-th root α of 1 to $(B_{\infty,\bar{K}/K}/p^m)^{\times}$ (1.1.5), we see that the image of $\sigma \longmapsto \frac{\sigma(v)}{v}$ in $H^1(K_n, J_{\infty,\bar{K}/K}/p^m)$ is represented by the cocycle

$$\sigma \longmapsto \log\left(\left(\frac{\sigma(\tilde{v})}{\tilde{v}}\right)^{p^m}\right) = \log\left(\frac{\sigma(\tilde{v}^{p^m}h^{-1})}{\tilde{v}^{p^m}h^{-1}}\right) = \sigma(\log(\tilde{v}^{p^m}h^{-1})) - \log(\tilde{v}^{p^m}h^{-1}).$$

Here $\log(\tilde{v}^{p^m}h^{-1}) \in J_{\bar{S}_n}$ is defined since $\tilde{v}^{p^m}h^{-1} \in (\bar{S}_n)^{\times}$ maps to $1 \in O_{\hat{K}}^{\times}$. By

$$d\log(\tilde{v}^{p^m}h^{-1}) \equiv -\frac{dh}{h} \mod p^m,$$

and by 2.2.9, the class of the last cocycle is equal to the image of $-\mu_{\xi,n}(\frac{dh}{h})$ in $H^1(K_n, J_{\infty,\bar{K}/K}/p^m)$.

§2.3. <u>The proof of Thm. 2.2.7.</u>

The case $r = 1$ of the following proof is due to M. Kurihara.

2.3.1. In the following, we regard $\mathrm{coLie}(G)^{\otimes r}$ as embedded in $H^0(K, T^{\otimes(-r)} \otimes_{O_F} J_{\infty,\bar{K}/K}/J_{\infty,\bar{K}/K}^{[r+1]})$ (2.2.4). We fix $n \geq 1$. For an integer c such that $(r-1)!^{-1}\pi^{-nr}c \in O_K$ and for $m \geq n$, we define the maps

$$\alpha_{c,m},\ \beta_{c,m},\ \tau_{c,m} :$$
$$S_m \otimes_{\mathcal{O}(G)} \Omega^1(G) \longrightarrow H^1(K_n,\ (T^{\otimes(-r)} \otimes_{O_F} J_{\infty,\bar{K}/K}/J^{[r+1]}_{\infty,\bar{K}/K})/\pi^m))$$

as follows. Let $\alpha_{c,m}$ be c times the composite map

$$S_m \otimes_{\mathcal{O}(G)} \Omega^1(G) \xrightarrow{\mu_{r,\xi,m}} H^1(K_m,\ (T^{\otimes(-r)} \otimes_{O_F} J_{\infty,\bar{K}/K}/J^{[r+1]}_{\infty,\bar{K}/K})/\pi^m)$$
$$\xrightarrow{\text{trace}} H^1(K_n,\ (T^{\otimes(-r)} \otimes_{O_F} J_{\infty,\bar{K}/K}/J^{[r+1]}_{\infty,\bar{K}/K})/\pi^m).$$

Let $\beta_{c,m}$ be the map defined by

$$\beta_{c,m}(\eta) = ((r-1)!^{-1}\pi^{-nr}c)\omega^{\otimes r} \otimes ((\tfrac{d}{\omega})^{r-1}(\tfrac{1}{\omega}\mathrm{Trace}_{m,n}(\eta)))(\xi_n) \cup \log(\chi_G)$$

where ω is any O_K-basis of coLie(G), $\mathrm{Trace}_{m,n}$ is the trace map $S_m \otimes_{\mathcal{O}(G)} \Omega^1(G) \longrightarrow S_n \otimes_{\mathcal{O}(G)} \Omega^1(G)$ defined below, and $\log(\chi_G)$ is the element of $H^1(K, O_F) = \mathrm{Hom}_{\mathrm{cont}}(\mathrm{Gal}(\bar{K}/K), O_F)$ defined as the composite

$$\mathrm{Gal}(\bar{K}/K) \xrightarrow{\chi_G} O_F^\times \xrightarrow{\log} O_F \quad \text{with } \chi_G \text{ the action on T.}$$

Let $\tau_{c,m} = \alpha_{c,m} + \beta_{c,m}$.

We give the definition of $\mathrm{Trace}_{m,n}$ above (and a little more). We have a commutative diagram of rings

$$\begin{array}{ccccc}
\mathcal{O}(G) & \longrightarrow & S_n & \xrightarrow{\xi_n} & O_{K_n} \\
{\scriptstyle [\pi^{m-n}]^*}\downarrow & & {\scriptstyle [\pi^{m-n}]^*}\downarrow & & \downarrow \\
\mathcal{O}(G) & \longrightarrow & S_m & \xrightarrow[\xi_m]{} & O_{K_m}
\end{array}$$

in which the squares are "push-outs", where $[\pi^{m-n}]^*$ are homomorphisms induced by $[\pi^{m-n}] : G \longrightarrow G$. We have trace maps associated to the vertical arrows, which we denote by $\mathrm{Trace}_{m,n}$, and a commutative diagram

$$\begin{array}{ccccc} \mathcal{O}(G) & \longrightarrow & S_m & \xrightarrow{\xi_m} & O_{K_m} \\ \text{Trace}_{m,n}\downarrow & & \text{Trace}_{m,n}\downarrow & & \text{Trace}_{m,n}\downarrow \\ \mathcal{O}(G) & \longrightarrow & S_n & \xrightarrow[\xi_n]{} & O_{K_n} . \end{array}$$

Then $\text{Trace}_{m,n}(\mathcal{O}(G)) \subset \pi^{m-n}\mathcal{O}(G)$ as is easily seen, and hence $\text{Trace}_{m,n}(S_m) \subset \pi^{m-n}S_n$. The desired map $\text{Trace}_{m,n} : S_m \otimes_{\mathcal{O}(G)} \Omega^1(G) \longrightarrow S_n \otimes_{\mathcal{O}(G)} \Omega^1(G)$ is defined by

$\text{Trace}_{m,n}(x \otimes \omega) = (\pi^{n-m}\text{Trace}_{m,n}(x)) \otimes \omega \quad (x \in S_m,\ \omega \in \text{coLie}(G))$.

(note $\pi^{n-m}\text{Trace}_{m,n}(x) \in S_n$). This definition is natural because we have the projection formula (note $[\pi^{m-n}]^*\omega = \pi^{m-n}\omega$)

$\text{Trace}_{m,n}(x \otimes [\pi^{m-n}]^*\omega) = \text{Trace}_{m,n}(x) \otimes \omega \quad (x \in S_n,\ \omega \in \text{coLie}(G))$.

Now Thm. 2.2.7 is reduced to the following Prop. 2.3.2 and the fact that $\log(\chi_G)$ and $\log(\chi_{\text{cyclo}})$ have the same image in $H^1(K, \mathbb{C}_p)$ (this fact follows from $V \otimes_F \mathbb{C}_p \cong \mathbb{C}_p(1)$).

<u>Proposition</u> 2.3.2. <u>There exists an integer</u> c <u>such that</u> $(r-1)!^{-1}\pi^{-nr}c \in O_K$ <u>and</u> $\tau_{c,m} = 0$ <u>for all</u> $m \geq n$. (c <u>can be taken independent of</u> m).

It is easily seen that $\alpha_{c,m}$ and $\beta_{c,m}$, and so $\tau_{c,m}$ vanish on $J^{[r]}_{S_m} \otimes_{\mathcal{O}(G)} \Omega^1(G)$. The following 2.3.3 is the key step in the proof of 2.3.2.

<u>Lemma</u> 2.3.3. <u>There exists a non-zero integer</u> c <u>such that</u> $(r-1)!^{-1}\pi^{-nr}c \in O_K$ <u>and such that</u> $\tau_{c,m}$ <u>kill</u> $\zeta^{r-1}_{\omega,\xi,m}S_m \otimes_{\mathcal{O}(G)} \Omega^1(G)$ <u>for all</u> $m \geq n$.

This will be deduced (see 2.3.8) from

<u>Lemma</u> 2.3.4. <u>The map</u>

$$S_m \otimes_{\mathcal{O}(G)} \Omega^1(G) \xrightarrow{\mu_{r,\xi,m}} H^1(K_m, (T^{\otimes(-r)} \otimes_{O_F} J_{\infty,\bar{K}/K}/J^{[r+1]}_{\infty,\bar{K}/K})/\pi^m)$$

<u>sends</u> $\ell^{r-1}_{\omega,\xi,m} h \otimes \omega \quad (h \in S_m,\ \omega \in \mathrm{coLie}(G))$ <u>to</u>

$$-\omega^{\otimes r} \otimes h(\xi_m) \cup \chi_m$$

<u>where</u> χ_m <u>is the element of</u> $H^1(K_m, O_F/\pi^m) = \mathrm{Hom}_{\mathrm{cont}}(\mathrm{Gal}(\bar{K}/K_m), O_F/\pi^m)$ <u>defined by</u>

$$\chi_m(\sigma) = \pi^{-m}(\chi_G(\sigma) - 1) \bmod \pi^m \quad (\sigma \in \mathrm{Gal}(\bar{K}/K_m)).$$

We prove 2.3.4 using a lemma 2.3.6 below.

<u>Lemma</u> 2.3.5. The connecting map of the exact sequence

$$0 \longrightarrow J_{\infty,\bar{K}/K} \xrightarrow{\pi^m} J_{\infty,\bar{K}/K} \longrightarrow J_{\infty,\bar{K}/K}/\pi^m \longrightarrow 0$$

sends $\ell_{\omega,\xi,m} \in H^0(K_m, J_{\infty,\bar{K}/K}/\pi^m)$ to $-\mu_{\xi,m}(\omega) \in H^1(K_m, J_{\infty,\bar{K}/K})$.

<u>Proof</u>. Take $x \in J_{\bar{S}_m}$ such that $dx = \omega$. Then, $\mu_{\xi,m}(\omega) \in H^1(K_m, J_{\infty,\bar{K}/K})$ is represented by the cocycle

$$\mathrm{Gal}(\bar{K}/K_m) \longrightarrow J_{\infty,\bar{K}/K} \quad ; \quad \sigma \longmapsto \sigma(x) - x \qquad (2.2.9).$$

Let $y = \ell_{\omega,\xi,m} - \pi^m x$. Then $dy = 0$ so $y \in J_{\infty,\bar{K}/K}$, and $y \bmod \pi^m = \ell_{\omega,\xi,m} \bmod \pi^m$. Hence the cocycle

$$\mathrm{Gal}(\bar{K}/K_m) \longrightarrow J_{\infty,\bar{K}/K} \quad ; \quad \sigma \longmapsto \pi^{-m}(\sigma(y) - y) \in J_{\infty,\bar{K}/K}$$

represents the image of $\ell_{\omega,\xi,m} \in H^0(K_m, J_{\infty,\bar{K}/K}/\pi^m)$ in $H^1(K_m, J_{\infty,\bar{K}/K})$ under the connecting map, but

$$\pi^{-m}(\sigma(y) - y) = x - \sigma(x).$$

<u>Lemma</u> 2.3.6. <u>In</u> $H^1(K_m, J_{\infty,\bar{K}/K}/\pi^m)$, <u>we have</u>

$$\ell_{\omega,\xi,m} \cup \chi_m = -\mu_{\xi,m}(\omega).$$

($\ell_{\omega,\xi,m}$ is regarded as an element of $H^0(K_m, J_{\infty,\bar{K}/K}/\pi^m)$.)

<u>Proof</u>. Let $X = \mathrm{coLie}(G) \otimes_{O_F} T$, $Y = J_{\infty,\bar{K}/K}$. Consider the commutative diagram of exact sequences

$$\begin{array}{ccccccccc} 0 & \longrightarrow & X/\pi^m & \xrightarrow{\pi^m} & X/\pi^{2m} & \longrightarrow & X/\pi^m & \longrightarrow & 0 \\ & & \downarrow & & \downarrow & & \downarrow & & \\ 0 & \longrightarrow & Y/\pi^m & \xrightarrow{\pi^m} & Y/\pi^{2m} & \longrightarrow & Y/\pi^m & \longrightarrow & 0 \end{array}$$

where the vertical arrows are the pairings $\omega \otimes \xi \longrightarrow (\ell_{\omega,\xi,*})_*$. We have a commutative diagram

$$\begin{array}{ccc} H^0(K_m, X/\pi^m) & \longrightarrow & H^1(K_m, X/\pi^m) \\ \downarrow & & \downarrow \\ H^0(K_m, Y/\pi^m) & \longrightarrow & H^1(K_m, Y/\pi^m) . \end{array}$$

The upper horizontal arrow sends $\omega \otimes \xi_m \in \mathrm{coLie}(G) \otimes H^0(K_m, T/\pi^m) = H^0(K_m, X/\pi^m)$ to $\omega \otimes \chi_m \cup \xi_m \in \mathrm{coLie}(G) \otimes H^1(K_m, T/\pi^m) = H^1(K_m, X/\pi^m)$ as is seen easily and the last element is sent to $\ell_{\omega,\xi,m} \cup \chi_m$ by the right vertical arrow. On the other hand the lower horizontal arrow sends $\ell_{\omega,\xi,m}$ to $-\mu_{\xi,m}(\omega)$ by 2.3.5. This proves 2.3.6.

2.3.7. Now we prove 2.3.4 by using 2.3.6. In $H^1(K_m, (T^{\otimes(-r)} \otimes_{O_F} J_{\infty,\bar{K}/K}/J^{[r+1]}_{\infty,\bar{K}/K})/\pi^m)$, for $h \in S_m$ and a basis ω of coLie(G), we have

$$\begin{aligned} &\mu_{r,\xi,m}(\ell^{r-1}_{\omega,\xi,m} h \otimes \omega) \\ &= (\xi^{\otimes r})^{-1} \otimes \ell^{r-1}_{\omega,\xi,m} h(\xi_m)\mu_{\xi,m}(\omega) \quad \text{by the definition of } \mu_{r,\xi,m} \\ &= -\omega^{\otimes r} \otimes h(\xi_m)\chi_m \quad \text{by 2.3.6.} \end{aligned}$$

2.3.8. We deduce 2.3.3 from 2.3.4.

We first consider $\alpha_{c,m}$. Since

$$\log(\chi_G(\sigma)) = (\chi_G(\sigma) - 1) + \frac{1}{2}(\chi_G(\sigma) - 1)^2 + \dots\dots ,$$

there exists a non-zero integer c, which is independent of m, such that

(*) $c\pi^{-m}\log(\chi_G(\sigma)) \in O_F$, $c\pi^{-m}\log(\chi_G(\sigma)) \equiv c\pi^{-m}(\chi_G(\sigma) - 1) \bmod \pi^m$

for all $m \geq n$ and all $\sigma \in \mathrm{Gal}(\bar{K}/K_m)$. For a non-zero integer c for which ($*$) holds, and for $h \in S_m$, we have by 2.3.4

$$\alpha_{c,m}(\ell^{r-1}_{\omega,\xi,m} h \otimes \omega) = -\omega^{\otimes r} \otimes \mathrm{Trace}_{m,n}(h(\xi_m)(c\pi^{-m}\log(\chi_G))) .$$

($\mathrm{Trace}_{m,n}$ here is the trace map $H^1(K_m, O_{\bar{K}}/\pi^m) \longrightarrow H^1(K_n, O_{\bar{K}}/\pi^m)$.) Now π times the last thing is equal to π times

$$-\omega^{\otimes r} \otimes (\pi^{n-m}\mathrm{Trace}_{m,n}(h(\xi_m)))(c\pi^{-n}\log(\chi_G)),$$

with $\mathrm{Trace}_{m,n}$ here the trace map $O_{K_m} \longrightarrow O_{K_n}$ whose image is contained in $\pi^{m-n}O_{K_n}$. This is because the trace $H^1(K_i, O_{\bar{K}}) \longrightarrow H^1(K_{i-1}, O_{\bar{K}})$ for $n < i \le m$ sends $(\pi^{i-m}\mathrm{Trace}_{m,i}(h(\xi_m)))(c\pi^{1-i}\log(\chi_G))$ to $(\pi^{i-m}\mathrm{Trace}_{m,i-1}(h(\xi_m)))(c\pi^{1-i}\log(\chi_G))$ (for $c\pi^{1-i}\log(\chi_G)$ comes from $H^1(K_{i-1}, \;))$.

Next we consider $\beta_{c,m}$. We have for $h \in S_m$

$$\mathrm{Trace}_{m,n}(\ell^{r-1}_{\omega,\xi,m} h \otimes \omega) = \ell^{r-1}_{\omega,\xi,n}(\pi^{n-m}\mathrm{Trace}_{m,n}(h)) \otimes \omega$$

where $\mathrm{Trace}_{m,n}$ on the right hand side is the trace map $S_m \longrightarrow S_n$ associated to $[\pi^{m-n}]^* : S_n \longrightarrow S_m$ (2.3.1). By $d\ell_{\omega,\xi,m} = \pi^n\omega$, we have for any $g \in S_n$

$$(\tfrac{d}{\omega})^{r-1}(\ell^{r-1}_{\omega,\xi,n} g) \equiv (r-1)!\,\pi^{n(r-1)} g \mod J_{S_n}.$$

These show

$$\{(\tfrac{d}{\omega})^{r-1}(\tfrac{1}{\omega}\mathrm{Trace}_{m,n}(\ell^{r-1}_{\omega,\xi,m} h \otimes \omega))\}(\xi_n)$$
$$= (r-1)!\,\pi^{n(r-1)}\pi^{n-m}\mathrm{Trace}_{m,n}(h(\xi_m)).$$

Hence 2.3.4 implies that for some non-zero integer c such that $(r-1)!^{-1}\pi^{-nr}c \in O_K$, both $-\alpha_{c,m}(\ell^{r-1}_{\omega,\xi,m} h \otimes \omega)$ and $\beta_{c,m}(\ell^{r-1}_{\omega,\xi,m} h \otimes \omega)$ coincides with

$$\omega^{\otimes r} \otimes \pi^{n-m}\mathrm{Trace}_{m,n}(h(\xi_m))(c\pi^{-n}\log(\chi_G)) .$$

<u>Lemma</u> 2.3.9. <u>There exists a non-zero integer</u> c <u>such that</u> $(r-1)!^{-1}\pi^{-nr}c \in O_K$ <u>and such that</u> $\tau_{c,m}$ <u>kill</u> $J^{[r-1]}_{S_m} \otimes_{\mathcal{O}(G)} \Omega^1(G)$ <u>for all</u> $m \ge n$.

This follows from 2.3.3 and the following

Lemma 2.3.10. *Let ω be a basis of* coLie(G). *There exists a non-zero integer* c *which kills* $J_{S_m}^{[r-1]}/(J_{S_m}^{[r]} + \ell^{r-1}_{\omega,\xi,m}S_m)$ *for all* $m \geq n$.

To prove 2.3.10, we note

Lemma 2.3.11. *For* $i \geq 0$, $J_{S_m}^{[i]}/J_{S_m}^{[i+1]}$ *is a free* $O_{K_m} = S_m/J_{S_m}$*-module of rank* 1. *If* $h \in S_m$ *is an element of* J_{S_m} *such that* $h \bmod J_{S_m}^{[2]}$ *is a basis of* $J_{S_m}/J_{S_m}^{[2]}$, $h^{[i]}$ *is a basis of* $J_{S_m}^{[i]}/J_{S_m}^{[i+1]}$ *for any* $i \geq 0$.

The proof of 2.3.11 is omitted.

Proof of 2.3.10. By 2.3.11, it is enough to show that there exists a non-zero integer which kills $J_{S_m}/(J_{S_m}^{[2]} + \ell_{\omega,\xi,m}S_m)$ for all $m \geq n$.

Let I be the augumentation ideal $\mathrm{Ker}(\mathcal{O}(G) \longrightarrow O_K)$ of $\mathcal{O}(G)$. Then $I/I^2 \xrightarrow{\simeq} J_{S_0}/J_{S_0}^{[2]}$. Since $\ell_\omega \bmod J_{S_0}^{[2]}$ is a basis of $J_{S_0}/J_{S_0}^{[2]}$, we can take a generator h of I such that $h \equiv \ell_\omega \bmod J_{S_0}^{[2]}$. Let $h_m = [\pi^m]^* h$ for $m \geq 0$. Then for $m \geq 1$, $g_m = h_m h_{m-1}^{-1}$ is a generator of the ideal $\mathrm{Ker}(\xi_m : \mathcal{O}(G) \longrightarrow O_{K_m})$. We have

$$g_m\mathcal{O}(G)/g_m^2\mathcal{O}(G) \xrightarrow{\simeq} J_{S_m}/J_{S_m}^{[2]}$$

and this isomorphism sends the class of h_m to that of $\ell_{\omega,\xi,m}$. So for $m \geq 1$,

$$\begin{aligned} J_{S_m}/(J_{S_m}^{[2]} + \ell_{\omega,\xi,m}S_m) &\simeq g_m\mathcal{O}(G)/(g_m^2\mathcal{O}(G) + h_m\mathcal{O}(G)) \\ &\simeq \mathcal{O}(G)/(g_m\mathcal{O}(G) + h_{m-1}\mathcal{O}(G)) \\ &\simeq [\pi^{m-1}]^*(\mathcal{O}(G)/(g_1\mathcal{O}(G) + h\mathcal{O}(G))). \end{aligned}$$

Since g_1 and h have no common prime divisor in $\mathcal{O}(G)$, $\mathcal{O}(G)/(g_1\mathcal{O}(G) + h\mathcal{O}(G))$ is killed by a non-zero integer, and this integer kills $J_{S_m}/(J_{S_m}^{[2]} + \ell_{\omega,\xi,m}S_m)$ for all m.

2.3.12. To proceed from 2.3.9 to Prop. 2.3.2, we consider the action $[a]^*: S_m \otimes_{\mathcal{O}(G)} \Omega^1(G) \longrightarrow S_m \otimes_{\mathcal{O}(G)} \Omega^1(G)$ of $a \in O_F$. For any integer c such that $(r-1)!^{-1}\pi^{-nr}c \in O_K$ and for any $a \in \mathrm{Ker}(O_F^\times \longrightarrow (O_F/\pi^n)^\times)$, we have

$$\alpha_{c,m}\circ[a]^* = a^r\alpha_{c,m}, \qquad \beta_{c,m}\circ[a]^* = a^r\beta_{c,m}$$

as is easily seen and hence

$$\tau_{c,m}\circ[a]^* = a^r\tau_{c,m}.$$

By this, Prop. 2.3.2 is reduced to 2.3.9 and the following

__Lemma__ 2.3.13. __Let__ $C_m = (S_m/J_{S_m}^{[r-1]}) \otimes_{\mathcal{O}(G)} \Omega^1(G)$, __and let__ $D_m \subset C_m$ __be the subgroup of__ C_m __generated by__ $([a]^* - a^r)C_m$ __for all__ $a \in \mathrm{Ker}(O_F^\times \longrightarrow (O_F/\pi^n)^\times)$. __Then there exists a non-zero integer which kills__ C_m/D_m __for all__ $m \geq n$.

The proof of 2.3.13 is reduced to

__Lemma__ 2.3.14. __Let__ $1 \leq i \leq r-1$. __Let__ $E_m = J_{S_m}^{[i-1]}/J_{S_m}^{[i]} \otimes_{\mathcal{O}(G)} \Omega^1(G)$ __and let__ F_m __be the subgroup of__ E_m __generated by__ $([a]^* - a^r)E_m$ __for all__ $a \in \mathrm{Ker}(O_F^\times \longrightarrow (O_F/\pi^n)^\times)$. __Then there exists a non-zero integer which kills__ E_m/F_m __for all__ $m \geq n$.

__Proof__. With g_m as in the proof of 2.3.10 and with ω a basis of coLie(G), consider the isomorphism of O_{K_m}-modules

$$f \;:\; O_{K_m} \xrightarrow{\cong} J_{S_m}^{[i-1]}/J_{S_m}^{[i]} \otimes_{\mathcal{O}(G)} \Omega^1(G) \quad ; \quad 1 \longmapsto g_m^{[i-1]} \otimes \omega.$$

Then

$$fa^i\sigma_a = [a]^*f \qquad \text{for any} \quad a \in \mathrm{Ker}(O_F^\times \longrightarrow (O_F/\pi^n)^\times)$$

where $\sigma_a \in \mathrm{Gal}((\bigcup_j K_j)/K_n)$ is the unique element whose action on T is a. The inverse image of F_m (2.3.14) under f is the subgroup of O_{K_m} generated by $(a^i\sigma_a - a^r)O_{K_m}$ for all $a \in \mathrm{Ker}(O_F^\times \longrightarrow (O_F/\pi^n)^\times)$. Thus we are reduced to

<u>Lemma</u> 2.3.15. <u>Let</u> r <u>be a non-zero integer</u>. <u>Let</u> A_m <u>be the subgroup of</u> O_{K_m} <u>generated by</u> $(\sigma_a - a^r)O_{K_m}$ <u>for all</u> $a \in \mathrm{Ker}(O_F^\times \longrightarrow (O_F/\pi^n)^\times)$. <u>Then there exists a non-zero integer which kills</u> O_{K_m}/A_m <u>for all</u> $m \geq n$.

<u>Proof</u>. Since A_m is an O_{K_n}-submodule of O_{K_m} containing $p^i O_{K_m}$ for some $i \geq 0$ and since

$$O_{K_m}/p^i \times O_{K_m}/p^i \longrightarrow O_{K_n}/p^i \quad ; \quad (x, y) \longmapsto \pi^{n-m}\mathrm{Trace}_{m,n}(xy)$$

is a perfect duality of finitely generated free O_{K_n}/p^i-modules, we are reduced, by considering the dual, to

<u>Lemma</u> 2.3.16. <u>Fix</u> $r \in \mathbb{Z}$, $r \neq 0$. <u>Then there exists a non-zero integer which kills</u>

$$\{x \in O_{K_m}/p^i \ ; \ \sigma_a(x) = a^r x \quad \text{for all } a \in \mathrm{Ker}(O_F^\times \longrightarrow (O_F/\pi^n)^\times)\}$$

<u>for all</u> $m \geq n$ <u>and all</u> $i \geq 0$.

<u>Proof</u>. The last group is isomorphic to $H^0(K_n, O_{K_m}/p^i \otimes_{O_F} T^{\otimes(-r)})$ which is embedded in $H^0(K_n, O_{\bar{K}}/p^i \otimes_{O_F} T^{\otimes(-r)})$. Hence it is sufficient to show that $H^q(K_n, \hat{O_{\bar{K}}} \otimes_{O_F} T^{\otimes(-r)})$ $(q = 0, 1)$ is killed by some power of p. Since

$$(\hat{O_{\bar{K}}} \otimes_{O_F} T^{\otimes(-r)}) \otimes \mathbf{Q} \cong (\mathrm{coLie}(G)^{\otimes r} \otimes_{O_K} \hat{O_{\bar{K}}}(-r)) \otimes \mathbf{Q},$$

we are reduced to the fact $H^q(K_n, \hat{O_{\bar{K}}}(-r))$ is killed by some power of p (2.2.4) (1).

Chapter III. Global subjects.

§1. Explicit reciprocity laws and zeta values.

We relate the explicit reciprocity laws in Chap. II §2 to values of partial Riemann zeta functions, and to values of some Hecke L-series of quadratic imaginary fields of class number 1. (Thm. 1.2.6.)

§1.1 is a preliminary review.

§1.1. The fundamental functions on $\mathbb{G}_m$ and on elliptic curves, and zeta values (review).

In this §1.1, let K be a field and let G be either the multiplicative group $\mathbb{G}_{m,K}$ over K or an elliptic curve over K. We denote by K(G) the field of rational functions on G.

1.1.1 For $a \in \mathrm{End}(G) \setminus \{0\}$, let $a^* : K(G) \longrightarrow K(G)$ be the pull back by a, let deg(a) be the degree of this homomorphism of fields, and let $N_a : K(G)^\times \longrightarrow K(G)^\times$ be the norm homomorphism for a^*. We have $N_a(a^*f) = f^{\deg(a)}$ for $f \in K(G)^\times$. For $a \in \mathbb{Z}$, we have $\deg(a) = |a|$ if $G = \mathbb{G}_{m,K}$ and $\deg(a) = a^2$ if G is an elliptic curve.

Definition 1.1.2. An element $\theta \in K(G)^\times$ is called the fundamental function on G if it satisfies the following two conditions.

(1) $N_a(\theta) = \theta$ for any $a \in \mathrm{End}(G) \setminus \{0\}$.

(2) θ does not have zero or pole on G except a zero of order 1 at the origin of G.

The fundamental function on G does not exist. (In the case G is an elliptic curve, this can be seen by the fact that the degree of a principal divisor is zero which contradicts the condition (2).)

Clearly the author has to explain why he puts an empty definition 1.1.2. He thinks it is very helpful for the study of some zeta values, to find "some truth" in the following wrong sentense: "The fundamental function exists and unique. In the case $G = \mathbb{G}_{m,K}$ or G is an elliptic curve with complex multiplication, some zeta values are obtained as the values of the iterated derivatives of the logarithm of the fundamental function at torsion points of G".

1.1.3. For $G = \mathbb{G}_{m,K} = \mathrm{Spec}(K[t, t^{-1}])$ (t is the standard coordinate of $\mathbb{G}_{m,K}$), the function $(1-t)(1-t^{-1})$ plays the role of θ^2. Indeed, it is the unique element of $K(G)^{\times} = K(t)^{\times}$ which is invariant under N_a for any $a \in \mathrm{End}(G) \smallsetminus \{0\} = \mathbb{Z} \smallsetminus \{0\}$ and which does not have zero or pole on G except a zero of order 2 at the origin.

The "some truth" in 1.1.2 for $G = \mathbb{G}_{m,K}$ is that the values of $(t\frac{d}{dt})^r(\log((1-t)(1-t^{-1})))$ $(r \geq 0)$ at torsion points of $\mathbb{G}_{m,K}$ are values of partial Riemann zeta functions. (Chap. I, §1.3, Chap. I, 3.3.6. Cf. 1.1.7 below.)

1.1.4. We denote the origin of G by e. For a, b ∈ End(G), we write (a, b) = 1 if $\mathrm{Ker}(a) \cap \mathrm{Ker}(b) = \{e\}$ as a scheme ({e} is endowed with the reduced scheme structure) and ab = ba. (If $G = \mathbb{G}_{m,K}$ or if char(K) = 0, End(G) is commutative and hence ab = ba is automatically satisfied.) The following 1.1.5 says that if G is an elliptic curve and a ∈ End(G) satisfying (a, 6) = 1, then there exists a function $\theta_a \in K(G)^{\times}$ which plays the role of $\theta^{a^2}(a^*\theta)^{-1}$.

<u>Proposition</u> 1.1.5. <u>Assume</u> G <u>is an elliptic curve</u>, <u>and let</u> a ∈ End(G), (a, 6) = 1. <u>Then there exists a unique rational function</u> $\theta_a \in K(G)^{\times}$ <u>satisfying the following two conditions</u> (1) (2).

(1) $N_b(\theta_a) = \theta_a$ <u>for any</u> $b \in \mathrm{End}(G)$ <u>such that</u> $(a, b) = 1$.

(2) <u>The divisor of</u> θ_a <u>is</u> $\deg(a)(e) - \mathrm{Ker}(a)$.

<u>This function</u> θ_a <u>satisfies</u>

(3) $(\theta_a)^{\deg(b)}(b^*\theta_a)^{-1} = (\theta_b)^{\deg(a)}(a^*\theta_b)^{-1}$

<u>for any</u> $b \in \mathrm{End}(G)$ <u>such that</u> $(b, 6) = 1$ <u>and</u> $ab = ba$.

The proof of 1.1.5 is given later in 1.1.13.

1.1.6. Now let K be a quadratic imaginary field of class number 1, and let E be an elliptic curve over K such that $\mathrm{End}(E) \longrightarrow \mathrm{End}(\mathrm{Lie}(E)) = K$ induces an isomorphism $\mathrm{End}(E) \xrightarrow{\cong} O_K$ which we regard as an identification. Then,

$$L(H^1(E), s) = \sum_{\mathfrak{a}} \overline{\nu(\mathfrak{a})} N(\mathfrak{a})^{-s} = \sum_{\mathfrak{a}} \nu(\mathfrak{a})^{-1} N(\mathfrak{a})^{-(s-1)}$$

where ν is a Hecke character of K such that $\nu(\mathfrak{a})$ is a generator of $\mathfrak{a}$ for any non-zero ideal $\mathfrak{a}$ of O_K which is prime to the conductor $\mathfrak{f}$ of ν, $\overline{?}$ is the complex conjugate of ?, and $\mathfrak{a}$ in the sum ranges over all ideals of O_K which are prime to $\mathfrak{f}$.

Let $\mathfrak{g}$ be a non-zero ideal of O_K such that $\mathfrak{g} \subset \mathfrak{f}$ and let L be the abelian extension of K corresponding to the kernel of $\mathrm{Gal}(\bar{K}/K) \longrightarrow \mathrm{Aut}_{O_K}({}_{\mathfrak{g}}E) \cong (O_K/\mathfrak{g})^{\times}$ (here ${}_{\mathfrak{g}}E$ denotes the subgroup scheme of E of $\mathfrak{g}$-torsion points of E).

We will consider for $r \in \mathbb{Z}$, $r \geq 0$ and for $\sigma \in \mathrm{Gal}(L/K)$, the partial L-function

$$L(\nu^{-r}, \sigma\text{-part}, s) = \sum_{(\mathfrak{a}, L/K) = \sigma} \nu(\mathfrak{a})^{-r} N(\mathfrak{a})^{-s}$$

where the sum is taken over all non-zero ideals $\mathfrak{a}$ of O_K which are prime to $\mathfrak{g}$ and whose Artin symbols $(\mathfrak{a}, L/K) \in \mathrm{Gal}(L/K)$ are equal to σ. This L-function absolutely converges if $\mathrm{Re}(s) > 1 - \frac{r}{2}$. If $r \geq 1$

(resp. $r = 0$), it is extended to the whole complex plane as a holomorphic (resp. meromorphic) function.

Fix an embedding $K \longrightarrow \mathbb{C}$. Let

$$\exp : \mathrm{Lie}(E) \otimes_K \mathbb{C} \longrightarrow E(\mathbb{C})$$

be the exponential map.

The following theorem is classical (cf. [Da][We$_1$][dS$_2$]) ((2) is Kronecker's "second limit formula").

Theorem. Let γ be a generator of the invertible O_K-module $g^{-1}H_1(E(\mathbb{C}), \mathbb{Z})$, and let $\alpha = \exp(\gamma) \in {}_gE(\mathbb{C}) = {}_gE(L)$. Let $\mathfrak{a}$ be a non-zero ideal of O_K which is prime to $6g$, and let $\sigma \in \mathrm{Gal}(L/K)$, $\tau \underset{\mathrm{def}}{=} (\mathfrak{a}, L/K) \in \mathrm{Gal}(L/K)$.

(1) Let $r \in \mathbb{Z}$, $r \geq 1$, and let $\omega \in \mathrm{coLie}(E) \setminus \{0\}$. Then, the value of

$$\left(\frac{d}{\omega}\right)^r \log(\theta_{\nu(\mathfrak{a})})$$

at $\sigma\alpha$ is equal to $(-1)^{r-1}(r-1)!\,(\int_\gamma \omega)^{-r}$ times

$$N(\mathfrak{a})L(\nu^{-r}, \sigma\text{-part}, 0) - \nu(\mathfrak{a})^r L(\nu^{-r}, \sigma\tau\text{-part}, 0).$$

(2) $\log(|\theta_{\nu(\mathfrak{a})}(\sigma\alpha)|)$

$$= -\lim_{s\to 0} s^{-1}(N(\mathfrak{a})L(1, \sigma\text{-part}, s) - L(1, \sigma\tau\text{-part}, s)).$$

1.1.7. The above theorem is an analogue of the following fact for partial Riemann zeta functions introduced in Chap. I, §1.3 and Chap. I, 3.3.6.

Let $\alpha = \exp(2\pi i a N^{-1})$ with $N, a \in \mathbb{Z}$, $N \geq 1$, and assume $\alpha \neq 1$. Then:

(1) For $r \in \mathbb{Z}$, $r \geq 1$, the value of

$$\frac{1}{2}\left(t\frac{d}{dt}\right)^r \log((1-t)(1-t^{-1}))$$

at α is equal to $(-1)^{r-1}(r-1)!\left(\frac{2\pi i}{N}\right)^{-r}$ times

$$(\zeta_{a(N)} + (-1)^r \zeta_{-a(N)})(r).$$

(2) $$\frac{1}{2}\log((1-\alpha)(1-\alpha^{-1})) = -\lim_{s\to 0} s^{-1}(\zeta_{a(N)}(s) + \zeta_{-a(N)}(s)).$$

Both in 1.1.6 and 1.1.7, zeta values appear each time we shake the magical stick $\frac{d}{\omega}$. Is there such magical stick for any motif?

1.1.8. We review briefly the proof of the case $r \geq 3$ of (1) of the the theorem in 1.1.6. (By the assumption $r \geq 3$, we can avoid the delicate argument concerning the convergence and analytic continuation. Cf. [We_1] for the whole proof of the theorem in 1.1.6.)

We use the following 1.1.9.

1.1.9. Let G be an elliptic curve over $\mathbb{C}$. Let $a \in \mathrm{End}(G) \setminus \{0\}$, $(a, 6) = 1$, let $r \in \mathbb{Z}$, $r \geq 3$, and let $\omega \in \mathrm{coLie}(G) \setminus \{0\}$. Then the composite meromorphic function

$$\mathrm{Lie}(G) \xrightarrow{\exp} G(\mathbb{C}) \xrightarrow{(*)} \mathbb{C} \quad ; \quad (*) = (\tfrac{d}{\omega})^r \log(\theta_a)$$

is equal to

$$z \longmapsto (-1)^{r-1}(r-1)!(\deg(a) E_{r,\omega}(z) - \mathrm{Lie}(a)^r E_{r,\omega}(az))$$

where $\mathrm{Lie}(a) \in \mathbb{C}$ is the effect of a on $\mathrm{Lie}(G)$ and

$$E_{r,\omega}(z) = \sum_{\gamma} (z+\gamma, \omega)^{-r}$$

in which γ ranges over all elements of

$$H_1(G(\mathbb{C}), \mathbb{Z}) = \mathrm{Ker}(\exp : \mathrm{Lie}(G) \longrightarrow G(\mathbb{C}))$$

and $(\ ,\ )$ is the canonical pairing $\mathrm{Lie}(G) \times \mathrm{coLie}(G) \longrightarrow \mathbb{C}$.

1.1.10. 1.1.9 is an analogue of the following fact: For $r \in \mathbb{Z}$, $r \geq 1$, the composite meromorphic function

$$\mathbb{C} \xrightarrow{\exp} \mathbb{C}^\times \xrightarrow{(*)} \mathbb{C} \quad ; \quad (*) = \frac{1}{2}\,(t\frac{d}{dt})^r \log((1-t)(1-t^{-1}))$$

is equal to

$$z \longmapsto (-1)^{r-1}(r-1)! \sum_{n\in\mathbb{Z}} (z - 2\pi i n)^{-r} \qquad \text{if } r \geq 2,$$

$$z \longmapsto \frac{1}{2}(-1)^{r-1}(r-1)! \sum_{n\in\mathbb{Z}} \{(z - 2\pi i n)^{-1} + (z + 2\pi i n)^{-1}\} \quad \text{if } r = 1.$$

1.1.11. We deduce the case $r \geq 3$ of (1) of the theorem in 1.1.6 from 1.1.9. We are reduced to showing

$$E_{r,\omega}(\sigma\alpha) = (\int_\gamma \omega)^{-r} L(\nu^{-r}, \sigma\text{-part}, 0).$$

(We used the fact $\deg(a) = a\bar{a} = N(\mathfrak{a})$ for $a \in O_K = \mathrm{End}(E)$ and for $\mathfrak{a} = (a)$.) Take $b \in O_K$ such that $\sigma\alpha = b\alpha$. Then,

$$E_{r,\omega}(\sigma\alpha) = \sum_{\delta} (b\gamma + \delta, \omega)^{-r} = (\sum_{x \in \mathfrak{g}} (b + x)^{-r}) (\int_\gamma \omega)^{-r}$$

where δ ranges over all elements of $H_1(G(\mathbb{C}), \mathbb{Z})$. But

$$\sum_{x \in \mathfrak{g}} (b + x)^{-r} = \sum_{(\mathfrak{a}, L/K) = \sigma} \nu(\mathfrak{a})^{-r} = L(\nu^{-r}, \sigma\text{-part}, 0)$$

where the first equation follows from the fact that the maps

$$\mathfrak{a} \longmapsto \nu(\mathfrak{a}), \quad (a) \longleftarrow a$$

are bijections inverse to each other, between the set of all ideals of O_K which are prime to $\mathfrak{g}$ such that $(\mathfrak{a}, L/K) = \sigma$ and the set of all elements of O_K which are $\equiv b \bmod \mathfrak{g}$.

1.1.12. Similarly we can deduce 1.1.7 from 1.1.10 and obtain another proof of Chap. I, 1.3.2 (2).

1.1.13. We give the proof of 1.1.5. By Abel's theorem, $\deg(a)(e) - \mathrm{Ker}(a)$ is a principal divisor. Take any $g \in K(G)^{\times}$ such that $\mathrm{div}(g) = \deg(a)(e) - \mathrm{Ker}(a)$. Then for any $b \in \mathrm{End}(G)$ such that $(a, b) = 1$, we have $\mathrm{div}(N_b(g)) = \mathrm{div}(g)$ and hence there exists $\lambda_b \in K^{\times}$ such that $N_b(g) = \lambda_b g$. Let $\theta_a = \lambda_3\lambda_2^{-3} g$. We show θ_a satisfies the conditions (1)(2). (2) is clear, so we prove (1). For any $b, c \in \mathrm{End}(G)$ such that $(a, b) = 1$, $(a, c) = 1$ and $bc = cb$, we obtain from $N_bN_c = N_cN_b$

$$\lambda_b\lambda_c^{\deg(b)} = \lambda_c\lambda_b^{\deg(c)}.$$

By considering the cases $c = 2, 3$ of this equation, we get $\lambda_b = (\lambda_2^3\lambda_3^{-1})^{(\deg(b)-1)}$ for any $b \in \mathrm{End}(G)$ such that $(a, b) = 1$. From this we see (1) holds.

The uniqueness of θ_a is shown as follows. By (1), the ratio α of two functions satisfying the conditions (1) (2) has the property $\alpha = N_b(\alpha) = \alpha^{\deg(b)}$ for any b as in (1). By taking b = 2, 3, we have $\alpha = 1$.

Finally we prove (3). The both sides of (3) has the same divisor

$$\deg(a)\deg(b)(e) - \deg(a)\mathrm{Ker}(b) - \deg(b)\mathrm{Ker}(a) + \mathrm{Ker}(ab) .$$

So the ratio of the both sides is a non-zero element of $K^{\times}$, and by applying N_2 and N_3 to the both sides of (3), we see this ratio is 1.

§1.2. Relation with explicit reciprocity laws.

1.2.1. In this §1.2, let p be a prime number, and let K be either

(1) $\mathbf{Q}$

or

(2) a quadratic imaginary field of class number one.

Let S be a finite set of non-zero prime ideals of O_K including all prime divisors of p, and let $X = \mathrm{Spec}(O_K[\frac{1}{S}])$.

In Chap. I §3.3, we tried to define zeta elements $z_{\Lambda}(X, \mathcal{F})$ for Λ as in Chap. I, 3.1.1 and for a smooth invertible Λ-sheaf $\mathcal{F}$ on X in the case (1) above. A similar story exists in the case (2) (1.2.3). As cyclotomic units played essential roles in the case (1), "elliptic units" play essential roles in the case (2). We show that in both the cases (1) and (2), our zeta elements are often related to zeta values (as is expected from Chap. I, Conj. 3.2.2, condition (v)) via the explicit reciprocity law in Chap. II, §2.

In the case (2), we fix in this §1.2 an elliptic curve E over K such that $\mathrm{End}(E) \cong O_K$. We assume E has good reductions outside S, and also good reductions at prime divisors of p. We identify O_K with End(E) in the way that the composite $O_K \longrightarrow \mathrm{End}(E) \longrightarrow \mathrm{End}_K(\mathrm{Lie}(E)) = K$ is the natural inclusion. In the case (2), fix also an embedding $K \longrightarrow \mathbf{C}$.

Both in the cases (1) and (2), fix an algebraic closure $\bar{K}$ of K, and let $K^{ab,S}$ be the union of all finite abelian extensions of K in $\bar{K}$ which are unramified outside S.

This §1.2 is written under the influence of the definitions of "elliptic elements in p-adic Galois cohomology" of Soulé [So_4].

1.2.2. Assume we are in the case 1.2.1 (2). We consider an analogue of Chap. I, 3.3.5. Let

$$\Lambda_0 = \mathbf{Z}_p[[\mathrm{Gal}(K^{ab,S}/K)]], \qquad \mathcal{F}_0 = \varprojlim_L (f_L)_* f_L^*(\mathbf{Z}_p(1)_X)$$

where L ranges over finite subextensions of $K^{ab,S}/K$ and f_L denotes the canonical morphism $\mathrm{Spec}(O_L[\frac{1}{S}]) \longrightarrow \mathrm{Spec}(O_K[\frac{1}{S}])$. Take a non-zero element a of O_K which is prime to 6 and to any elements of S. In this 1.2.2, we define an element

$$z_{a,\Lambda_0}(X, \mathcal{F}_0) \in H^1(X, \mathcal{F}_0) \otimes_{\Lambda_0} H^0(\mathbf{C}, \mathcal{F}_0(-1))^{-1}.$$

The values of θ_a (§1.1) at torsion points of E (called elliptic units) play essential roles in the definition. This is analogous to the fact that the values of $(1-t)(1-t^{-1})$ at torsion points of $\mathbf{G}_{m,K} = \mathrm{Spec}(K[t, t^{-1}])$ (called cyclotomic units) played essential roles in the definition of $z_{\Lambda_+}(X; \mathcal{F}_+)$ in Chap. I, 3.3.5.

Let $\mathfrak{f}$ be the conductor of the Hecke character associated to E (§1.1). For a non-zero ideal $\mathfrak{g}$ of O_K such that $\mathfrak{g} \subset \mathfrak{f}$ and such that S coincides with the set of prime divisors of $\mathfrak{g}$, let $K(\mathfrak{g}) \subset K^{ab,S}$ be the finite abelian extension of K corresponding to the kernel of $\mathrm{Gal}(K^{ab,S}/K) \longrightarrow \mathrm{Aut}({}_{\mathfrak{g}}E)$, where ${}_{\mathfrak{g}}E$ is the subgroup scheme of E of $\mathfrak{g}$-torsion points of E. Then $K^{ab,S} = \bigcup_{\mathfrak{g}} K(\mathfrak{g})$. Let

$$(1)_{a,\mathfrak{g}} \quad H^0(K(\mathfrak{g}) \otimes_K \mathbf{C}, \mathbf{Z}_p) \cong \bigoplus_{\iota} H^0(\mathbf{C}, \mathbf{Z}_p) \longrightarrow H^1(O_{K(\mathfrak{g})}[\tfrac{1}{S}], \mathbf{Z}_p(1))$$

be the homomorphism $(a_\iota)_\iota \longmapsto \sum_\iota a_\iota(\iota^{-1}(\theta_a(x_g)))$ where ι ranges over all embeddings $K(g) \longrightarrow \mathbb{C}$ whose restriction to K coincides with the given embedding of K into $\mathbb{C}$, and where $x_g = \exp(\gamma_g)$ with γ_g an O_K-basis of $g^{-1}H_1(E(\mathbb{C}), \mathbb{Z})$. (Then $\theta_a(x_g)$ is independent of γ_g because $\theta_a \circ [u] = \theta_a$ for any $u \in (O_K)^\times$ by the propery 1.1.5 (1) of θ_a. It is known that $\theta_a(x_g) \in (O_{K(g)}[\frac{1}{S}])^\times$ ([dS$_2$] Chap. II §2). This map $(1)_{a,g}$ forms an inverse system when g varies. The inverse limit

$$H^0(\mathbb{C}, \mathcal{F}_0(-1)) \longrightarrow H^1(X, \mathcal{F}_0)$$

corresponds to an element of $H^1(X, \mathcal{F}_0) \otimes_{\Lambda_0} H^0(\mathbb{C}, \mathcal{F}_0(-1))^{-1}$, which we denote by $z_{a,\Lambda_0}(X, \mathcal{F}_0)$.

1.2.3. Assume we are in the case 1.2.1 (2). We consider an analogue of Chap. I, 3.3.7. We try to define a canonial Λ-basis $z_\Lambda(X, \mathcal{F})$ of $\Delta_\Lambda(X, \mathcal{F})$ for any Λ as in Chap. I, 3.1.1 and for any smooth invertible Λ-sheaf $\mathcal{F}$ on X.

Define $z_{\Lambda_0}(X, \mathcal{F}_0) \in H^1(X, \mathcal{F}_0) \otimes_{\Lambda_0} H^0(\mathbb{C}, \mathcal{F}_0(-1)) \otimes_{\Lambda_0} Q(\Lambda_0)$ by

$$z_{\Lambda_0}(X, \mathcal{F}_0) = (N(\mathfrak{a}) - \sigma_{\mathfrak{a}})^{-1} z_{\nu(\mathfrak{a}),\Lambda_0}(X, \mathcal{F}_0)$$

where $\mathfrak{a}$ is an ideal of O_K such that $(\mathfrak{a}, 6) = 1$, $\mathfrak{a} \neq O_K$ and $\mathfrak{a}$ is prime to any element of S. Then, $z_{\Lambda_0}(X, \mathcal{F}_0)$ is independent of the choice of $\mathfrak{a}$. We want to regard $z_{\Lambda_0}(X, \mathcal{F}_0)$ as a Λ_0-basis of $\Delta_{\Lambda_0}(X, \mathcal{F}_0)$. If this is done, we get $z_\Lambda(X, \mathcal{F})$ for any $(\Lambda, \mathcal{F})$ as above, by the same method as in Chap. I, 3.3.2. The problem is reduced to the following conjecture which is analogous to Chap. I, 3.3.8. Let g be an ideal of O_K such that the set of prime divisors of gp coincides with S, and let $\Lambda = \mathbb{Z}_p[[G_{gp^\infty}]]$ with $G_{gp^\infty} = \varprojlim_n \mathrm{Gal}(K(gp^n)/K)$, $\mathcal{F}_{gp^\infty} = \mathcal{F}_0 \otimes_{\Lambda_0} \Lambda_{gp^\infty}$. Then by the same reason as in Chap. I, 3.3.7,

$H^i(X, \mathcal{F}_{gp^\infty}) \otimes_{\Lambda_{gp^\infty}} Q(\Lambda_{gp^\infty})$ is zero if $i \neq 1$, and is an invertible $Q(\Lambda_{gp^\infty})$-module if $i = 1$.

<u>Conjecture</u>. The image of $z_{\Lambda_0}(X, \mathcal{F}_0)$ in

$$\Delta_{\Lambda_{gp^\infty}}(X, \mathcal{F}_{gp^\infty}) \otimes_{\Lambda_{gp^\infty}} Q(\Lambda_{gp^\infty})$$

$$H^1(X, \mathcal{F}_{gp^\infty}) \otimes_{\Lambda_{gp^\infty}} H^0(C, \mathcal{F}_{gp^\infty}(-1)) \otimes_{\Lambda_{gp^\infty}} Q(\Lambda_{gp^\infty})$$

is a Λ_{gp^∞}-basis of $\Delta_{\Lambda_{gp^\infty}}(X, \mathcal{F}_{gp^\infty})$.

For various S, this conjecture is proved as a consequence of the solution of the "main conjecture for the Iwasawa theory of quadratic imaginary fields" proved by Rubin [Ru]. (See Chap. I, 3.3.9 about the delicate difference between the above conjecture and the "main conjecture".)

1.2.4. The following 1.2.4 and 1.2.5 are preparations of Thm. 1.2.6. Let Λ be a finite product of finite extensions of K, let $\Lambda = A \otimes Q_p$, and let $\mathcal{F}$ be an invertible smooth Λ-sheaf on X such that there exist an integer $r \geq 1$ and a homomorphism $\lambda : \mathrm{Gal}(K^{ab,S}/K) \longrightarrow A^\times$ which is continuous for the discrete topology of $A^\times$ satisfying the following condition: In the case 1.2.1 (1) (resp. (2)),

$$\chi_{\mathcal{F}}(\sigma) = \chi_{cyclo}(\sigma)^{1-r}\lambda(\sigma) \quad \text{for all } \sigma \in \mathrm{Gal}(Q^{ab,S}/Q) \quad \text{and}$$

$$\chi_{\mathcal{F}}(\tau) = -1 \quad \text{where } \tau \in \mathrm{Gal}(Q^{ab,S}/Q) \text{ is the complex conjugation.}$$

(resp.

$$\chi_{\mathcal{F}}(\sigma) = \chi_{cyclo}(\sigma)\chi_E(\sigma)^{-r}\lambda(\sigma) \quad \text{for all } \sigma \in \mathrm{Gal}(K^{ab,S}/K)$$

where $\chi_E : \mathrm{Gal}(K^{ab,S}/K) \longrightarrow (O_K \otimes Z_p)^\times \subset \Lambda^\times$ is the action on the p-adic Tate module T_pE).

We define an element

$$z_\Lambda(X, \mathcal{F}) \in H^1(X, \mathcal{F}) \otimes_\Lambda H^0(K \otimes R, \mathcal{F}(-1))^{-1}$$

by using the element $z_{\Lambda_+}(X, \mathcal{F}_+)$ in Chap. I, 3.3.5 in which we assumed $p \neq 2$ but p can be 2 in this section (resp. by using $z_{a,\Lambda_0}(X, \mathcal{F}_0)$ in 1.2.2 where a is an element of O_K which is prime to 6 and to any element of S). If we denote Λ_+ and $\mathcal{F}_+$ (resp. Λ_0 and $\mathcal{F}_0$) by Λ' and $\mathcal{F}'$, respectively, we have a non-canonical Λ-isomorphism

(1) $\quad \mathcal{F} \cong \mathcal{F}' \otimes_{\Lambda'} \Lambda$

where $\Lambda' \longrightarrow \Lambda$; $\sigma \longmapsto \chi_{cyclo}(\sigma)\chi_{\mathcal{F}}(\sigma)^{-1}$,

and (1) induces a homomorphism

(2) $\quad H^1(X, \mathcal{F}') \otimes_{\Lambda'} H^0(K \otimes \mathbb{R}, \mathcal{F}'(-1))^{-1}$

$\longrightarrow H^1(X, \mathcal{F}) \otimes_{\Lambda} H^0(K \otimes \mathbb{R}, \mathcal{F}(-1))^{-1}$

which is independent of the choice of the Λ-isomorphism (1).

Case 1.2.1 (1) with $p \neq 2$: We define $z_{\Lambda}(X, \mathcal{F})$ as the image of $z_{\Lambda_+}(X, \mathcal{F}_+)$ under the map (2).

Case 1.2.1 (1) with $p = 2$. We define

$$z_{\Lambda_+}(X, \mathcal{F}_+) \in H^1(X, \mathcal{F}_+) \otimes H^0(K \otimes \mathbb{R}, \mathcal{F}_+(-1)) \otimes \mathbb{Q}$$

in the same way as in the case $p \neq 2$ (2^{-1} exists in the definition of $z_+(X, \mathcal{F}_+)$ (Chap. I, 3.3.5) but this does not matter after $\otimes\, \mathbb{Q}$), and define $z_{\Lambda}(X, \mathcal{F})$ as the image of $z_{\Lambda_+}(X, \mathcal{F}_+)$ under the map (2).

Case 1.2.1 (2): We define $z_{a,\Lambda}(X, \mathcal{F})$ as the image of $z_{a,\Lambda_0}(X, \mathcal{F}_0)$ under the map (2). Choose a non-zero ideal $\mathfrak{a}$ of O_K such that $(\mathfrak{a}, 6g) = 1$ and such that $N(\mathfrak{a})\nu(\mathfrak{a})^{-r}$ is not a root of 1.
Let $\sigma_{\mathfrak{a}} = (\mathfrak{a}, K^{ab,S}/K)$ be the Artin symbol and define

$$z_{\Lambda}(X, \mathcal{F}) = (N(\mathfrak{a}) - N(\mathfrak{a})\chi_{\mathcal{F}}(\sigma_{\mathfrak{a}})^{-1})^{-1} z_{\nu(\mathfrak{a}),\Lambda}(X, \mathcal{F}).$$

$(N(\mathfrak{a}) - N(\mathfrak{a})\chi_{\mathcal{F}}(\sigma_{\mathfrak{a}})^{-1}$ is invertible in Λ because $\chi_{\mathcal{F}}(\sigma_{\mathfrak{a}}) = N(\mathfrak{a})\nu(\mathfrak{a})^{-r}\alpha$ for a root α of 1 in Λ). Then $z_{\Lambda}(X, \mathcal{F})$ is independent of the choice of Ω.

1.2.5. Let A and $\mathcal{F}$ be as in 1.2.4. We define invertible A-modules Σ and Ω as follows. Take a finite subextension L/K of $K^{ab,S}/K$ such that λ (1.2.4) factors through Gal(L/K).

In the case 1.2.1 (1) (resp. (2)), let

$$\Sigma' = \bigoplus_{\iota} \mathbf{Q}(2\pi i)^{-r}, \qquad \Omega' = L$$

(resp. $\Sigma' = \bigoplus_{\iota} H_1(E(\mathbb{C}), \mathbf{Q})^{\otimes(-r)}, \qquad \Omega' = \mathrm{coLie}(E)^{\otimes r} \otimes_K L$)

where r is as in 1.2.4, ι ranges over all embeddings $L \longrightarrow \mathbb{C}$ (resp. all embeddings $L \longrightarrow \mathbb{C}$ whose restriction to K coinicides with the given embedding of K into $\mathbb{C}$, and $\otimes(-r)$ is taken as an invertible K-module). Let

$$\Sigma = \Sigma' \otimes_{K[\mathrm{Gal}(L/K)]} A, \qquad \Omega = \Omega' \otimes_{K[\mathrm{Gal}(L/K)]} A.$$

We have canonical isomorphisms of Λ-modules

$$\Sigma \otimes \mathbf{Q}_p \cong H^0(K \otimes \mathbb{R}, \mathcal{F}(-1)),$$

$$\Omega \otimes \mathbf{Q}_p \cong D_{dR}(\mathcal{F}).$$

We define an isomorphism of $A \otimes \mathbb{R}$-modules

$$a \;:\; \Omega \otimes \mathbb{R} \xrightarrow{\cong} \Sigma \otimes \mathbb{R}$$

or equivalently an isomorphism of $A \otimes \mathbb{R}$-module

$$(\Omega \otimes_A \Sigma^{-1}) \otimes \mathbb{R} \cong A \otimes \mathbb{R}.$$

We define a to be the map induced by $a' : \Omega' \otimes \mathbb{R} \xrightarrow{\cong} \Sigma' \otimes \mathbb{R}$ defined as follows.

In the case 1.2.1 (1), let a' be the composite map

$$\Omega' \otimes \mathbb{R} = L \otimes \mathbb{R} \longrightarrow \bigoplus_{\iota} \mathbb{C} \longrightarrow \bigoplus_{\iota} \mathbb{C}/(\mathbb{R}(2\pi i)^{1-r}) \xleftarrow{\cong} \bigoplus_{\iota} \mathbb{R}(2\pi i)^{-r} = \Sigma' \otimes \mathbb{R}$$

where the first arrow is induced by the embeddings $\iota : L \longrightarrow \mathbb{C}$. In the case 1.2.1 (2), let a' be the composite map

$$\Omega' \otimes \mathbb{R} = \mathrm{coLie}(E)^{\otimes r} \otimes_K L \otimes_{\mathbf{Q}} \mathbb{R} \xrightarrow{\cong} \bigoplus_{\iota} \mathrm{coLie}(E)^{\otimes r} \otimes_K \mathbb{C}$$

$$\xrightarrow{\cong} \bigoplus_{\iota} \mathrm{Hom}_K(H_1(E(\mathbb{C}),\ \mathbb{Q})^{\otimes r},\ \mathbb{C}) = \bigoplus_{\iota} H_1(E(\mathbb{C}),\ \mathbb{Q})^{\otimes(-r)} \otimes_K \mathbb{C} \cong \Sigma' \otimes \mathbb{R}$$

where the second arrow of the first line is induced by the embeddings $\iota : L \longrightarrow \mathbb{C}$ and the first arrow of the second line is given by

$$\omega^{\otimes r} \longmapsto (\gamma^{\otimes r} \longmapsto (\textstyle\int_\gamma \omega)^r)$$

$$(\omega \in \mathrm{coLie}(E) \setminus \{0\},\quad \gamma \in H_1(E(\mathbb{C}),\ \mathbb{Q}) \setminus \{0\}) .$$

In the following Thm. 1.2.6, we use the notations $z_\Lambda(X,\ \mathcal{F})$, Σ, Ω which were sometime used assuming some conjectures. However, $z_\Lambda(X,\ \mathcal{F})$ (resp. Σ and Ω) in 1.2.6 was (resp. were) defined in 1.2.4 (resp. 1.2.5) without any conjecture, and <u>we do not assume any conjecture in Thm. 1.2.6</u>.

<u>Theorem</u> 1.2.6. <u>Let</u> A, $\mathcal{F}$ <u>and</u> r <u>be as in</u> 1.2.4. <u>Then, the element</u> $z_\Lambda(X,\ \mathcal{F})$ <u>defined in</u> 1.2.4 <u>is sent by the map</u>

$$\exp^* \otimes \mathrm{id.} \quad : \quad H^1(X,\ \mathcal{F}) \otimes_A \Sigma^{-1} \longrightarrow (\Omega \otimes_A \Sigma^{-1}) \otimes \mathbb{Q}_p$$

<u>to an element of</u> $\Omega \otimes_A \Sigma^{-1}$ <u>whose image in</u> $A \otimes \mathbb{R}$ (1.2.5) <u>coinicdes with</u> $(-1)^{r-1} L_{A,S}(\mathcal{F}^*(1),\ 0)$.

(For the notations Σ and Ω, cf. 1.2.5.)

The case r = 1 of 1.2.6 is a known result, and is related (in the situation 1.2.1 (2)) to the work of Coates-Wiles [CW] as is explained in [dS_2] chap. IV §2. A part of 1.2.6 was obtained in [BK] (7.13).

We give here the proof of 1.2.6 in the case 1.2.1 (2). We omit the proof for the case 1.2.1 (1) which is similar and easier.

Let p be a prime ideal of O_K lying over p, K_p the local field of K at p, and let O_p be the valuation ring of K_p. Since we have assumed (1.2.1) that E has a good reduction at p, there exists an elliptic curve $\mathfrak{E}$ over O_p such that $\mathfrak{E} \otimes_{O_p} K_p = E \otimes_K K_p$. Then the formal completion of $\mathfrak{E}$ is identified with the Lubin-Tate group over O_p

corresponding to the pair $(K_{\mathfrak{p}}, \pi)$ where $\pi = \nu(\mathfrak{p})$. Let $\mathfrak{h}$ be an ideal of O_K such that $\mathfrak{h} \subset \mathfrak{f}$ and such that the set of prime divisors of $\mathfrak{h}$ coinsides with $S \setminus \{\mathfrak{p}\}$. Let $K'_{\mathfrak{p}}$ be the composite field $K(\mathfrak{h})K_{\mathfrak{p}}$ and regard the pair $(K_{\mathfrak{p}}, K'_{\mathfrak{p}})$ as the pair (F, K) in Chap. II §2. Then we have

$$(K'_{\mathfrak{p}})_n = K(\mathfrak{h}\mathfrak{p}^n)K_{\mathfrak{p}} \quad \text{for any } n \geq 0.$$

Now take an O_K-basis δ of $H_1(E(\mathbb{C}), \mathbb{Z})$. Take also an embedding $K^{ab,S} \longrightarrow \mathbb{C}$, and denote by ι_n the restriction of this embedding to $K(\mathfrak{h}\mathfrak{p}^n)$ for each $n \geq 0$. Consider the map $(1)_{a,\mathfrak{g}}$ in 1.2.3 for $\mathfrak{g} = \mathfrak{h}\mathfrak{p}^n$, and denote by the same letter ι_n the element of

$$H^0(K(\mathfrak{h}\mathfrak{p}^n) \otimes \mathbb{R}, \mathbb{Z}_p) = \bigoplus_{\iota} H^0(\mathbb{C}, \mathbb{Z}_p)$$

whose ι-component is 1 if $\iota = \iota_n$ and is 0 if $\iota \neq \iota_n$. Then when n varies, $(\iota_n)_n$ forms a projective system, and the image of $(\iota_n)_n$ under

$$H^0(K(\mathfrak{g}) \otimes \mathbb{R}, \mathbb{Z}_p) \xrightarrow{(1)_{a,\mathfrak{g}}} H^1(O_{K(\mathfrak{g})}[\tfrac{1}{S}], \mathbb{Z}_p(1))$$
$$\longrightarrow H^1((K'_{\mathfrak{p}})_n, \mathbb{Z}_p(1)) \quad \text{with } \mathfrak{g} = \mathfrak{h}\mathfrak{p}^n \ (n \geq 1)$$

coincides with the norm compatible system $(u_n)_n$ with

$$u_n = \iota_n^{-1}(\theta_a(\exp(\pi^{-n}h^{-1}\delta))),$$

where we fixed a generator h of the ideal $\mathfrak{h}$. Let $T = T_{\mathfrak{p}}E$. We define an $O_{\mathfrak{p}}$-basis ξ of $H^0(K^{ab,S}, T)$ as the image of $h^{-1}\delta$ under the isomorphisms

$$H_1(E(\mathbb{C}), \mathbb{Z}) \otimes \mathbb{Z}_p \cong H^0(\mathbb{C}, T) \cong H^0(K^{ab,S}, T),$$

where the last isomorphism is induced by the above embedding of $K^{ab,S}$ into $\mathbb{C}$. Then the Coleman power series $g_{u,\xi}$ associated to $u = (u_n)_n$ and ξ is the function

$$z \longmapsto \theta_a(z + \iota^{-1}(\exp(h^{-1}\delta)))$$

on the Lubin-Tate group.

Now we prove 1.2.6. In 1.2.5, let $L = K(\mathfrak{h}p^n)$, $n \geq 1$. It is sufficient to prove 1.2.6 in the case where $A = \mathbb{Q}[\mathrm{Gal}(L/K)]$ and $\mathcal{F} = f_* f^*(T^{\otimes(-r)}(1))$ for some $r \geq 1$ where f is the canonical morphism $\mathrm{Spec}(O_L[\frac{1}{S}]) \longrightarrow \mathrm{Spec}(O_K[\frac{1}{S}])$. (This is because any $(A, \mathcal{F})$ is obtained as $(?, \mathcal{F} \otimes_A ?)$ with $(A, \mathcal{F})$ as in this special case, for some ring ? over A which is a finite product of finite extensions of K.) In this case, we have

$$L_{A,S}(X, \mathcal{F}^*(1), s) = \sum_{\mathfrak{a}} \nu(\mathfrak{a})^{-r} N(\mathfrak{a})^{-s} (\mathfrak{a}, L/K)^{-1}$$

as a function with values in $\mathbb{C}[\mathrm{Gal}(L/K)])$ where $\mathfrak{a}$ ranges over all non-zero ideals of O_K which are prime to S and $(\mathfrak{a}, L/K) \in \mathrm{Gal}(L/K)$ denotes the Artin symbol. Consider the composite map

$$\Sigma \longrightarrow H^1(O_K[\tfrac{1}{S}], \mathcal{F}) \cong H^1(O_L[\tfrac{1}{S}], T^{\otimes(-r)}(1))$$
$$\longrightarrow H^1((K'_p)_n, T^{\otimes(-r)}(1))$$

where the first arrow is defined by the element $z_{a,\Lambda}(X, \mathcal{F})$ in 1.2.4. Consider the element of Σ whose ι-component is $\delta^{\otimes(-r)}$ if $\iota = \iota_n$ and is 0 if $\iota \neq \iota_n$. Then, this element is sent to

$$h^{-r} \cdot s_r(u, \xi) \in H^1((K'_p)_n, T^{\otimes(-r)}(1))$$

by the above composite map. Note that we have computed the Coleman power series $g_{u,\xi}$.

Now 1.2.6 follows from this by the theorem introduced in 1.1.6 and by Chap. II, 2.1.7.

We finish this section by some comments which are a little funny.

1.2.7. Zeta values are defined in $\mathbb{C}$ and explicit reciprocity laws live in the deep inside of the p-adic world, and the distance between them is very big. Their relationship is so surprising to me that I wonder whether such relationship arised before the universe started or after.

1.2.8. As we have seen in this section and in Chap. I §3, cyclotomic units and elliptic units are related to zeta values in many (mysteriously many!) ways. I am surprized by this, and especially by the fact that they express zeta values via the explicit reciprocity laws. In the p-adic world, it is a very hard work to express values of complex zeta functions. However such difficulty does not matter for great units who seem to have a strong wish to express zeta values. In the p-adic world, they seem not to forget that they were great and expressed zeta values in the archimedean world.

So, I feel that cyclotomic units and elliptic units are "incarnations" of zeta values. In [Ki], the crane entered the home of Yohyo transforming herself into a girl, to express her thanks to him. (Cf. the summary of [Ki] given in Bibliography.) Similarly, zeta values enter explicit reciprocity laws transforming themselves into cyclotomic units and elliptic units.

I respect this sincere wish of the crane.

1.2.9. Mysterious properties of zeta values seem to tell us (in a not so loud voice) that our universe has the same properties: The universe is not explained just by real numbers. It has p-adic properties (as is claimed by some people in physics) and it is related to profound objects which we call for simplicity the crane, the galaxy train, and the homeland of zeta values. We ourselves may have the same properties.

Are there physical meanings of zeta elements?

Bibliography.

[Be_1] Beilinson, A., Higher regulators and values of L-functions, J. Soviet Math. 30 (1985) 2036-2070.

[Be_2] Beilinson, A., Polylogarithm and cyclotomic elements, preprint.

[Bl] Bloch, S., Lectures on algebraic cycles, Duke Univ. Math. Series (1980).

[BK] Bloch, S. and Kato, K., L-functions and Tamagawa numbers motives, in The Grothendieck Festscherift, Vol. I (1980) 334-400.

[Bo] Borel, A., Stable real cohomology of arithmetic groups, Ann. Sci. École Norm. Sup. 7 (1974) 235-272.

[CP] Coates, J. and Perrin-Riou, B., On p-adic L-functions attached to motives over $\mathbb{Q}$, in Advanced Studies in Pure Math. 17 (1989) 23-54.

[CW] Coates J. and Wiles, A., On the conjecture of Birch and Swinnerton-Dyer, Invent. Math. 39 (1977) 223-251.

[Co] Coleman, R., Division values in local fields, Inv. Math. 53 (1979) 91-116.

[Da] Damerell, R. M., L-functions of elliptic curves with complex multiplication, I, Acta Arith. 17 (1970) 287-301, II, ibid. 19 (1971) 311-317.

[De_1] Deligne, P., Théorie de Hodge II, Publ. Math. IHES 40 (1972) 5-57.

[De_2] Deligne, P., Théorème de finitude en cohomologie ℓ-adique, in Lecture Notes in Math. 569 (SGA $4\frac{1}{2}$), Springer (1977) 233-261.

[De_3] Valeurs de fonctions L et périodes d'intégrales, Proc. Symp. Pure Math., vol. 33, Part 2, AMS (1979) 313-349.

[De_4] Deligne, P., La conjecture de Weil II, Publ. Math. IHES 52 (1981).

[De_5] Deligne, P., Le groupe fondamental de la droite projective moins trois points, in Galois groups over $\mathbb{Q}$, Springer (1989) 79-298.

[dS_1] de Shalit, E., The explicit reciprocity law in local class field theory, Duke Math. J. (1986) 163-176.

[dS_2] de Shalit, E., Iwasawa theory of elliptic curves with complex multiplication, Academic Press (1987).

[dS_3] de Shalit, E., The explicit reciprocity law of Bloch and Kato, preprint.

[Fa] Faltings, G., Crystalline cohomology and p-adic Galois representations, in Algebraic Analysis, Geometry, and Number Theory, Johns Hopkins Univ. (1989) 25-80.

[Fo_1] Fontaine, J.-M., Sur certains types de représentations p-adiques du groupe de Galois d'un corps local: construction d'un anneau de Barsotti-Tate, Ann. of Math. 115 (1982) 547-608.

[Fo_2] Fontaine, J.-M., Formes différentielles et modules de Tate des variétés abeliennes sur les corps locaux, Invent. Math. 65 (1982) 379-409.

[Fo_3] Fontaine, J.-M., Cohomologie crystalline et représentations p-adiques, in Lecture Notes in Math. 1016, Springer (1983) 86-108.

[FM] Fontaine, J.-M. and Messing, W., p-adic periods and p-adic étale cohomology, Contemporary Math. 67 (1987) 179-207.

[FP_1] Fontaine, J.-M. and Perrin-Riou, B., Autour des conjectures de Bloch et Kato, I. C. R. Acad. Sci. Paris, t. 313, Série I (1991) 189-196, II, ibid., 349-356, III, ibid., 421-428.

[FP_2] Fontaine, J.-M. and Perrin-Riou, B., Autour des conjectures de Bloch et Kato, cohomologie galoisienne et valeurs de fonctions L, preprint.

[Gr] Grothendieck, A., Formule de Lefschetz et rationalité des fonctions L, in Sém. Bourbaki, vol. 1965/66, Benjamin (1966) exposé 306

[Il] Illusie, L., Cohomologie de de Rham et cohomologie étale p-adique (Sém. Bourbaki exposé 726), in Astérisque (1990) 325-374.

[Ja] Jannsen, U., On the ℓ-adic cohomology of varieties over number fields and its Galois cohomology, in Galois groups over Q, Springer (1989) 315-360.

[Ka_1] Kato, K., The explicit reciprocity law and the cohomology of Fontaine-Messing, Bull. Soc. Math. France 119 (1991) 397-441.

[Ka_2] Kato, K., Iwasawa theory and p-adic Hodge theory, preprint.

[Ka_3] Kato, K., in preparation.

[Ki] Kinoshita, J., The twilight-crane (1949). (A drama basing on a Japanese legend.)

A poor young man Yohyo saved the life of a crane which was shot by someone. On a night after, a girl came to his home. Living with him, she produced beautiful cloths, but she asked him not to look at her when she was weaving. They were married and happy. However Yohyo became interested only in the money which he earned by selling her productions. One night he finally looked her weaving, to find a crane producing a cloth by using her own feathers, and finished their happy days (she had to return to the sky).

[KM] Knudsen, F. and Mumford, D., The projectivity of the moduli space of stable curves I, Math. Scand. 39, 1 (1976) 19-55.

[Ko] Kolyvagin, V. A., Euler systems, The Grothendieck Festschrift, vol. 2, Birkhaüser (1990) 435-483.

[Ma] Mazur, B., Notes on the étale cohomology of number fields, Ann. Sci. Ec. Norm. Sup. 6 (1973) 521-556.

[MW] Mazur, B and Wiles, A., Class fields of abelian extensions of Q, Invent. Math. 76 (1984) 179-330.

[Mi] Miyazawa K. (a Japanese poet), A night on the galaxy train (written around 1924).

A boy travels the night sky by a train (in his dream) with his friend. They visit constellations; the Northern Cross, the Scorpion, the Southern Cross, ... along the Milky Way. At the end of the story, it is found that this friend had been dead and it is suggested that the train was bringing this friend to the heaven.

From the mathematical point of view, the boy wakes from the dream without reaching the homeland of zeta values but he learned many important things from his travel (i.e. from his study of zeta values).

[Qu] Quillen, D., Higher algebraic K-theory, I., in Lecture Notes in Math. 341, Springer (1973) 85-147.

[Ra] Rapoport, M., Schappacher, N. and Schneider, P. (ed.), Beilinson's conjectures on special values of L-functions, Academic Press (1988).

[Ru] Rubin, K., The main conjectures" of Iwasawa theory for imaginary quadratic fields, Invent. math. 103 (1991) 25-68.

[Se] Serre, J.-P., Cohomologie Galoisienne, Lecture Notes in Math. 5, Springer (1965).

[So_1] Soulé, C., K-théorie des anneaux d'entiers de corps de nombres et cohomologie étale, Invent. Math. 55 (1979) 251-295.

[So_2] Soulé, C., On higher p-adic regulators, in Lecture Notes in Math. 854, Springer (1981) 371-401.

[So_3] Soulé, C., The rank of étale cohomology of varieties over p-adic or number fields, Comp. Math. 53 (1984) 113-131.

[So_4] Soulé, C., p-adic K-theory of elliptic curves, Duke Math. J. 54 (1987) 249-269.

[Ta_1] Tate, J., On the conjecture of Birch and Swinnerton-Dyer and a geometric analog, in Sém. Bourbaki, vol. 1965/66, Benjamin (1966) exposé 306.

[Ta_2] Tate, J., p-divisible groups, Proceedings of a conference on local fields, Driebergen, 1966, Springer (1967) 158-183.

[Wa] Washington, L. C., Introduction to cyclotomic fields, Springer (1982).

[We_1] Weil, A., Elliptic functions according to Eisenstein and Kronecker, Springer (1976).

[We_2] Weil, A., Number theory: An approach through history; From Hammurapi to Legendre, Birkhäuser (1983).

[Wi_1] Wiles, A., Higher explicit reciprocity laws, Ann. Math. 107 (1978) 235-254.

[Wi_2] Wiles, A., The Iwasawa conjecture for totally real fields, Ann. of Math. 131 (1990) 493-540.

[Wo] Wolfgang, K. S., λ-rings and Adams operators in algebraic K-theory, included in [Ra], 93-102.

[SGA4] Artin, M. and Grothendieck, and Verdier, J. L., Théorie des topos et cohomologie étale des schémas, Lecture Notes in Math. 269, 270, 305, Springer (1972/73).

Tokyo Institute of Technology

Applications of Arithmetic Algebraic Geometry to Diophantine Approximations

Paul Vojta*
Department of Mathematics
University of California
Berkeley, CA 94720 USA

Contents

Let us start by recalling the statement of Mordell's conjecture, first proved by Faltings in 1983.

Theorem 0.1. *Let C be a curve of genus > 1 defined over a number field k. Then $C(k)$ is finite.*

In this series of lectures I will describe an application of arithmetic algebraic geometry to obtain a proof of this result using the methods of diophantine approximations (instead of moduli spaces of abelian varieties). I obtained this proof in 1989 [**V 4**]; it was followed in that same year by an adaptation due to Faltings, giving the following more general theorem, originally conjectured by Lang [**L 1**]:

*Partially supported by NSF grant DMS-9001372 .

Theorem 0.2 ([**F 1**]). *Let X be a subvariety of an abelian variety A, and let k be a number field over which both of them are defined. Suppose that there is no nontrivial translated abelian subvariety of $A \times_k \bar{k}$ contained in $X \times_k \bar{k}$. Then the set $X(k)$ of k-rational points on X is finite.*

In 1990, Bombieri [Bo] also found a simplification of the proof [V 4]. While it does not prove any more general finiteness statements, it does provide for a very elementary exposition, and can be more readily used to obtain explicit bounds on the number of rational points.

Early in 1991, Faltings succeeded in dropping the assumption in Theorem 0.2 that $X \times_k \bar{k}$ not contain any translated abelian subvarieties of A, obtaining another conjecture of Lang ([L 2], p. 29).

Theorem 0.3 ([**F 2**]). *Let X be a subvariety of an abelian variety A, both assumed to be defined over a number field k. Then the set $X(k)$ is contained in a finite union $\bigcup_i B_i(k)$, where each B_i is a translated abelian subvariety of A contained in X.*

The problem of extending this to the case of integral points on subvarieties of semiabelian varieties is still open. One may also rephrase this problem as showing finiteness for the intersection of X with a finitely generated subgroup Γ of $A(\overline{\mathbb{Q}})$. The same sort of finiteness question can then be posed for the division group

$$\{g \in A(\overline{\mathbb{Q}}) \mid mg \in \Gamma \text{ for some } m \in \mathbb{N}\};$$

this has recently been solved by M. McQuillan (unpublished); see also [Ra].

Despite the fact that arithmetic algebraic geometry is a very new set of techniques, the history of this subject goes back to a 1909 paper of A. Thue. Recall that a **Thue equation** is an equation

$$f(x, y) = c, \qquad x, y \in \mathbb{Z}$$

where $c \in \mathbb{Z}$ and $f \in \mathbb{Z}[X, Y]$ is irreducible and homogeneous, of degree at least three. Thue proved that such equations have only finitely many solutions.

The lectures start, therefore, by recalling some very classical results. These include a lemma of Siegel which constructs small solutions of systems of linear equations and, later, Minkowski's theorem on successive minima.

Next follows a brief sketch of the proof of Roth's theorem. It is this proof (or, more precisely, a slightly earlier proof due to Dyson) which motivated the new proof of Mordell's conjecture.

After that, we will consider how to apply the language of arithmetic intersection theory to this proof, and prove Mordell's conjecture using some of the methods of Bombieri. This will be followed by the original (1989) proof using the Gillet-Soulé Riemann-Roch theorem. These proofs will only be sketched, as they are written in detail elsewhere, and newer methods are available.

Finally, we give in detail Faltings' proof of Theorem 0.3, with a few minor simplifications.

In this paper, places v of a number field k will be taken in the classical sense, so that places corresponding to complex conjugate embeddings into $\mathbb{C}$ will be identified. Also, absolute values $\|\cdot\|_v$ will be normalized so that $\|x\|_v = |\sigma(x)|$ if v corresponds to a real embedding $\sigma\colon k \hookrightarrow \mathbb{R}$; $\|x\|_v = |\sigma(x)|^2$ if v corresponds to a complex embedding, and $\|p\|_v = p^{-ef}$ if v is $\mathfrak{p}$-adic, where $\mathfrak{p}$ is ramified to order e over a rational prime p and f is the degree of the residue field extension. With these normalizations, the product formula reads

$$\prod_v \|x\|_v = 1, \qquad x \in k,\ x \neq 0. \tag{0.4}$$

A **line sheaf** on a scheme X means a sheaf which is locally isomorphic to $\mathcal{O}_X$; *i.e.*, an invertible sheaf. Similarly a **vector sheaf** is a locally free sheaf.

More notations appear in Definition 2.3 and in Section 5.

§1. History; integral and rational points

In its earliest form, the study of diophantine approximations concerns trying to prove that, given an algebraic number α, there are only finitely many $p/q \in \mathbb{Q}$ (written in lowest terms) satisfying an inequality of the form

$$\left|\frac{p}{q} - \alpha\right| < \frac{c}{|q|^\kappa}$$

for some value of κ and some constant $c > 0$. It took many decades to obtain the best value of κ: letting $d = [\mathbb{Q}(\alpha) : \mathbb{Q}]$, the progress is as follows:

$\kappa = d$, c computable	Liouville, 1844
$\kappa = \frac{d+1}{2} + \epsilon$	Thue, 1909
$\kappa = \min\{\frac{d}{s} + s - 1 \mid s = 2, \dots, d\} + \epsilon$	Siegel, 1921
$\kappa = \sqrt{2d} + \epsilon$	Dyson, Gel'fond (independently), 1947
$\kappa = 2 + \epsilon$	Roth, 1955

Of course, stronger approximations may be conjectured; *e.g.*,

$$\left|\frac{p}{q} - \alpha\right| < c|q|^{-2}(\log q)^{-1-\epsilon}.$$

See ([L 3], p. 71).

Beginning with Thue's work, these approximation results can be used to prove finiteness results for certain diophantine equations, as the following example illustrates.

Example 1.1. The (Thue) equation

$$x^3 - 2y^3 = 1, \qquad x, y \in \mathbb{Z} \tag{1.2}$$

has only finitely many solutions.

Indeed, this equation may be rewritten

$$\frac{x}{y} - \sqrt[3]{2} = \frac{1}{y(x^2 + \sqrt[3]{2}xy + \sqrt[3]{4}y^2)}.$$

But for $|y|$ large the absolute value of the right-hand side is dominated by some multiple of $1/|y|^3$; if (1.2) had infinitely many solutions, then the inequalities of Thue, *et al.* would be contradicted.

For a second example, consider a particular case of Mordell's conjecture (Theorem 0.1).

Example 1.3. The equation

$$x^4 + y^4 = z^4, \qquad x, y, z \in \mathbb{Q} \tag{1.4}$$

in projective coordinates (or $x^4 + y^4 = 1$ in affine coordinates) has only finitely many solutions.

The intent of these lectures is to show that Theorem 0.1 can be proved by the methods of diophantine approximations. At first glance this does not seem likely, since it is no longer true that solutions must go off toward infinity. But let us start by considering how, in the language of schemes, these two problems are very similar.

In the first example, let $W = \operatorname{Spec} \mathbb{Z}[X,Y]/(X^3 - 2Y^3 - 1)$ and $B = \operatorname{Spec} \mathbb{Z}$ be schemes, and let $\pi: W \to B$ be the morphism corresponding to the injection

$$\mathbb{Z} \hookrightarrow \mathbb{Z}[X,Y]/(X^3 - 2Y^3 - 1).$$

Then solutions (x, y) to the equation (1.2) correspond bijectively to sections $s: B \to W$ of π since they correspond to homomorphisms

$$\mathbb{Z}[X,Y]/(X^3 - 2Y^3 - 1) \to \mathbb{Z}, \qquad X \mapsto x, Y \mapsto y$$

and the composition of these two ring maps gives the identity map on $\mathbb{Z}$.

In the second example, let $W = \operatorname{Proj} \mathbb{Z}[X,Y,Z]/(X^4 + Y^4 - Z^4)$ and $B = \operatorname{Spec} \mathbb{Z}$. Then sections $s: B \to W$ of π correspond bijectively to closed points on the generic fiber of π with residue field $\mathbb{Q}$. In one direction this is the valuative criterion of properness, and in the other direction the bijection is given by taking the closure in W. These closed points correspond bijectively to rational solutions of (1.4).

Thus, in both cases, solutions correspond bijectively to sections of $\pi: W \to B$. The difference between integral and rational points is accounted for by the fact that in the first case π is an affine map, and in the second it is projective.

Note that, in the second example, any ring with fraction field $\mathbb{Q}$ can be used in place of $\mathbb{Z}$ as the affine ring of B (by the valuative criterion of properness). But, in the case of integral points, localizations of $\mathbb{Z}$ make a difference: using $B = \operatorname{Spec} Z[\frac{1}{2}]$, for example, allows solutions in which x and y may have powers of 2 in the denominator.

§2. Siegel's lemma

Siegel's lemma is a corollary of the "pigeonhole principle." Actually, the idea dates back to Thue, but he did not state it explicitly as a separate lemma.

Lemma 2.1 (Siegel's lemma). *Let A be an $M \times N$ matrix with $M < N$ and having entries in $\mathbb{Z}$ of absolute value at most Q. Then there exists a nonzero vector $\mathbf{x} = (x_1, \ldots, x_n) \in \mathbb{Z}^N$ with $A\mathbf{x} = 0$, such that*

$$|x_i| \leq \left[(NQ)^{M/(N-M)}\right] =: Z, \qquad i = 1, \ldots, N.$$

Proof. The number of integer points in the box

$$0 \leq x_i \leq Z, \qquad i = 1, \ldots, N \tag{2.2}$$

is $(Z+1)^N$. On the other hand, for all $j = 1, \ldots, N$ and for each such $\mathbf{x}$, the jth coordinate y_j of the vector $\mathbf{y} := A\mathbf{x}$ lies in the interval $[-n_j QZ, (N - n_j)QZ]$, where n_j is the number of negative entries in the jth row of A. Therefore there are at most

$$(NQZ + 1)^M < (Z+1)^N$$

possible values of $A\mathbf{x}$. Hence there must exist vectors $\mathbf{x}_1 \neq \mathbf{x}_2$ satisfying (2.2) and such that $A\mathbf{x}_1 = A\mathbf{x}_2$. Then $\mathbf{x} = \mathbf{x}_1 - \mathbf{x}_2$ satisfies the conditions of the lemma. □

In order to further emphasize the arithmetic-geometric nature of the subject, all results will be done in the context of a ring R_S, obtained by localizing the ring R of integers of a number field k away from primes in a finite set S of places of k. We also will always assume that S contains the set of archimedean places of k.

In the case of Siegel's lemma, the generalization to number fields is sufficient; this was proved at least as early as LeVeque ([LeV], proof of Thm. 4.14). The form that we will use is due to Bombieri and Vaaler. First, however, we need to define some heights (cf. also Section 5). Recall that, classically, the height of an element $x \in k \setminus \{0\}$ is defined as

$$h(x) = \frac{1}{[k:\mathbb{Q}]} \sum_v \log \max(\|x\|_v, 1).$$

Definition 2.3.

(a). For vectors $\mathbf{x} \in k^N$, $\mathbf{x} \neq 0$,

$$h(\mathbf{x}) = \frac{1}{[k:\mathbb{Q}]} \sum_v \log \max_{1 \leq i \leq N} \|x_i\|_v.$$

(b). For $P \in \mathbb{P}^n(k)$ with homogeneous coordinates $[x_0 : \cdots : x_n]$, $h(P)$ is defined as the height of the vector $(x_0, \ldots, x_n) \in k^{n+1}$. By the product formula (0.4), it is independent of the choice of homogeneous coordinates.

(c). For $x \in k$, $h(x) = h([1:x])$, the height of the corresponding point in $\mathbb{P}^1$.

(d). For an $M \times N$ matrix A of rank M, $h(A)$ is defined as the height of the vector consisting of all $M \times M$ minors of A.

For a number field k, let $D_{k/\mathbb{Q}}$ denote the discriminant and s the number of complex places. Then the generalization of Siegel's lemma to number fields is the following.

Theorem 2.4 (Siegel-Bombieri-Vaaler, ([B-V], Theorem 9)). *Let A be an $M \times N$ matrix of rank M with entries in k. Then there exists a basis $\{\mathbf{x}_1, \dots, \mathbf{x}_{N-M}\}$ of the kernel of A (regarded as a linear transformation from k^N to k^M) such that*

$$\sum_{i=1}^{N-M} h(\mathbf{x}_i) \le h(A) + \frac{N-M}{[k:\mathbb{Q}]} \log\left(\left(\frac{2}{\pi}\right)^s \sqrt{|D_k|}\right).$$

The exact value of the constant will not be needed here; it has been included only for reference.

Note that, in addition to allowing arbitrary number fields, this result gives information on all generators of the kernel of A; this will be used briefly when discussing the proof of Theorem 0.2 (Section 18).

§3. The index

Let $Q(X,Y)$ be a nonzero polynomial in two variables. Then recall that the **multiplicity of Q at 0** is the smallest integer t such that $a_{ij}X^iY^j$ is a nonzero monomial in Q with $i+j=t$. This definition treats the two variables symmetrically, whereas here it will be necessary to treat them with weights which may vary. Therefore we will define a multiplicity using weighted variables, which is called the **index**.

Definition 3.1. Let

$$Q(X_1, \dots, X_n) = \sum_{\ell_1,\dots,\ell_n \ge 0} a_{\ell_1,\dots,\ell_n} X_1^{\ell_1} \cdots X_n^{\ell_n} =: \sum_{(\ell)\ge 0} a_{(\ell)} X^{(\ell)}$$

be a nonzero polynomial in n variables, and let $d_1, \dots, d_n$ be positive real numbers. Then the **index of Q at 0 with weights $d_1, \dots, d_n$** is

$$t(Q,(0,\dots,0),d_1,\dots,d_n) = \min\left\{ \sum_{i=1}^{n} \frac{\ell_i}{d_i} \;\middle|\; a_{(\ell)} \ne 0 \right\}$$

Often the notation will be shortened to $t(Q,(0,\dots,0))$ when $d_1, \dots, d_n$ are clear from the context.

Note that, although stated for polynomials, the above definition applies equally well to power series. Moreover, replacing some X_i with a power series $b_1X_i' + b_2X_i'^2 + \dots$ ($b_1 \ne 0$) does not change the value of the index. Likewise, the index is preserved if Q is multiplied by some power series with nonzero constant term (*i.e.*, a unit). And finally, there is no reason why one cannot allow several variables in place of each X_i. Thus the index can be defined more generally for sections of line sheaves on products of varieties.

Definition 3.2. Let γ be a rational section of a line sheaf $\mathscr{L}$, on a product $X_1 \times \cdots \times X_n$ of varieties. Let $P = (P_1, \dots, P_n)$ be a regular point on $\prod X_i$, and suppose that γ is regular at P. Let γ_0 be a section which generates $\mathscr{L}$ in a neighborhood of P, and for each $i = 1, \dots, n$ let z_{ij}, $j = 1, \dots, \dim X_i$, be a system of local parameters at P_i, with $z_{ij}(P_i) = 0$ for all j. Then γ/γ_0 is a regular function in a neighborhood of P, so it can be written as a power series

$$\gamma/\gamma_0 = \sum_{(\ell)\geq 0} a_{(\ell)} X^{(\ell)}.$$

Here $(\ell) = (\ell_{ij})$, $i = 1, \dots, n$, $j = 1, \dots, \dim X_i$ is a $(\dim X_1 + \cdots + \dim X_n)$-tuple. Also let $d_1, \dots, d_n$ be positive real numbers. Then the **index of γ at P with weights** $d_1, \dots, d_n$, denoted $t(\gamma, P, d_1, \dots, d_n)$ or just $t(\gamma, P)$, is

$$\min\left\{\sum_{i=1}^{n}\sum_{j=1}^{\dim X_i} \frac{\ell_{ij}}{d_i} \,\middle|\, a_{(\ell)} \neq 0\right\}.$$

As noted already, this definition does not depend on the choices of γ_0 or of local parameters z_{ij}.

As a special case of this definition, if $n = 2$, if all X_i are taken to be $\mathbb{P}^1$, and if no P_i is ∞, then γ is just a polynomial in two variables and this definition specializes to the preceding definition after P is translated to the origin.

§4. Sketch of the proof of Roth's theorem

In a nutshell, the proof of Roth's theorem amounts to a complicated system of inequalities involving the index. First we state the theorem, in general.

Theorem 4.1 (Roth [**Ro**]). *Fix $\epsilon > 0$, a finite set of places S of k, and $\alpha_v \in \overline{\mathbb{Q}}$ for each $v \in S$. Then for almost all $x \in k$,*

$$\frac{1}{[k:\mathbb{Q}]}\sum_{v\in S} -\log\min(\|x-\alpha_v\|_v, 1) \leq (2+\epsilon)h(x). \tag{4.2}$$

If $k = \mathbb{Q}$ and $S = \{\infty\}$, then this statement reduces to that of Section 1.

We will only sketch the proof here; complete expositions can be found in [**L 5**] and [**Sch**], as well as [**Ro**].

First, we may first assume that all α_v lie in k. Otherwise, let k' be some finite extension field of k containing all α_v, let S' be the set of places w of k' lying over $v \in S$, and for each $w \mid v$ let α_w be a certain conjugate of α_v. (In order to write $\|x - \alpha_v\|_v$ when $\alpha_v \notin k$, some extension of $\|\cdot\|_v$ to $k(\alpha_v)$ must be chosen; then the α_w should be chosen correspondingly.) With proper choices of α_w, the left-hand side will remain unchanged when k is replaced by k', as will the right-hand side.

The basic idea of the proof is to assume that there are infinitely many counterexamples to (4.2), and derive a contradiction. In particular, we choose n good approximations which satisfy certain additional constraints outlined below.

The proof will be split into five steps, although the often the first two steps are merged, or the last two steps.

The first two steps construct an auxiliary polynomial Q with certain properties. Let $\alpha_1, \dots, \alpha_m$ be the distinct values taken on by all α_v, $v \in S$. Also let $d_1, \dots, d_n$ be positive integers. These will be taken large; independently of everything else in the proof, they may be taken arbitrarily large if their ratios are fixed.

The polynomial should be a nonzero polynomial in n variables $X_1, \ldots, X_n$, and its degree in each X_i should be at most d_i, for each i. The first requirement is that the polynomial should have index $\geq n(\frac{1}{2} - \epsilon_1)$ at each point $(\alpha_i, \ldots, \alpha_i)$, $i = 1, \ldots, m$. Such polynomials can be constructed by solving a linear algebra problem in which the variables are the coefficients of Q and the linear equations are given by the vanishing of various derivatives of Q at the chosen points $(\alpha_i, \ldots, \alpha_i)$. A nonzero solution exists if the number of linear equations is less than the number of variables (coefficients of Q). Step 1 consists of showing that this is the case; the exact inequalities on ϵ_1 and m will be omitted, however, since they will not be needed for this exposition.

Thus, step 1 is geometrical in nature.

Step 2 involves applying Siegel's lemma to show that such a polynomial can be constructed with coefficients in R_S and with bounded height. (Here we let the height of a polynomial be the height of its vector of coefficients.) From step 1, we know the values of M and N for Siegel's lemma; the height will then be bounded by

$$h(Q) \leq c_1 \sum_{i=1}^{n} d_i. \tag{4.3}$$

Here the constant c_1 (not a Chern class!) depends on k, S, n, and $\alpha_1, \ldots, \alpha_m$. This bound holds because $M/(N-M)$ will be bounded from above, and coefficients of constraints will be powers of α, multiplied by certain binomial coefficients.

Step 3 is independent of the first two steps; for this step, we choose elements $x_1, \ldots, x_n \in k$ not satisfying (4.2). Further, the vectors

$$\left(-\log \min(\|x_i - \alpha_v\|_v, 1)\right)_{v \in S} \in \mathbb{R}^{\#S}$$

all need to point in approximately the same direction, for $i = 1, \ldots, n$. This is easy to accomplish by a pigeonhole argument, since the vectors lie in a finite dimensional space. To be precise, there must exist real numbers κ_v, $v \in S$ such that

$$-\log \min(\|x_i - \alpha_v\|_v, 1) \geq \kappa_v h(x_i), \qquad v \in S,\ i = 1, \ldots, n$$

and such that

$$\frac{1}{[k:\mathbb{Q}]} \sum_{v \in S} \kappa_v = 2 + \frac{\epsilon}{2}. \tag{4.4}$$

The points must also satisfy the conditions

$$h(x_1) \geq c_2$$

and

$$h(x_{i+1})/h(x_i) \geq r, \qquad i = 1, \ldots, n-1.$$

These conditions are easily satisfied, since by assumption there are infinitely many $x \in k$ not satisfying (4.2), and the heights of these x go to infinity.

Having chosen $x_1, \ldots, x_n$, let d be a large integer and let d_i be integers close to $d/h(x_i)$, $i = 1, \ldots, n$. For step 4, we want to obtain a lower bound for the index of

Q at the point $(x_1, \ldots, x_n)$. This is done by a Taylor series argument: if $v \in S$, then write

$$Q(X_1, \ldots, X_n) = \sum_{(\ell)\geq 0} b_{v,(\ell)}(X - \alpha_v)^{(\ell)}.$$

Bounds on the sizes $\|b_{v,(\ell)}\|_v$ can be obtained from (4.3) in Step 2; moreover $b_{v,(\ell)} = 0$ if

$$\frac{\ell_1}{d_1} + \cdots + \frac{\ell_n}{d_n} < n\left(\frac{1}{2} - \epsilon_1\right),$$

by the index condition in step 1. Then the only terms with nonzero coefficients are those with high powers of some factors $X_i - \alpha_v$, so the falsehood of (4.2) implies a quite good bound on $\|Q(x_1, \ldots, x_n)\|_v$ for $v \in S$. Indeed, the nonzero terms in this Taylor series are bounded by

$$\begin{aligned}\prod_{i=1}^{n} &\exp(-\kappa_v \ell_i h(x_i)) \cdot \text{other factors}\\ &\leq \exp\left(-\kappa_v d \sum_{i=1}^{n} \frac{\ell_i}{d_i}\right) \cdot \text{other factors}\\ &\leq \exp(-\kappa_v d n(\tfrac{1}{2} - \epsilon_1)) \cdot \text{other factors}\end{aligned}$$

At $v \notin S$, we also have bounds on $\|Q(x_1, \ldots, x_n)\|_v$, depending on the denominators in $x_1, \ldots, x_n$. If these bounds are good enough, then the product formula is contradicted, implying that $Q(x_1, \ldots, x_n) = 0$. Indeed, taking the product over all v, the other factors come out to roughly $\exp([k : \mathbb{Q}]dn)$; by (4.4) this gives

$$\prod_v \|Q(x_1, \ldots, x_n)\|_v \leq \exp(-[k : \mathbb{Q}](2 + \tfrac{\epsilon}{2})dn(\tfrac{1}{2} - \epsilon_1)) \cdot \exp([k : \mathbb{Q}]dn) < 1.$$

Applying the same argument to certain partial derivatives of Q similarly gives vanishing, so we obtain a lower bound for the index of Q at $(x_1, \ldots, x_n)$.

Note that the choice of the d_i counterbalances the varying heights of the x_i, so in fact each x_i has roughly equal effect on the estimates in this step.

Finally, in step 5 we show that this lower bound contradicts certain other properties of Q. One possibility is to use the height $h(Q)$ from step 2.

Lemma 4.5 (Roth, [Ro]; see also [Bo]). *Let $Q(X_1, \ldots, X_n) \not\equiv 0$ be a polynomial in n variables, of degree at most d_i in X_i, with algebraic coefficients. Let $x_1, \ldots, x_n$ be algebraic numbers, and let $t = t(Q, (x_1, \ldots, x_n), d_1, \ldots, d_n)$ be the index of Q at $(x_1, \ldots, x_n)$. Suppose that $\epsilon_2 > 0$ is such that*

$$\frac{d_{i+1}}{d_i} \leq \epsilon_2^{2^{n-1}}, \qquad i = 1, \ldots, n-1$$

and

$$d_i h(x_i) \geq \epsilon_2^{-2^{n-1}}(h(Q) + 2nd_1), \qquad i = 1, \ldots, n. \tag{4.6}$$

Then

$$\frac{t}{n} \leq 2\epsilon_2.$$

Another approach is to use the index information from step 1. The following is a version which has been simplified, to cut down on extra notation. It is true more generally either in n variables, or on products of two curves.

Lemma 4.7 (Dyson, [D]). *Let* $\operatorname{Vol}(t)$ *be the area of the set*

$$\{(x_1, x_2) \in [0,1]^2 \mid x_1 + x_2 \le t\},$$

so that $\operatorname{Vol}(t) = t^2/2$ *if* $t \le 1$. *Let* $\xi_1, \dots, \xi_m$ *be* m *points in* $\mathbb{C}^2$ *with distinct first coordinates and distinct second coordinates. Let* Q *be a polynomial in* $\mathbb{C}[X_1, X_2]$ *of degree at most* d_1 *in* X_1 *and* d_2 *in* X_2. *Then*

$$\sum_{i=1}^{m} \operatorname{Vol}(t(Q, \xi_i, d_1, d_2)) \le 1 + \frac{d_2}{2d_1} \max(m-2, 0).$$

Historically, Dyson's approach was the earlier of the two, but it has been revived in recent years by Bombieri. I prefer it for aesthetic reasons, although the method of using Roth's lemma is much quicker. In particular, it was the form of Dyson's lemma which suggested the particular line sheaf to use in Step 1 of the Mordell proof.

Roth's innovation in this area was the use of n good approximations instead of two; but for now we will just use two good approximations—it sufficed for Thue's work on integral points.

§5. Notation

The rest of this paper will make heavy use of the language and results of Gillet and Soulé extending Arakelov theory to higher dimensions. For general references on this topic, see [So 2], [G-S 1], and [G-S 2].

In Arakelov theory it is traditional to regard distinct but complex conjugate embeddings of k as giving rise to distinct local archimedean fibers. Here, however, we will follow the much older convention of general algebraic number theory, that complex conjugate embeddings be identified, and therefore give rise to a single archimedean fiber. This is possible to do in the Gillet-Soulé theory, because all objects at complex conjugate places are assumed to be taken into each other by complex conjugation.

Then, if $\mathscr{X}$ is an arithmetic variety and v is an archimedean place, let $\mathscr{X}_v$ denote the set $\mathscr{X}(\bar{k}_v)$, identified with a complex manifold via some fixed embedding of $\bar{k}_v$ into $\mathbb{C}$.

The notation also differs from Gillet's and Soulé's in another respect. Namely, instead of using a pair $\overline{D} = (D, g_D)$ to denote, say, an arithmetic divisor (*i.e.*, an element of $\hat{Z}^1(\mathscr{X})$), we will use the single letter D to refer to the tuple, and D_{fin} to refer to its first component: $D = (D_{\text{fin}}, g_D)$. Likewise, $\mathscr{L}$ will usually refer to a metrized line sheaf whose corresponding non-metrized line sheaf is denoted $\mathscr{L}_{\text{fin}}$, etc. This notation is probably closer to that in Arakelov's original work than the more recent work of Gillet and Soulé and others. Also, I feel that the objects with the additional structure at infinity are the more natural objects to be considering, and the notation should reflect this fact.

In the theory of the Gillet-Soulé Riemann theorem (cf. [G-S 3] and [G-S 4]), it is natural to use the L^2 norm to assign a metric to a global section γ of a metrized line sheaf; however, in this theory it is more convenient to use the supremum norm instead:

$$\|\gamma\|_{\sup,v} := \sup_{P\in X(\bar{k}_v)} \|\gamma(P)\|.$$

Here the norm on the right is the norm of $\mathscr{L}$ on $\mathscr{X}_v$, since $P \in \mathscr{X}(\bar{k}_v)$. If instead we had $P \in X(k)$, we would write $\|\gamma(P)\|_v$ to specify the norm at the point (denoted P_v) on $\mathscr{X}_v$ corresponding to $P \in X(k)$. Also, if $E \subseteq \mathscr{X}$ is the image of the section corresponding to P, then also let E_v equal P_v, as a special case of the notation $\mathscr{X}_v$.

For future reference, we note here that often we will be considering $\mathbb{Q}$-divisors or $\mathbb{Q}$-divisor classes; these are divisors with rational coefficients (note that we do *not* tensor with $\mathbb{Q}$: this kills torsion, which leads to technical difficulties when converting to a line sheaf). Unless otherwise specified, divisors will always be assumed to be Cartier divisors. If L is such a $\mathbb{Q}$-divisor class, then writing $\mathscr{O}(dL)$ will implicitly imply an assumption that d is sufficiently divisible so as to cancel all denominators in L. Also, the notations $\Gamma(X,L)$ and $h^i(X,L)$ will mean $\Gamma(X,\mathscr{O}(L))$ and $h^i(X,\mathscr{O}(L))$, respectively.

And finally, given any sort of product, let pr_i denote the projection morphism to the ith factor.

§6. Derivatives

In adapting the proof of Roth's theorem to prove Mordell's conjecture, we replace $\mathbb{P}^1$ with an arbitrary curve C. Therefore instead of dealing with polynomials, we need to consider something more intrinsic on C^n, namely, sections of certain line sheaves. Step 4 of Roth's proof therefore needs some notion of partial derivatives of sections of line sheaves at a point.

To begin, let $\pi\colon X \to B$ be an arithmetic surface corresponding to the curve C. By a theorem of Abhyankar [Ab] or [Ar], we may assume that X is a regular surface. Let W be some arithmetic variety which for now we will assume to be $X \times_B \cdots \times_B X$ (but actually it will be a slight modification of that variety); it is then a model for C^n. Let γ be a section of a metrized line sheaf $\mathscr{L}$ on W, let $(P_1,\ldots,P_n)$ be a rational point on C^n, and let $E \subseteq W$ be the corresponding arithmetic curve, so that $E \cong B$ via the restriction of $q\colon W \to B$.

Now the morphism $\pi\colon X \to B$ is not necessarily smooth, but that is not a problem here, since we are dealing with rational points. Indeed, for $i = 1,\ldots,n$ let E_i denote the arithmetic curve in X corresponding to P_i; then the intersection number of E_i with any fiber is 1 (or rather $\log q_v$), which means that E_i can only meet one local branch of the fiber and can meet that branch only at a smooth point. Thus the completed local ring $\widehat{\mathscr{O}}_{X,E_{i,v}}$ is generated over $\widehat{\mathscr{O}}_{B,v}$ by the local generator of the divisor E_i; hence π is smooth in a neighborhood of E_i; likewise q is smooth in a neighborhood of E.

Let x be a closed point in E, and let $\widehat{\mathscr{O}}_{B,q(x)}$ be the completed local ring of B at $q(x)$. For $i = 1,\ldots,n$ let z_i be local equations representing the divisors E_i. Then, as noted above, the completed local ring $\widehat{\mathscr{O}}_{W,x}$ of W at x can be written

$$\widehat{\mathscr{O}}_{W,x} \cong \widehat{\mathscr{O}}_{B,q(x)}[[z_1,\ldots,z_n]].$$

If γ_0 is a local generator for $\mathscr{L}$ at x, then γ/γ_0 is an element of this local ring, which can then be written

$$\frac{\gamma}{\gamma_0} = \sum_{(\ell)\geq 0} b_{(\ell)} z^{(\ell)}.$$

Then each term $\gamma_0 b_{(\ell)} z^{(\ell)}$ lies in the subsheaf

$$\mathscr{L} \otimes \mathscr{O}(-\ell_1 \operatorname{pr}_1^{-1} E_1 - \cdots - \ell_n \operatorname{pr}_n^{-1} E_n) \subseteq \mathscr{L}.$$

In general this element depends on the choices of γ_0 and $z_1, \dots, z_n$. However, if $b_{(\ell')} = 0$ for all tuples $(\ell') \neq (\ell)$ satisfying $\ell'_1 \leq i_1, \dots, \ell'_n \leq i_n$ (*i.e.*, if (ℓ) is a **leading term**), then the restriction to E of this term is independent of the above choices. This is the definition of the **partial derivative**

$$D_{(\ell)}\gamma(P_1, \dots, P_n) \in \left(\mathscr{L} \otimes \mathscr{O}(-\ell_1 \operatorname{pr}_1^{-1} E_1 - \cdots - \ell_n \operatorname{pr}_n^{-1} E_n)\right)\Big|_E.$$

In particular, this derivative is a *regular section* of the above sheaf. We will also need the corresponding fact for the points on E at archimedean places. This translates into an upper bound on the metric of the above section at archimedean places, depending on choices of metrics for $\mathscr{O}(E_i)$.

To metrize $\mathscr{O}(E_i)$ in a uniform way, for each archimedean place v choose a metric for $\mathscr{O}(\Delta)$ on $C(\bar{k}_v) \times C(\bar{k}_v)$. For example, one could use the Green's functions of Arakelov theory, but any smooth metric will suffice. Then $\mathscr{O}(E_i)$ will be taken as the restriction of this metric to $E_{i,v} \times C(\bar{k}_v)$.

The next lemma takes place entirely on the local fiber of an achimedean place v; *i.e.*, on a complex manifold. Therefore for this lemma let C be a compact Riemann surface.

By a compactness argument there exists a constant $\rho = \rho(C) > 0$ such that for all $P \in C$ there exists a neighborhood $U_P \subseteq C$ of P with coordinate $z_P \colon U_P \xrightarrow{\sim} \mathbb{D}_\rho$ such that $z_P(P) = 0$ and such that the norm at P of z_P, regarded as a section of $\mathscr{O}(-P)$, satisfies the inequality

$$|z_P(P)| \leq 1. \tag{6.1}$$

Lemma 6.2. *Fix a point $P_0 = (P_1, \dots, P_n) \in C^n$. Let $\mathscr{L}$ be a metrized line sheaf on W, and let $\gamma \in \Gamma(W, \mathscr{L})$. Let γ_0 be a section of $\mathscr{L}$ on $U := U_{P_1} \times \cdots \times U_{P_n}$ which is nonzero at P_0. Suppose $(\ell) = (\ell_1, \dots, \ell_n)$ is a tuple for which the derivative $D_{(\ell)} f^*\gamma(P_1, \dots, P_n)$ is defined. Then*

$$-\log \|D_{(\ell)}\gamma(P_0)\| \geq -\log \|\gamma\|_{\sup} + \sum_{i=1}^{n} \ell_i \log \rho + \log \inf_{P \in U} \frac{\|\gamma_0(P)\|}{\|\gamma_0(P_0)\|}.$$

Proof. Writing z_{P_0} for the tuple of functions $(\operatorname{pr}_1^* z_{P_1}, \dots, \operatorname{pr}_n^* z_{P_n})$ and writing

$$\gamma = \gamma_0 \sum_{(i)\geq 0} a_{(i)} z_{P_0}^{(i)}$$

gives

$$\|\gamma\|_{\sup} \geq \sup_{P\in U}\left(\|\gamma_0(P)\|\left|\sum_{(i)\geq 0} a_{(i)} z_{P_0}^{(i)}(P)\right|\right)$$

$$\geq \inf_{P\in U}\|\gamma_0(P)\|\cdot \sup_{z\in \mathbf{D}_\rho^n}\left|\sum_{(i)\geq 0} a_{(i)} z^{(i)}\right|$$

$$\geq \|\gamma_0(P_0)\| \inf_{P\in U}\frac{\|\gamma_0(P)\|}{\|\gamma_0(P_0)\|}|a_{(\ell)}|\rho^{\ell_1+\dots+\ell_n}$$

by Cauchy's inequalities in several variables,

$$|a_{(\ell)}|\rho^{\ell_1+\dots+\ell_n} \leq \sup_{z\in \mathbf{D}_\rho^n}\left|\sum_{(i)\geq 0} a_{(i)} z^{(i)}\right|$$

Thus by (6.1),

$$\|(\gamma_0 a_{(\ell)} z_{P_0}^{(\ell)})(P_0)\|\rho^{\ell_1+\dots+\ell_n} \inf_{P\in U}\frac{\|\gamma_0(P)\|}{\|\gamma_0(P_0)\|} \leq \|\gamma\|_{\sup}.$$

□

Of course, one can now transport this result into the arithmetic setting by using the formula $\|\cdot\|_v = |\cdot|^{[k_v:\mathbb{R}]}$. Also, if $P \in C(k)$ and $v \mid \infty$, we shall write $U_{P,v}$ and $z_{P,v}$ in place of U_{P_v} and z_{P_v}, respectively.

Corollary 6.3. *Let γ be a global section of a metrized line sheaf $\mathscr{L}$ on W. Let $E \subseteq W$ correspond to some $P_0 = (P_1, \dots, P_n) \in C^n$. Let $d_1, \dots, d_n$ be positive real numbers. For each $v \mid \infty$ let $\gamma_{0,v}$ be a local generator of $\mathscr{L}\big|_{U_v}$, where $U_v = U_{P_1,v} \times \dots \times U_{P_n,v}$. Then the index $t = t(\gamma, P_0, d_1, \dots, d_n)$ is at least*

$$t \geq \frac{-\deg \mathscr{L}\big|_E - \sum_{v|\infty} \log \|\gamma\|_{\sup,v} + \sum_{v|\infty} \log \inf_{P\in U_v} \frac{\|\gamma_{0,v}(P)\|}{\|\gamma_{0,v}(P_{0,v})\|}}{\max_{1\leq i\leq n} \deg \mathscr{O}(-d_i E_i)\big|_{E_i} - [k:\mathbb{Q}]\log\rho \max_{1\leq i\leq n} d_i}.$$

Proof. Let (ℓ) be a multi-index for which $D_{(\ell)}\gamma(P_0)$ is defined and nonzero. Then $D_{(\ell)}\gamma(P_0)$ is a nonzero section of

$$\mathscr{L} \otimes \mathscr{O}(-\ell_1 \operatorname{pr}_1^* E_1 - \dots - \ell_n \operatorname{pr}_n^* E_n)\big|_E.$$

We obtain the inequality by computing the degree δ of this line sheaf in two ways. First, its degree is

$$\delta = \deg \mathscr{L}\big|_E + \sum_{i=1}^n \ell_i \deg \mathscr{O}(-E_i)\big|_{E_i}$$

$$\leq \deg \mathscr{L}\big|_E + \sum_{i=1}^n \frac{\ell_i}{d_i} \cdot \max_{1\leq i\leq n} \deg \mathscr{O}(-d_i E_i)\big|_{E_i}.$$

On the other hand, δ can be computed from the degree of the arithmetic divisor $D_{(\ell)}\gamma(P_0)$ on E, via Lemma 6.2:

$$\begin{aligned}\delta &\geq \sum_{v|\infty} -\log \|D_{(\ell)}\gamma(P_0)\|_v \\ &\geq \sum_{v|\infty} -\log \|\gamma\|_{\sup,v} + \sum_{v|\infty} \log \inf_{P\in U_v} \frac{\|\gamma_{0,v}(P)\|}{\|\gamma_{0,v}(P_{0,v})\|} + [k:\mathbb{Q}] \log \rho \sum_{i=1}^{n} \ell_i \\ &\geq \sum_{v|\infty} -\log \|\gamma\|_{\sup,v} + \sum_{v|\infty} \log \inf_{P\in U_v} \frac{\|\gamma_{0,v}(P)\|}{\|\gamma_{0,v}(P_{0,v})\|} + \sum_{i=1}^{n} \frac{\ell_i}{d_i} \cdot [k:\mathbb{Q}] \log \rho \max_{1\leq i\leq n} d_i\end{aligned}$$

(here we assume $\rho < 1$). Combining these two inequalities gives the corollary, since we may choose (ℓ) such that

$$t = \sum_{i=1}^{n} \frac{\ell_i}{d_i}.$$

□

§7. Proof of Mordell, with some simplifications by Bombieri

To introduce the various proofs of Mordell's conjecture, etc., we start with a sketch of a proof using ideas from Bombieri's proof, since the methods do not involve as much machinery. More details on this proof can be found in [Bo].

First, we need some notation. Let C be the curve, and let g be its genus; then $g > 1$. Let K_C be a canonical divisor on C, and let $F = K_C/(2g-2)$, so that F has degree 1. Here F is a $\mathbb{Q}$-divisor—a divisor with rational coefficients. On $C \times C$, let pr_1 and pr_2 be the projections to the factors, let $F_i = \mathrm{pr}_i^* F$ for $i = 1, 2$, and let Δ be the diagonal on $C \times C$. Let

$$\Delta' = \Delta - F_1 - F_2.$$

Let $\delta > 0$ and $r > 1$ be rational; we require that $a_1(r) := \sqrt{(g+\delta)r}$ also be rational. Then $a_2(r) := \sqrt{(g+\delta)/r}$ is also rational. The goal of the first two steps of the proof, then, will be to construct a certain global section of the line sheaf $\mathcal{O}(dY)$, where d is a large sufficiently divisible positive integer and

$$Y = Y_r = \Delta' + a_1 F_1 + a_2 F_2.$$

In Bombieri's proof, Step 1 is rather easy: by duality, $h^2(C \times C, dY) = 0$ for d sufficiently large; therefore by the Hirzebruch Riemann-Roch theorem,

$$\begin{aligned}h^0(C \times C, dY) &\geq \frac{d^2 Y^2}{2} \\ &\geq \delta d^2,\end{aligned}$$

since $\Delta'^2 = -2g$, $\Delta' F_1 = \Delta' F_2 = F_1^2 = F_2^2 = 0$, and $F_1 \cdot F_2 = 1$.

Step 2 requires a bit more cleverness. First, fix $N > 0$ such that NF is very ample, and fix global sections $x_0, \ldots, x_n$ of $\mathcal{O}(NF)$ giving a corresponding embedding into $\mathbb{P}^n$. We also regard these as sections of $\mathcal{O}(NF_1)$ on $C \times C$ via pr_1 and let $x'_0, \ldots, x'_n$ be the corresponding sections of $\mathcal{O}(NF_2)$ defined via pr_2.

Also take s to be a sufficiently large integer such that

$$B := sF_1 + sF_2 - \Delta'$$

is very ample, and let $y_0, \dots, y_m$ be global sections of $\mathcal{O}(B)$ on $C \times C$ giving a corresponding embedding into $\mathbb{P}^m$.

Then one of Bombieri's key ideas is to write

$$Y = \delta_1 NF_1 + \delta_2 NF_2 - B,$$

where $\delta_i = (a_i + s)/N$. Then, for d sufficiently large, a section γ of $\mathcal{O}(dY)$ will have the property that γy_i^d will be represented by polynomials Φ_i of degree $d\delta_1$ in $x_0, \dots, x_n$ and of degree $d\delta_2$ in $x'_0, \dots, x'_n$. Indeed, let $\mathcal{I}$ be the ideal sheaf of the image of $C \times C$ in $\mathbb{P}^n \times \mathbb{P}^n$. Since $\mathcal{O}(\delta_1, \delta_2)$ (defined as $\mathrm{pr}_1^* \mathcal{O}(\delta_1) \otimes \mathrm{pr}_2^* \mathcal{O}(\delta_2)$) is ample, for d sufficiently large we will have $H^1(\mathbb{P}^n \times \mathbb{P}^n, \mathcal{I} \otimes \mathcal{O}(d\delta_1, d\delta_2)) = 0$, and by the long exact sequence in cohomology, the map

$$\begin{aligned} H^0(\mathbb{P}^n \times \mathbb{P}^n, \mathcal{O}(d\delta_1, d\delta_2)) &\to H^0(C \times C, \mathcal{O}(d\delta_1, d\delta_2)\big|_{C\times C}) \\ &= H^0(C \times C, \mathcal{O}(d\Delta_1 NF_1 + d\Delta_2 NF_2)) \end{aligned}$$

will be surjective. Of course, it is not *injective:* instead, one chooses a subspace of $H^0(\mathbb{P}^n \times \mathbb{P}^n, \mathcal{O}(d\delta_1, d\delta_2))$ for which the map is injective, and for which the cokernel is sufficiently small. If the coordinates $x_0, \dots, x_n$ are chosen suitably generically, then these conditions hold for the subspace spanned by

$$x_0^{d\delta_1} (x'_0)^{d\delta_2} \cdot \left(\frac{x_1}{x_0}\right)^a \left(\frac{x_2}{x_0}\right)^b \left(\frac{x'_1}{x'_0}\right)^{a'} \left(\frac{x'_2}{x'_0}\right)^{b'},$$
$$0 \le a + b \le d\delta_1,\ 0 \le b \le N,\ 0 \le a' + b' \le d\delta_2,\ 0 \le b' \le N.$$

Then, apply Siegel's lemma to tuples $(\Phi_1, \dots, \Phi_m)$ in the above subspace, subject to the linear constraints

$$\Phi_i y_j^d = \Phi_j y_i^d, \qquad i, j \in \{0, \dots, m\}.$$

A solution to this system then yields local sections Φ_i / y_i^d which patch together to give a global section

$$\gamma \in \Gamma(C \times C, \mathcal{O}(dY))$$

with

$$h(\gamma) \le cda_1 + o(d). \tag{7.1}$$

Here the height $h(\gamma)$ is defined as the height of the vector of coefficients of the polynomials $\Phi_1, \dots, \Phi_m$.

From this point Bombieri continues to proceed very classically, using Weil's theory of heights instead of arithmetic intersection theory. But instead we will continue the proof in more modern language.

Note, first of all, that one can fix metrics on F and Δ, giving a metric on Y. Then we obtain a bound for the sup norm on Y, in terms of the sizes of the coefficients of the polynomials Φ_i.

Step 3 now uses an idea from a 1965 paper of Mumford [M]. As in Roth's theorem, it uses a pigeonhole argument, but this time the argument takes place in $J(k) \otimes_{\mathbb{Z}} \mathbb{R}$; here J denotes the Jacobian of C. By the Mordell-Weil theorem ([L 5], Ch. 6), this is again a finite dimensional vector space.

Fixing a point $P_0 \in C$, let $\phi_0 \colon C \to J$ be the map given by $P \mapsto \mathcal{O}(P - P_0)$. By ([M], pp. 1011–1012),

$$\mathcal{O}(\Delta - \{P_0\} \times C - C \times \{P_0\}) \simeq \mathrm{pr}_1^* \phi_0^* \Theta_0 + \mathrm{pr}_2^* \phi_0^* \Theta_0 - (\phi_0 \times \phi_0)^* (\mathrm{pr}_1 + \mathrm{pr}_2)^* \Theta_0$$

where Θ_0 is the divisor class of the theta divisor on J. But we need Θ_0 to be a symmetric divisor class. Therefore fix $a \in J$ such that $(2g-2)a = \mathcal{O}((2g-2)P_0 - K_C)$, let $\phi(P) = \phi_0(P) + a$, and let Θ be the class of the theta divisor defined relative to the map ϕ. Then Θ is a symmetric divisor class, and

$$\Delta' = -(\phi \times \phi)^* \mathcal{P} \tag{7.2}$$

on $C \times C$, up to $(2g-2)$-torsion, where

$$\mathcal{P} := (\mathrm{pr}_1 + \mathrm{pr}_2)^* \Theta - \mathrm{pr}_1^* \Theta - \mathrm{pr}_2^* \Theta$$

is the Poincaré divisor class. See also ([L 5], Ch. 5, § 5).

Let $\hat{h}_\Theta$ denote the Néron-Tate canonical height on J relative to Θ (cf. ([Sil], Thm. 4.3) or ([L 5], Ch. 5, §§3, 6, 7)); it is quadratic in the group law and therefore defines á bilinear pairing

$$(P_1, P_2)_\Theta = \hat{h}_\Theta(P_1 + P_2) - \hat{h}_\Theta(P_1) - \hat{h}_\Theta(P_2)$$

which gives the vector space $J(k) \otimes_{\mathbb{Z}} \mathbb{R}$ a dot product structure. The canonical height also satisfies

$$\hat{h}_\Theta(P) = h_\Theta(P) + O(1). \tag{7.3}$$

Let $|P|^2 = (P, P)_\Theta$, so that $|P|^2 = 2\hat{h}_\Theta(P)$. Then, assuming that $C(k)$ is infinite, we may find an infinite subsequence such that all points in the subsequence point in approximately the same direction: all P_1 and P_2 in this subsequence satisfy

$$(P_1, P_2)_\Theta \geq (\cos\theta)\sqrt{|P_1|^2 |P_2|^2} \tag{7.4}$$

for a given $\theta \in (0, \pi)$. Also choose P_1 and P_2 so that $h(P_1)$ is large, and so that $h(P_2)/h(P_1)$ is large.

Mumford shows that $\phi^* \Theta = gF$; thus by (7.2), (7.3), and (7.4),

$$h_{\Delta'}(P_1, P_2) \leq -2g(\cos\theta)\sqrt{h_F(P_1) h_F(P_2)} + O(1).$$

Expressing this in terms of Y, and rewriting it in terms of metrized line sheaves on $W := X \times_B X$ gives

$$\frac{1}{[k : \mathbb{Q}]} \deg \mathcal{O}(dY)|_E \leq d\big(a_1 h_F(P_1) + a_2 h_F(P_2) - 2g(\cos\theta)\sqrt{h_F(P_1) h_F(P_2)} + O(1)\big).$$

Taking r close to $h_F(P_2)/h_F(P_1)$ and recalling the definitions of a_1 and a_2 then gives

$$\frac{1}{[k:\mathbb{Q}]}\deg \mathcal{O}(dY)\big|_E \le 2d\big((\sqrt{g+\delta}-g\cos\theta)\sqrt{h_F(P_1)h_F(P_2)}+O(1)\big).$$

For θ small enough, this is negative. If it is sufficiently small, then by the product formula (0.4), the restriction to E of any sufficiently small global section of $\mathcal{O}(dY)$ must vanish, so γ must vanish along E.

As was the case in Roth's proof, Step 4 consists of applying a variant of the above argument to certain partial derivatives, giving a lower bound for the index of γ at (P_1, P_2). But the work has already been done: take $d_i = da_i$ for $i = 1,2$ and note that the inequality of Corollary 6.3 gives:

$$t \ge \frac{2d(\sqrt{g+\delta}-g\cos\theta)\sqrt{h_F(P_1)h_F(P_2)} - cda_1 + o(d)}{(2g-2)da_1h_F(P_1)+cda_1}.$$

Indeed, by (7.1) the term $\sum_{v|\infty}\log\|\gamma\|_{\sup,v}$ in the numerator of the expression in Corollary 6.3 is bounded by $cda_1+o(d)$, and the generators $\gamma_{0,v}$ for various $\mathcal{O}(dY_r)$ may be taken uniformly in d and r, so the term $\sum_{v|\infty}\log\inf_{P\in U_v}\|\gamma_{0,v}(P)\|/\|\gamma_{0,v}(P_{0,v})\|$ also is bounded by cda_1.

But now note that $a_1h_F(P_1)$ is approximately $\sqrt{g+\delta}\sqrt{h_F(P_1)h_F(P_2)}$. Thus the first terms in the numerator and denominator are dominant as the $h(P_i)$ become large. This gives a lower bound for the index.

One can then project $C\times C$ down to $\mathbb{P}^1\times\mathbb{P}^1$, take the norm of γ to get a polynomial, and apply Roth's lemma to obtain a contradiction. We omit the details because they will appear in more generality in Section 18.

§8. Proof using Gillet-Soulé Riemann-Roch

In this case, we still use the same notations δ, r, a_1, a_2, F_1, F_2, Δ, Δ', and Y as before. However, Step 1 is a little more complicated, in that we prove that if r is sufficiently large, then Y is ample. For details on this and other parts of the proof, see [V 3] and [V 4].

Step 2 is the part which I wish to emphasize—this is where the Gillet-Soulé Riemann-Roch theorem is used. First, we assume C has semistable reduction over k, and let X be the regular semistable model for C over $B(=\operatorname{Spec} R)$ ([L 7], Ch. V, §5). Then $X\times_B X$ is regular except at points above nodes on the fibers of each factor. At such points, though, the singularity is known explicitly and can be resolved by replacing it with a projective line. Let $q\colon W\to B$ be the resulting model for $C\times C$.

The divisors F and Δ on the generic fiber need to be extended to X and W, respectively. To extend F, we take $\omega_{X/B}$ at finite places, and fix a choice of metrics with positive curvature. The (Arakelov) canonical metric is one possible choice, but it is not required. To extend Δ, we take its closure on W, and choose a metric for it. Again, the Arakelov Green's function is one possible choice. Then F_1, F_2, Δ', and Y become arithmetic divisors on W as well. By the Gillet-Soulé Riemann-Roch theorem, then,

$$\sum_{i=0}^{2}(-1)^i\deg R^iq_*\mathcal{O}(dY) = \frac{d^3Y^3}{6}+O(d^2).$$

We want a lower bound for $\deg q_* \mathcal{O}(dY)$; this is obtained as follows. First, since Y is ample, the free parts of $R^i q_* \mathcal{O}(dY)$ vanish for $i > 0$ if d is sufficiently large. The torsion part of the $R^1 q_*$ term is nonnegative, and the torsion part in the $R^2 q_*$ term is zero by a duality argument. Much longer arguments in a similar vein (but with their own analytic character) show that the same is true for analytic torsion, up to $O(d^2 \log d)$. Thus we find that

$$\deg q_* \mathcal{O}(dY) \geq \frac{d^3 Y^3}{6} - O(d^2 \log d).$$

Here Y^3 grows like $-O(\sqrt{r})$. Since the rank of $O(dY)$ is approximately δd^2, the ratio $\deg q_* \mathcal{O}(dY) / \operatorname{rank} q_* \mathcal{O}(dY)$ is approximately $O(-d\sqrt{r})$. Then it follows by a geometry of numbers argument that there exists a global section γ of $\mathcal{O}(dY)$ with

$$\prod_{v|\infty} \|\gamma\|_{L^2,v} \leq \exp(cd\sqrt{r}).$$

But we really need an inequality involving the sup norm. Of course, the trivial inequality

$$\|\gamma\|_{L^2,v} \leq \|1\|_{L^2,v} \|\gamma\|_{\sup,v}$$

holds, but this goes in the wrong direction.

It is slightly more difficult to prove an inequality in the opposite direction.

Lemma 8.1. *Fix a measure ν on a complex manifold X of dimension n. Then for each metrized line sheaf $\mathcal{L}$ on X there exists a constant $c_{\mathcal{L}} > 0$ such that for all $\gamma \in \Gamma(X, \mathcal{L})$,*

$$\|\gamma\|_{L^2} \geq c_{\mathcal{L}} \|\gamma\|_{\sup}.$$

Moreover, if $\mathcal{L} \cong \mathcal{L}_1^{\otimes i_1} \otimes \cdots \otimes \mathcal{L}_m^{\otimes i_m}$, then we may take $c_{\mathcal{L}} = c_{\mathcal{L}_1}^{i_1} \cdots c_{\mathcal{L}_m}^{i_m}$.

Proof. By a compactness argument, there exists a constant $\rho > 0$ and for each $P \in X$ a local coordinate system on a neighborhood U_P of P, $z_P \colon U_P \xrightarrow{\sim} \mathbb{D}_\rho^n$, such that $z_P(P) = 0$ and

$$dd^c|z_1|^2 \wedge \cdots \wedge dd^c|z_n|^2 \leq \nu.$$

Also, for each P and each $\mathcal{L}$ there exist local holomorphic sections $\gamma_{0,P}$ of $\mathcal{L}|_{U_P}$ such that $\|\gamma_{0,P}(P)\| = 1$ and

$$c'_{\mathcal{L}} := \inf_{\substack{P \in X \\ Q \in U_P}} \|\gamma_{0,P}(Q)\|^2$$

is strictly positive. This may require shrinking ρ, depending on $\mathcal{L}$. Then, letting P be the point where γ attains its maximum,

$$\begin{aligned} \|\gamma\|_{L^2}^2 &\geq c'_{\mathcal{L}} \int_{\mathbb{D}_\rho^n} \left| \left(\frac{\gamma}{\gamma_{0,P}} \right)(Q) \right|^2 dd^c|z_1|^2 \wedge \cdots \wedge dd^c|z_n|^2 \\ &\geq c_{\mathcal{L}}^2 \|\gamma(P)\|^2 \end{aligned}$$

for some suitable $c_{\mathcal{L}} > 0$, by Parseval's inequality (or harmonicity).

The last statement follows by choosing the sections $\gamma_{0,P}$ for $\mathcal{L}$ compatibly with those chosen for the $\mathcal{L}_i$. □

Sharper bounds are possible (cf. ([V 4], 3.9)), but the above bound is sufficient for our purposes.

This proof of Mordell can then conclude with Steps 3–5 as before. Or, in either case, instead of Roth's lemma, we can use Dyson's lemma on a product of two curves.

Lemma 8.2 ([V 2]). *Let $\xi_1, \dots, \xi_m$ be m points on C^2 with distinct first coordinates and distinct second coordinates. Let γ be a global section of a line sheaf $\mathscr{L}$ on $C \times C$, and assume that $(\mathscr{L} . F_1) \ge d_2$ and $(\mathscr{L} . F_2) \ge d_1$. Then, recalling the notation* Vol() *from Lemma 4.7,*

$$\sum_{i=1}^{m} \mathrm{Vol}(t(\gamma, \xi_i, d_1, d_2)) \le \frac{(\mathscr{L}^2)}{2d_1 d_2} + \frac{(\mathscr{L} . F_1)}{2d_1} \max(2g - 2 + m, 0).$$

In this case, $d_i = da_i$ as before, and $\mathscr{L} = \mathscr{O}(dY)$. Then it follows that the first term on the right is $\delta/(g+\delta)$ and the second term is $(2g-1)/r$. Both can be made smaller than $t^2/2$ on the left, obtaining a contradiction.

It was this part of the argument that first led to some insight on the problem: instead of making certain terms on the left large, one could make Y^2 on the right small. This is how one can prove finiteness for diophantine equations without using diophantine approximation *per se*.

§9. The Faltings complex

The remainder of these lectures will be devoted to proving Faltings' generalization of Mordell's conjecture, Theorem 0.3. This will be done in detail. See also [F 1] and [F 2].

As a first step towards generalizing the technique to more general subvarieties of abelian varieties, recall the result of Mumford (7.2):

$$\Delta' = (j \times j)^*(\mathrm{pr}_1^* \Theta + \mathrm{pr}_2^* \Theta - (\mathrm{pr}_1 + \mathrm{pr}_2)^* \Theta),$$

where $\mathrm{pr}_1 + \mathrm{pr}_2$ in the last term refers to the sum under the group law on the Jacobian. Then one can replace Θ with any symmetric ample divisor class L on a general abelian variety A, and let the Poincaré divisor class

$$\mathscr{P} := (\mathrm{pr}_1 + \mathrm{pr}_2)^* L - \mathrm{pr}_1^* L - \mathrm{pr}_2^* L$$

play the role of (minus) Δ'. But now the theorem of the cube implies that for $a, b \in \mathbb{Z}$,

$$(a \cdot \mathrm{pr}_1 + b \cdot \mathrm{pr}_2)^* L = a^2 \mathrm{pr}_1^* L + b^2 \mathrm{pr}_2^* L + ab\mathscr{P}. \tag{9.1}$$

Then it follows that dY can be written (approximately) in the form

$$dY = (s_1 \cdot \mathrm{pr}_1 - s_2 \cdot \mathrm{pr}_2)^* L - \epsilon s_1^2 \mathrm{pr}_1^* L - \epsilon s_2^2 \mathrm{pr}_2^* L.$$

In this case, however, it will be necessary to work on a product of n copies of A, so let us define

$$L_{\delta,s} = \sum_{i<j} (s_i \cdot \mathrm{pr}_i - s_j \cdot \mathrm{pr}_j)^* L + \delta \sum_{i=1}^{n} s_i^2 \mathrm{pr}_i^* L \tag{9.2}$$

for rational δ and $s \in \mathbb{N}^n$. By the theorem of the cube, this expression is homogeneous of degree two in $s_1, \dots, s_n$, so we also extend this definition to $s \in \mathbb{Q}_{>0}^n$ by homogeneity.

One aspect of the expression (9.2) is that it clearly points out a key idea in the whole theory. Namely, on $\prod X_i$ the first term is large (ample, in the case of Theorem 0.2), and the second term is small (δ is taken negative but close to zero); however, on the arithmetic curve corresponding to our point $(P_1, \dots, P_n)$, the first term is small and the second term then dominates.

Another benefit of this expression is that, by the theorem of the cube,

$$(a \cdot \mathrm{pr}_i - b \cdot \mathrm{pr}_j)^* L + (a \cdot \mathrm{pr}_i + b \cdot \mathrm{pr}_j)^* L = 2a^2 \mathrm{pr}_i^* L + 2b^2 \mathrm{pr}_j^* L.$$

Thus, choosing global sections $\gamma_1, \dots, \gamma_m \in \Gamma(A, \mathcal{O}(L))$ which generate $\mathcal{O}(L)$ over the generic fiber A, for any $X_1, \dots, X_n$ we can form an injection

$$0 \to \Gamma\left(\prod X_i, dL_{\delta,s}\right) \to \Gamma\left(\prod X_i,\ d(2n-2+\delta)\sum_{i=1}^{n} s_i^2 \,\mathrm{pr}_i^* L\right)^a$$

by tensoring with products of terms of the form $(as_i \cdot \mathrm{pr}_i + as_j \cdot \mathrm{pr}_j)^* \gamma_{\ell_{ij}}^b$, where $ba^2 = d$ and a is sufficiently divisible. Here the tuples $(\ell)_{ij}$ vary over $\{1, \dots, m\}^{n(n-1)/2}$. Likewise, one can extend this sequence to an exact sequence

(9.3)
$$0 \to \Gamma\left(\prod X_i, dL_{\delta,s}\right) \to \Gamma\left(\prod X_i,\ d(2n-2+\delta)\sum_{i=1}^{n} s_i^2 \,\mathrm{pr}_i^* L\right)^a$$
$$\to \Gamma\left(\prod X_i,\ d\sum_{i<j}(s_i \cdot \mathrm{pr}_i - s_j \cdot \mathrm{pr}_j)^* L + d(2n-2+\delta)\sum_{i=1}^{n} s_i^2 \,\mathrm{pr}_i^* L\right)^{b'}$$

and embed this last term to obtain the **Faltings complex**

$$0 \to \Gamma\left(\prod X_i, dL_{\delta,s}\right) \to \Gamma\left(\prod X_i,\ d(2n-2+\delta)\sum_{i=1}^{n} s_i^2 \,\mathrm{pr}_i^* L\right)^a \tag{9.4}$$
$$\to \Gamma\left(\prod X_i,\ d(4n-4+\delta)\sum_{i=1}^{n} s_i^2 \,\mathrm{pr}_i^* L\right)^b$$

for some integer b. This integer will be quite large, but depends only on m and n.

§10. Overall plan

Now let us discuss the overall plan of the proof of Theorem 0.3. First of all, we may assume that X is closed in A.

Next, recall some standard facts about subvarieties of abelian varieties. For now, let $X \subseteq A$ be defined over $\mathbb{C}$.

Definition 10.1. Let $B(X)$ be the identity component of the algebraic group

$$\{a \in A \mid a + X = X\}.$$

Then the restriction to X of the quotient map $A \to A/B(X)$ gives a fibration $X \to X/B(X)$ whose fibers are all isomorphic to $B(X)$. This fibration is called the **Ueno fibration of** X. It is **trivial** when $B(X)$ is a point.

Theorem 10.2 (Ueno ([U], Thm. 10.9)). *If $B(X)$ is trivial, then X is of general type (and conversely).*

Thus, Theorem 0.3 gives an affirmative answer to a special case of a question posed by Bombieri [N]: if X is a variety of general type defined over a number field k, is it true that $X(k)$ is not dense in the Zariski topology? Furthermore, Lang has conjectured in ([L 6], Conj. 5.8) that the higher dimensional part of the closure of $X(k)$ should be geometric; *i.e.*, independent of k. That conjecture is also answered in this special case, by the following theorem.

Theorem 10.3 (Kawamata structure theorem ([Ka], Theorem 4)). *Let $Z(X)$ denote the union of all nontrivial translated abelian subvarieties of A contained in X. Then $Z(X)$ is a Zariski-closed subset, and each irreducible component of it has nontrivial Ueno fibration.*

In particular, if X has trivial Ueno fibration, then $Z(X) \neq X$ (and conversely).

Returning to the situation over k, we note that $B(X)$ and $Z(X)$ are defined over k, by simple Galois-theoretic arguments. (In general, $Z(X)$ is a scheme: it may be reducible.)

For the proof of Theorem 0.3, we note first that it will suffice to assume that $B(X)$ is trivial; otherwise the theorem on X follows from the theorem on $X/B(X)$.

Next, it will suffice to prove that $X(k) \setminus Z(k)$ is finite. Indeed, points in $Z(k)$ can be handled by Noetherian induction.

Now, for the main part of the proof, fix a very ample symmetric divisor class L on A, and an associated projective embedding. Degrees and heights of subspaces of X will refer to this embedding. In particular, for heights, we use the definition from ([B-G-S], Section 1): the height of a subvariety of projective space is the intersection number with a "standard" linear subspace of complementary dimension in $\mathbb{P}^n$; a "standard" linear subspace is one which is obtained by setting some of the coordinate functions equal to zero.

In the proof, we will fix

$$n = \dim X + 1$$

and work with points $P_1, \dots, P_n \in X(k) \setminus Z(k)$. Let $h_L(P_i)$ denote the heights of the P_i. These points will be chosen (later) so as to satisfy the usual Step 3 conditions, for certain constants c_1, $c_2 \geq 1$, and ϵ_1:

(10.4.1). $h_L(P_1) \geq c_1$;

(10.4.2). $h_L(P_{i+1})/h_L(P_i) \geq c_2$, $i = 1, \dots, n-1$;

(10.4.3). $P_1, \dots, P_n$ all point in roughly the same direction in $A(k) \otimes_{\mathbb{Z}} \mathbb{R}$: let $\hat{h}_L(P)$ denote the Néron-Tate canonical height associated to h_L and let

$$(P, Q)_L = \hat{h}_L(P+Q) - \hat{h}_L(P) - \hat{h}_L(Q)$$

be the associated bilinear form; then the assumption is that

$$(P_i, P_j)_L \geq (1-\epsilon_1)\sqrt{(P_i,P_i)(P_j,P_j)} \qquad \text{for all } i,j\,.$$

We will also call these conditions $C_P(c_1, c_2, \epsilon_1)$.

The proof also uses subvarieties $X_1, \dots, X_n$ of X satisfying the following conditions, denoted $C_X(c_3, c_4, P_1, \dots, P_n)$:

(**10.5.1**). Each X_i contains P_i.
(**10.5.2**). The X_i are geometrically irreducible and defined over k.
(**10.5.3**). The degrees $\deg X_i$ satisfy $\deg X_i \leq c_3$.
(**10.5.4**). The heights $h(X_i)$ are bounded by the formula

$$\sum_{i=1}^{n} \frac{h(X_i)}{h_L(P_i)} < c_4 \sum_{i=1}^{n} \frac{1}{h_L(P_i)}.$$

Here and from now on, constants c and c_i will depend on A, X, k, the projective embedding associated to L, and sometimes the tuple $(\dim X_1, \dots, \dim X_n)$. They will not depend on X_i, P_i, or (s). Also, they may vary from line to line.

The overall plan of the proof, then, is to construct subvarieties $X_1, \dots, X_n$ of X satisfying the conditions (10.5). We start with $X_1 = \dots = X_n = X$ and successively create smaller tuples of subvarieties, until reaching the point where $\dim X_j = 0$ for some j. In that case $X_j = P_j$, and $h(X_j) = h_L(P_j)$. Then, by (10.5.4),

$$1 = \frac{h(X_j)}{h_L(P_j)} \leq \sum_{i=1}^{n} \frac{h(X_i)}{h_L(P_i)} \leq c_4 \sum_{i=1}^{n} \frac{1}{h_L(P_i)} \leq \frac{c_4 n}{h_L(P_1)} \tag{10.6}$$

and thus

$$h_L(P_1) \leq c_4 n. \tag{10.7}$$

This contradicts (10.4.1) if c_1 is taken large enough.

The inductive step of the proof takes place if all X_i are positive dimensional. For $i = 1, \dots, n$ let s_i be rational numbers close to $1/\sqrt{h_L(P_i)}$. Let d be a positive integer. This is usually taken large and highly divisible, and may depend on practically everything else. We shall construct a small section of $\mathcal{O}(dL_{-\epsilon,s})$ for some $\epsilon > 0$ (depending on the $\dim X_i$). If the points $P_1, \dots, P_n$ were chosen suitably, then it is possible to construct subvarieties X_i' of each X_i such that the X_i' also satisfy (10.5) (possibly with different constants), and such that some X_i' is strictly smaller.

To help clarify this step, the dependence of the constants, etc. can be written symbolically as follows.

(10.8)

$$\begin{aligned}
&\forall\, c_3, c_4 \text{ and } \forall\, \delta_1, \dots, \delta_n \in \mathbb{N} \\
&\exists\, c_1, c_2, \epsilon_1, c_3', c_4' \text{ such that} \\
&\quad \forall\, P_1, \dots, P_n \in (X \setminus Z(X))(k) \text{ satisfying } C_P(c_1, c_2, \epsilon_1) \text{ and} \\
&\quad \forall\, X_1, \dots, X_n \subseteq X \text{ satisfying } C_X(c_3, c_4, P_1, \dots, P_n) \text{ and } \dim X_i = \delta_i\ \forall\, i \\
&\qquad \exists\, X_1', \dots, X_n' \text{ with } X_i' \subseteq X_i\ \forall\, i \text{ and } X_i' \neq X_i \text{ for some } i, \\
&\qquad\quad \text{and satisfying } C_X(c_3', c_4', P_1, \dots, P_n).
\end{aligned}$$

To conclude the proof, we now assume that $X(k) \setminus Z(k)$ is infinite. Then it is possible to choose $P_1, \dots, P_n$ in $X(k) \setminus Z(k)$ to satisfy Conditions (10.4) for all c_1, c_2, and ϵ_1 which occur in the finitely many times that the above main step can take place. This leads to a contradiction, so $X(k) \setminus Z(k)$ must be finite, and the theorem is proved.

For future reference, let

$$d_i = ds_i^2. \tag{10.9}$$

§11. Lower bound on the space of sections

Recall that $X_1, \dots, X_n$ are positive dimensional subvarieties of X, not lying in the Kawamata locus of X.

Lemma 11.1. *If $n \geq \dim X + 1$, then the morphism $f\colon \prod X_i \to A^{n(n-1)/2}$ given by $(x_1, \dots, x_n) \mapsto (x_i - x_j)_{i<j}$ is generically finite over its image.*

Proof. If $P \in X_{i,\text{reg}}$, then the tangent space $T_{X_i,P}$ may be identified with a linear subspace of the tangent space $T_{A,0}$ at the origin of A via translation. Then we may choose points $P_i \in X_{i,\text{reg}}$ such that f is smooth at $P := (P_1, \dots, P_n)$, and such that

$$\bigcap_{i=1}^{n} T_{X_i,P_i} = (0).$$

Then any tangent to the fiber of f at P must be zero, so f is a finite map there. □

Corollary 11.2. *If $n \geq \dim X + 1$, then the intersection number*

$$\left(\left(L_{0,1}\big|_{\prod X_i} \right)^{\sum \dim X_i} \right) > 0.$$

Proof. The $\mathbb{Q}$-divisor class $L_{0,1}$ is the pull-back to $\prod X_i$ of an ample divisor class on $A^{n(n-1)/2}$ via a generically finite morphism. □

For the remainder of the proof of Theorem 0.3, fix $n = \dim X + 1$.

Next we prove a homogeneity result in s.

Lemma 11.3. *Fix an embedding of k into $\mathbb{C}$. Then the cohomology class in $H^{1,1}_{\bar\partial}(A^n)$ corresponding to the divisor class*

$$\mathscr{P}_{ij} := (\mathrm{pr}_i + \mathrm{pr}_j)^* L - \mathrm{pr}_i^* L - \mathrm{pr}_j^* L$$

is represented over $A(\mathbb{C})^n$ by a form in

$$\mathrm{pr}_i^* \mathscr{E}^{1,0}(A) \otimes \mathrm{pr}_j^* \mathscr{E}^{0,1}(A) + \mathrm{pr}_i^* \mathscr{E}^{0,1}(A) \otimes \mathrm{pr}_j^* \mathscr{E}^{1,0}(A) \subseteq \mathscr{E}^{1,1}(A^n).$$

Proof. Let $d = \dim A$, and let A be given local coordinates $z_1, \dots, z_d$ obtained from the representation of A as $\mathbb{C}^d$ modulo a lattice. Let $\mathscr{O}(L)$ be given the metric with translation invariant curvature, which can then be written

$$\sum_{\alpha,\beta=1}^{d} a_{\alpha\beta} dz_\alpha \wedge d\bar z_\beta,$$

where $a_{\alpha\beta}$ are *constants.* For $\alpha = 1, \dots, d$ let $u_\alpha = \mathrm{pr}_i^* z_\alpha$ and $v_\alpha = \mathrm{pr}_j^* z_\alpha$. Using the above choice of metric on L, the curvature of $\mathscr{P}_{ij}$ is

$$\sum_{\alpha,\beta=1}^{d} a_{\alpha,\beta}(du_\alpha \wedge d\bar v_\beta + dv_\alpha \wedge d\bar u_\beta).$$

□

By counting degrees we immediately obtain the main homogeneity lemma:

Corollary 11.4. *Any intersection product*

$$\prod_{i<j} \mathscr{P}_{ij}^{e_{ij}} \cdot \prod_{i=1}^{n} \mathrm{pr}_i^* L^{e_i}$$

of maximal codimension on $\prod X_i$ vanishes unless

$$2e_i + \sum_{j<i} e_{ji} + \sum_{j>i} e_{ij} = 2 \dim X_i, \qquad i = 1, \dots, n.$$

Consequently, since

$$(s_i \cdot \mathrm{pr}_i - s_j \cdot \mathrm{pr}_j)^* L = s_i^2 \,\mathrm{pr}_i^* L + s_j^2 \,\mathrm{pr}_j^* L + s_i s_j \mathscr{P}_{ij},$$

it follows that the highest self-intersection number of $L_{\delta,s}$ on $\prod X_i$ is homogeneous of degree $2 \dim X_i$ in each s_i.

The next proposition is the main result of this section. The proof is due to M. Nakamaye and G. Faltings, independently.

Proposition 11.5. *There exist constants $c > 0$ and $\epsilon > 0$, depending only on X, A, L, $\dim X_1, \dots, \dim X_n$, and the bounds on $\deg X_i$, such that for all tuples $s = (s_1, \dots, s_n)$ of positive rational numbers,*

$$h^0\left(\prod X_i, dL_{-\epsilon,s}\right) > c d^{\sum \dim X_i} \prod_{i=1}^{n} s_i^{2 \dim X_i}$$

for all sufficiently large d (depending on s).

Proof. By Seshadri's criterion ([H 1], Ch. I, §7), $L_{\delta,s}$ is ample for $\delta > 0$. By Riemann-Roch, it follows that

$$h^0\left(\prod X_i, dL_{\delta,s}\right) = d^{\sum \dim X_i} \frac{(L_{\delta,s}^{\sum \dim X_i})}{(\sum \dim X_i)!}(1+o(1)).$$

To shorten notation, let $N = \sum \dim X_i$ for the remainder of the proof. For each index j, let H_j be the subscheme of $\prod X_i$ cut out by some section of $\mathrm{pr}_j^* \mathcal{O}(L)$. Then, as above,

$$h^0\left(H_j, dL_{\delta,s}\big|_{H_j}\right) = d^{N-1}\frac{(\mathrm{pr}_j^* L \,.\, L_{\delta,s}^{N-1})}{(N-1)!}(1+o(1)).$$

For each i (including j), $\mathrm{pr}_i^* \mathcal{O}(L)$ is represented by an effective divisor on H_j, so

$$h^0\left(H_j, \left(dL_{\delta,s} - \sum m_i \,\mathrm{pr}_i^* L\right)\Big|_{H_j}\right) \le d^{N-1}\frac{(\mathrm{pr}_j^* L \,.\, L_{\delta,s}^{N-1})}{(N-1)!}(1+o(1))$$

for each tuple $m_1, \dots, m_n$ of nonnegative integers. Note that the $o(1)$ term does not depend on $m_1, \dots, m_n$.

The exact sequence in cohomology attached to the short exact sequence

$$0 \to \mathcal{O}\left(dL_{\delta,s} - \sum m_i \,\mathrm{pr}_i^* L - \mathrm{pr}_j^* L\right) \to \mathcal{O}\left(dL_{\delta,s} - \sum m_i \,\mathrm{pr}_i^* L\right)$$
$$\to \mathcal{O}\left(dL_{\delta,s} - \sum m_i \,\mathrm{pr}_i^* L\right)\Big|_{H_j} \to 0$$

gives the inequality

$$\begin{aligned} &h^0\left(\prod X_i,\ dL_{\delta,s} - \sum m_i \,\mathrm{pr}_i^* L - \mathrm{pr}_j^* L\right) \\ &\quad \ge h^0\left(\prod X_i,\ dL_{\delta,s} - \sum m_i \,\mathrm{pr}_i^* L\right) - h^0\left(H_j,\ dL_{\delta,s} - \sum m_i \,\mathrm{pr}_i^* L\right) \\ &\quad \ge h^0\left(\prod X_i,\ dL_{\delta,s} - \sum m_i \,\mathrm{pr}_i^* L\right) - d^{N-1}\frac{(\mathrm{pr}_j^* L \,.\, L_{\delta,s}^{N-1})}{(N-1)!}(1+o(1)). \end{aligned}$$

Therefore, since $dL_{-\epsilon,s} = dL_{\delta,s} - d(\delta+\epsilon)\sum_i s_i^2 H_i$, this estimate gives

$$\begin{aligned} h^0\left(\prod X_i, dL_{-\epsilon,s}\right) &\ge h^0\left(\prod X_i, dL_{\delta,s}\right) - d(\delta+\epsilon)\sum_{i=1}^n s_i^2 h^0\left(H_i,\ dL_{\delta,s}\big|_{H_i}\right) \\ &\ge d^N\left(\frac{(L_{\delta,s}^N)}{N!} - (\delta+\epsilon)\sum_{i=1}^n s_i^2 \frac{(\mathrm{pr}_i^* L \,.\, L_{\delta,s}^{N-1})}{(N-1)!}\right)(1+o(1)). \end{aligned}$$

By Corollary 11.4, this lower bound equals

$$d^N \prod_{i=1}^n s_i^{2\dim X_i}\left(\frac{(L_{\delta,1}^N)}{N!} - (\delta+\epsilon)\sum_{i=1}^n \frac{(\mathrm{pr}_i^* L \,.\, L_{\delta,1}^{N-1})}{(N-1)!}\right)(1+o(1)).$$

The quantity inside the parentheses is a polynomial in δ and ϵ whose constant term is positive, by Corollary 11.2. Therefore we may take sufficiently small $\delta > 0$, $\epsilon > 0$, and $c > 0$ such that

$$0 < c < \frac{(L_{\delta,1}^N)}{N!} - (\delta+\epsilon)\sum_{i=1}^{n} \frac{(\mathrm{pr}_i^* L \,.\, L_{\delta,1}^{N-1})}{(N-1)!}$$

The fact that these intersection numbers are taken on varying X_i is not a serious problem: since the degrees of the X_i are bounded, they can lie in only finitely many numerical equivalence classes. □

§12. More geometry of numbers

Let Γ be a metrized, finitely generated lattice over the ring of integers R of k. For all archimedean places v of k, let the completion Γ_v of Γ at v be given a measure such that, for each set $\gamma_1, \dots, \gamma_\delta$ of R-linearly independent vectors in Γ which generate Γ modulo torsion, the product over all v of the covolumes of the lattices generated over $\mathbb{Z}$ (if $k_v \cong \mathbb{R}$) or $\mathbb{Z}[i]$ (if $k_v \cong \mathbb{C}$) by $\gamma_1, \dots, \gamma_\delta$ equals $(\Gamma : \sum_{i=1}^{\delta} R\gamma_i)$. Also let $V(\Gamma)$ be the product, over all v, of the volumes (relative to the above measures) of the unit balls (relative to $\|\cdot\|_v$) in Γ_v. Define a length function $\ell(\gamma) = \prod_v \|\gamma\|_v$, and for $i = 1, \dots, \delta$ define successive minima λ_i to be the minimum λ such that there exist R-linearly independent elements $\gamma_1, \dots, \gamma_i \in \Gamma$ such that $\ell(\gamma_j) \le \lambda$ for all $j = 1, \dots, i$.

Lemma 12.1. *In this situation,*

(a). *There exist constants c_1 and c_2 depending only on k such that*

$$(c_2\delta)^{-\delta} \le \lambda_1 \cdots \lambda_\delta V(\Gamma) \le c_3^\delta.$$

(b). *Let $\beta\colon \Gamma_1 \to \Gamma_2$ be a homomorphism of metrized R-modules. Let δ_0 and δ_2 be the ranks (over R) of the kernel and image of β, respectively. Also assume that C is a constant such that*

$$\prod_{v|\infty} \|\beta(\gamma)\|_v \le C \prod_{v|\infty} \|\gamma\|_v \quad \textit{for all } \gamma \in \Gamma_1. \tag{12.2}$$

Then

$$V(\Gamma_1) \le 2^{[k:\mathbb{Q}]\delta_0} V(\operatorname{Ker}\beta) C^{\delta_2} V(\operatorname{Im}\beta). \tag{12.3}$$

Proof. Part (a) is a generalization of Minkowski's theory of convex bodies. It follows from ([V 1], 6.1.11). There, it is proved only for parallelepipeds, but the proof holds more generally for convex length functions. The proof of part (b) is straightforward. □

We note that this lemma can be used to prove a Siegel lemma over number fields, although it is not as sharp as Theorem 2.4. Indeed, metrize $\Gamma_1 := R^N$ by a metric $\|\mathbf{x}\|^2 = |x_1|^2 + \cdots + |x_N|^2$ at real places, and $\|\mathbf{x}\| = |x_1|^2 + \cdots + |x_N|^2$ at complex places. Then $h(\mathbf{x}_i) = (1/[k:\mathbb{Q}]) \log \lambda_i$, in the language of Theorem 2.4. Also, in (12.2), let $C = \prod_{v|\infty} N^{[k_v:\mathbb{R}]} \max_{i,j} \|a_{ij}\|$; this will play the role of Q in Lemma 2.1

or $\exp(h(A))$ in Theorem 2.4. Then $V(\Gamma_1)$ and $V(\operatorname{Im}\beta)$ are functions only of the respective ranks, so (12.3) gives a lower bound for $V(\operatorname{Ker}\beta)$, depending only on C and the various ranks. Then part (a) of the lemma gives an upper bound on $\sum_{i=1}^{N-M} h(\mathbf{x}_i)$.

We will use Lemma 12.1 to determine a lower bound for $V(\operatorname{Ker}\beta)$. It will then follow from part (a) that a global section γ exists, whose metric in Γ_1 is small (bounded by $\exp(cd\sum s_i^2)$), provided a suitable lower bound on δ_0 can be shown. This will be done in Section 14.

§13. Arithmetic of the Faltings complex

To obtain a small section of $\mathscr{O}(dL_{-\epsilon,s})$, one would like to extend the Faltings complex to a model $\mathscr{A}$ for A over the integers, and apply Siegel's lemma. However, when discussing the arithmetic of sections of $(a \cdot \mathrm{pr}_i + b \cdot \mathrm{pr}_j)^* \mathscr{O}(L)$, some care is needed: usually the morphism $a \cdot \mathrm{pr}_i + b \cdot \mathrm{pr}_j$ does not extend to a morphism of arithmetic schemes. Therefore, we use the theorem of the cube in order to convert such sections into sections of $\mathscr{O}(a^2 \cdot \mathrm{pr}_i^* L + b^2 \cdot \mathrm{pr}_j^* L + ab\mathscr{P}_{ij})$; the isomorphism is to be extended over $\operatorname{Spec} R$ so as to give an isomorphism over the zero section. Here, as in Section 9, $\mathscr{P}_{ij}$ denotes the Poincaré class

$$\mathscr{P}_{ij} := (\mathrm{pr}_i + \mathrm{pr}_j)^* L - \mathrm{pr}_i^* L - \mathrm{pr}_j^* L.$$

Then integrality of a section of $(a \cdot \mathrm{pr}_i + b \cdot \mathrm{pr}_j)^* \mathscr{O}(L)$ is to be understood via the corresponding isomorphism

$$(a \cdot \mathrm{pr}_i + b \cdot \mathrm{pr}_j)^* \mathscr{O}(L) \cong \mathscr{O}(a^2 \cdot \mathrm{pr}_i^* L + b^2 \cdot \mathrm{pr}_j^* L + ab\mathscr{P}_{ij}). \tag{13.1}$$

Here and in what follows, let $\mathscr{W}_0$ be some model for A^n such that $\mathscr{P}_{ij}$ is defined as a Cartier divisor class, for all i and j.

Lemma 13.2 ([F 1], Lemma 5.1). *Fix a model $\mathscr{A}$ for A over B, an arithmetic divisor class L on $\mathscr{A}$, and sections $\gamma_1, \dots, \gamma_m \in \Gamma(\mathscr{A}, \mathscr{O}(L))$ which generate $\mathscr{O}(L)$ over the generic fiber A. Let $\mathscr{A}^2$ be some model for $A \times A$ (as above). Then there exist constants $c_{5,v}$, with $c_{5,v} = 0$ for almost all v, satisfying the following property. Let $a, b \in \mathbb{Z}$, not both zero. Then for all archimedean places v the sections*

$$(a \cdot \mathrm{pr}_1 + b \cdot \mathrm{pr}_2)^* \gamma_i \in \mathscr{O}(a^2 \, \mathrm{pr}_1^* L + b^2 \, \mathrm{pr}_2^* L + ab\mathscr{P}_{12})$$

satisfy the following conditions.

(a). *For all $i = 1, \dots, m$ and for all $P \in A^2(\bar{k}_v)$,*

$$\log \|((a \cdot \mathrm{pr}_1 + b \cdot \mathrm{pr}_2)^* \gamma_i)(P)\|_v \le c_{5,v}(a^2 + b^2).$$

(b). *For all $P \in A^2(\bar{k}_v)$,*

$$\max_{1 \le i \le m} \log \|((a \cdot \mathrm{pr}_1 + b \cdot \mathrm{pr}_2)^* \gamma_i)(P)\|_v \ge -c_{5,v}(a^2 + b^2).$$

Furthermore, let X be a closed integral subscheme of A^2 and let $\mathscr{X}$ be its closure in $\mathscr{A}^2$. Let v be a non-archimedean place, let F be a component of the fiber of $\mathscr{X}$ at v, and let e be the multiplicity with which it occurs in the fiber at v. Then for all such $\mathscr{X}$, v, and F,

(c). *For all $i = 1,\dots,m$ the multiplicity at F of the divisor $((a\cdot \mathrm{pr}_1 + b\cdot \mathrm{pr}_2)^*\gamma_i)$ is bounded from below by $-c_{5,v}e(a^2+b^2)$.*

(d). *The minimum over $i = 1,\dots,m$ of the multiplicities at F of the divisors $((a\cdot \mathrm{pr}_1 + b\cdot \mathrm{pr}_2)^*\gamma_i)$ is at most $c_{5,v}e(a^2+b^2)$.*

Proof. First, note that if v is a place of good reduction, then (c) and (d) hold with $c_{5,v} = 0$, by the theorem of the cube on the fiber. Also note that the constants are independent of X.

We prove this result by stating two assertions about Weil functions, which simultaneously imply (a) and (c), and (b) and (d), respectively. For references on Weil functions, see ([L 5], Ch. 10). Indeed, let $D_i = ((a\cdot \mathrm{pr}_1 + b\cdot \mathrm{pr}_2)^*\gamma_i)$ for $i = 1,\dots,m$. Then the assertions of (a) and (c) follow from the inequality

$$\lambda_{D_i}(P) \ge -c_{5,v}(a^2+b^2) \qquad \text{for all } i = 1,\dots,m \text{ and all } P \in A^2(\bar{k}_v) \tag{13.3}$$

and the assertions of (b) and (d) follow from

$$\min_{1\le i\le m} \lambda_{D_i}(P) \le c_{5,v}(a^2+b^2) \qquad \text{for all } P \in A^2(\bar{k}_v). \tag{13.4}$$

As in (13.1), these Weil functions are to be derived from fixed Weil functions on certain representations of L and $\mathscr{P}$ by the theorem of the cube. The non-archimedean cases (c) and (d) follow from the above statements by choosing a point $P \in X(\bar{k}_v)$ whose section crosses F transversally.

To prove these assertions, we make canonical choices of Weil functions, called Néron functions, using the following result:

Theorem 13.5. *Let A be an abelian variety defined over k_v. To each divisor D on A which is not supported at 0, there exists a Weil function $\hat{\lambda}_D$, uniquely determined by the following properties.*

(1). $\hat{\lambda}_D(0) = 0$.

(2). *If D and D' are divisors on A not supported at 0, then*

$$\hat{\lambda}_{D+D'} = \hat{\lambda}_D + \hat{\lambda}_{D'}.$$

(3). *If $D = (f)$ is principal, then $\hat{\lambda}_D(P) = v(f(P)/f(0))$.*

(4). *We have $\hat{\lambda}_{[2]^*D} = \hat{\lambda}_D \circ [2]$.*

Such functions $\hat{\lambda}_D$ also satisfy the property that if $\phi\colon B \to A$ is a homomorphism of abelian varieties defined over k_v, then

$$\hat{\lambda}_{\phi^*D} = \hat{\lambda}_D \circ \phi. \tag{13.6}$$

Proof. See ([L 5], Ch. 11, Thm. 1.1). The normalization (1) eliminates the constants used there. □

Since the difference between λ_D and $\hat{\lambda}_D$ is bounded in absolute value for any fixed divisor D, it will suffice to prove (13.3) and (13.4) for $\hat{\lambda}$. This requires that the D_i not pass through 0, which we may assume after translating by a fixed point in $A(k_v)$, if necessary.

Then, by (13.6), $\hat{\lambda}_{D_i}$ is the same whether it is computed using (13.1) or directly from the definition of $(a \cdot \mathrm{pr}_i + b \cdot \mathrm{pr}_j)^*$. Since the divisors (γ_i) are all effective, we can choose $c_{5,v}$ such that

$$\hat{\lambda}_{(\gamma_i)}(P) \geq -c_{5,v}$$

for all P and all i, which implies (13.3). Also, since these divisors have no geometric point in common on the generic fiber, we can choose $c_{5,v}$ such that

$$\min_{1 \leq i \leq m} \hat{\lambda}_{(\gamma_i)}(P) \leq c_{5,v}$$

for all P, implying (13.4). □

This result is also proved in a weaker form in ([V 5], Lemma 6.5). In this case a stronger statement is needed, however, since the subscheme X may vary in parts (c) and (d).

Applying this lemma to the Faltings complex (9.4), parts (c) and (d) of Lemma 13.2 give the following corollary.

Corollary 13.7. *The Faltings complex can be extended to a sequence of R-modules by multiplying by integers of size at most* $\exp\left(cd \sum s_i^2\right)$, *and the cohomology of the complex will then be torsion, annihilated by similarly bounded integers.*

We also include here two lemmas which will also be used in the next section.

Lemma 13.8 ([F 1]). *Let $X_1, \dots, X_n$ be subvarieties of X, and let $\mathscr{X}_1, \dots, \mathscr{X}_n$ be their closures in $\mathscr{A}$. Then for all sufficiently large natural numbers $d_1, \dots, d_n$,*

$$V\left(\Gamma\left(\prod \mathscr{X}_i, \sum d_i \,\mathrm{pr}_i^* L\right)\right) \geq \exp\left(-c \cdot h^0\left(\prod \mathscr{X}_i, \sum d_i \,\mathrm{pr}_i^* L\right) \sum d_i\right)$$

for some constant c independent of $X_1, \dots, X_n$ and $d_1, \dots, d_n$.

Proof. Embed $\mathscr{A}$ into projective space by a very ample multiple of L. Then it follows that the direct sum $\bigoplus_{d \geq 0} \Gamma(\mathscr{X}, dL)$ is finitely generated over the ring of homogeneous polynomials. Therefore

$$V(\Gamma(\mathscr{A}, dL)) \geq \exp(-c \cdot h^0(\mathscr{X}, dL) \cdot d).$$

By ampleness of L (on the generic fiber), restricting to $\mathscr{X}_i$ will not increase this c. The lemma then follows, since $\Gamma\left(\prod \mathscr{X}_i, \sum d_i \,\mathrm{pr}_i^* L\right)$ is a direct sum of tensor products of such modules. □

Lemma 13.9. *Let $X_1, \dots, X_n$, $\mathscr{X}_1, \dots, \mathscr{X}_n$, and $d_1, \dots, d_n$ be as above, and assume also that the degrees of the X_i are bounded. Then for all sufficiently large $d_1, \dots, d_n$ and all nonzero $\gamma \in \Gamma\left(\prod \mathscr{X}_i, \mathscr{O}\left(\sum d_i \,\mathrm{pr}_i^* L\right)\right)$,*

$$\prod_{v \mid \infty} \|\gamma\|_{\sup,v} \geq \exp\left(-c \sum d_i h(X_i) - c' \sum d_i\right)$$

for some constants c and c′ independent of γ, $X_1, \ldots, X_n$, and $d_1, \ldots, d_n$.

Proof. As in (18.5) (below), there exist projections $\pi_i\colon \mathscr{X}_i \to \mathbb{P}^{m_i}_{\operatorname{Spec} R}$ ($m_i = \dim X_i$) of degree N_i such that the norm of γ is an integral section γ' of $\mathscr{O}(e_1, \ldots, e_n)$ with

$$e_i = d_i \prod_{\substack{j=1 \\ j \neq i}}^{n} N_j, \tag{13.10}$$

with norms at infinity derived from the Fubini-Study metric, and such that for all archimedean places v of k and all $P \in (\mathbb{P}^{m_1} \times \cdots \times \mathbb{P}^{m_n})(\bar{k}_v)$ over which $\pi_1 \times \cdots \times \pi_n$ is finite,

$$\prod_{Q \in (\pi_1 \times \cdots \times \pi_n)^{-1}(P)} \|\gamma(Q)\| \geq \|\gamma'(P)\| \exp\Big(c \sum d_i h(X_i) + c'' \sum d_i\Big). \tag{13.11}$$

The set of points in $\mathbb{P}^m$ whose homogeneous coordinates are all roots of unity is dense in the Zariski topology. Therefore in (13.11) we take $P = (P_1, \ldots, P_n)$ such that each P_i is of this form, and let

$$E \subseteq \mathbb{P}^{m_1}_{\operatorname{Spec} R} \times_{\operatorname{Spec} R} \cdots \times_{\operatorname{Spec} R} \mathbb{P}^{m_n}_{\operatorname{Spec} R}$$

denote the arithmetic curve corresponding to P. Then

$$\begin{aligned}
\frac{1}{[k:\mathbb{Q}]} \sum_{v \mid \infty} -\log \|\gamma\|_{\sup,v} &\leq \frac{1}{[k(P):\mathbb{Q}]} \sum_{w \mid \infty} -\log \|\gamma'(P)\|_v + c \sum d_i h(X_i) + c'' \sum d_i \\
&\leq \frac{1}{[k(P):\mathbb{Q}]} \sum_{w \mid \infty} \deg \mathscr{O}(e_1, \ldots, e_n)\big|_E \\
&\qquad + c \sum d_i h(X_i) + c'' \sum d_i \\
&\leq \sum_{i=1}^{n} e_i (h(P_i) + c''') + c \sum d_i h(X_i) + c'' \sum d_i \\
&\leq c \sum d_i h(X_i) + c' \sum d_i
\end{aligned}$$

This last step follows from (13.10), together with the fact that (naïve) heights of all the P_i vanish. Also, the places w range over archimedean places of $k(P)$. □

We also note that in the application of this lemma, the full version of (18.5) can be used (with the Poincaré divisors), and then (9.3) can be used in place of (9.4) for the Faltings complex.

§14. Construction of a global section

These past three sections now provide all the tools needed to construct a small global section.

Proposition 14.1. *Let (s) and $X_1, \ldots, X_n$ satisfy the conditions (10.5) and let ϵ be as in Proposition 11.5. Let $\mathscr{W}$ denote the closure of $X_1 \times \cdots \times X_n$ in $\mathscr{W}_0$. Then for all sufficiently large $d \in \mathbb{N}$ (depending on s), there exists an integral section*

$$\gamma \in \Gamma(\mathscr{W}, dL_{-\epsilon,s})$$

such that

$$\prod_{v|\infty} \|\gamma\|_{\sup,v} \leq \exp\left(cd \sum s_i^2\right). \tag{14.2}$$

Proof. Let $\beta: \Gamma_1 \to \Gamma_2$ be the last arrow in the Faltings complex (9.4). We have now extended the picture to schemes over $\operatorname{Spec} R$, so that for example

$$\Gamma_1 = \Gamma\left(\mathscr{W},\ d(2n-2+\epsilon)\sum s_i^2 \operatorname{pr}_i^* L\right)^a.$$

This is metrized by taking the largest of the sup norms of its components. Also let

$$\delta_0 = \operatorname{rank}(\operatorname{Ker}\beta);$$
$$\delta_1 = \operatorname{rank}(\Gamma_1);$$
$$\delta_2 = \operatorname{rank}(\operatorname{Im}\beta).$$

The proof will follow by applying Lemma 12.1 to β.

First, we will need to replace $\prod \mathscr{X}_i$ with $\mathscr{W}$ in Lemmas 13.8 and 13.9. This is easy to do, since for any fixed divisor class F on $\prod \mathscr{X}_i$, there is an injection of $\Gamma(\prod \mathscr{X}_i, dF)$ into $\Gamma(\mathscr{W}, dF)$ whose cokernel is annihilated by an integer independent of d. Then Lemma 13.8 implies that

$$V(\Gamma_1) \geq \exp\left(-\delta_1 cd \sum s_i^2\right). \tag{14.3}$$

By Lemma 13.9 and (10.5.4) (and 10.9), for all $\gamma \notin \operatorname{Ker}\beta$,

$$\prod_{v|\infty} \|\beta(\gamma)\| \geq \exp\left(-cd \sum s_i^2\right)$$

and therefore

$$V(\operatorname{Im}\beta) \leq \exp\left(cd\delta_2 \sum s_i^2\right).$$

Combining this with (14.3) and Lemma 12.1b gives the bound

$$V(\operatorname{Ker}\beta) \geq \exp\left(-cd\delta_1 \sum s_i^2\right).$$

Clearly $\delta_1 \leq cd^{\sum \dim X_i} \prod s_i^{2 \dim X_i}$; thus, by Proposition 11.5, δ_1/δ_0 is bounded; hence Lemma 12.1a gives a section $\gamma \in \operatorname{Ker}\beta$ with

$$\prod_{v|\infty} \|\gamma\|_v \leq \exp\left(cd. \sum s_i^2\right). \tag{14.4}$$

Here we used Lemma 13.2a to give (12.2). By Lemma 13.2b, then, the bound (14.4) holds also for the (sup) norms of γ as an element of $\Gamma(\mathscr{W}, dL_{-\epsilon,s})$. □

§15. Some analysis

This section contains some analytic results which will be needed for working with partial derivatives in the next section. Because of technical difficulties, the analytic approach of Lemma 6.2 is necessary at *all* places of k.

For archimedean places v, let $\mathbb{C}_v = \bar{k}_v$; after taking the algebraic closure the field is still complete. At non-archimedean places, this is no longer the case: only after taking the completion $\mathbb{C}_v$ of $\bar{k}_v$ is the resulting field both complete and algebraically closed. On such fields, functions defined by power series often behave similarly to functions on $\mathbb{C}^m$. There are also differences, though: for example $\mathbb{C}_v$ is totally disconnected, so it is often necessary to specify that a given function is defined by the *same* power series on all of the domain. This set of ideas is often referred to as rigid analysis. For more details on p-adic analysis, see [Ko]; for rigid analysis, see [T] or, for another perspective, [Be].

In the non-archimedean case, the absolute value $\|\cdot\|_v$ extends to an absolute value on $\mathbb{C}_v$; when discussing analysis on $\mathbb{C}_v$, we often write the absolute value as $|\cdot|$, omitting the subscript v. In the archimedean case, let $|\cdot|$ be the usual absolute value. For vectors $\mathbf{x} = (x_1, \dots, x_m) \in \mathbb{C}_v^m$ let $|\mathbf{x}|^2 = |x_1|^2 + \cdots + |x_m|^2$ if v is archimedean and $|\mathbf{x}| = \max(|x_1|, \dots, |x_m|)$ if it is non-archimedean. Let the open and closed discs $\mathbb{D}_\rho$ and $\overline{\mathbb{D}}_\rho$ be the sets $\{z \in \mathbb{C}_v^m \mid |z| < \rho\}$ and $\{z \in \mathbb{C}_v^m \mid |z| \le \rho\}$, respectively. This notation is inherited from the archimedean case, and is slightly misleading because in the non-archimedean case both sets are simultaneously open and closed. Also note that in the non-archimedean case, there is no difference between polydiscs and balls.

For non-archimedean v, Cauchy's inequalities still hold:

Lemma 15.1. *Let v be a non-archimedean place and assume that the multivariable power series*

$$f(z) = \sum_{(i)\ge 0} a_{(i)} z^{(i)}$$

converges (absolutely) in an open disc $\mathbb{D}_\rho$ of radius $\rho > 0$. Then for all tuples $(\ell) \ge 0$,

$$|a_{(\ell)}|\rho^{\ell_1+\cdots+\ell_m} \le \sup_{z\in\mathbb{D}_\rho} |f(z)|.$$

Proof. As in the archimedean case, it will suffice to prove the inequalities in the one-variable case. Also, it will suffice to show that $|a_\ell|(\rho')^\ell \le \sup_{z\in\overline{\mathbb{D}}_{\rho'}} |f(z)|$ for all $\rho' < \rho$; the lemma then follows by taking the limit. By dilation, we may assume $\rho' = 1$; then the power series converges at $z = 1$, so $|a_i| \to 0$. Multiplying by a constant, we assume $\max_i |a_i| = 1$; it will then suffice to show that $|f(z)| = 1$ for some z in the closed unit disc. But by the theory of the Newton polygon ([Ko], Ch. IV, §3), together with the Weierstrass Preparation Theorem, this holds for any z in the boundary of the disc which is not close to any of the roots of the Weierstrass polynomial corresponding to f. □

We will also use Newton's method and its p-adic corollary, Hensel's lemma. This will follow Lang ([L 4], Ch. 2, §2, Prop. 2), but will be done in the context of several variables. To begin, let K be a field with non-archimedean valuation $|\cdot|$, and let $A := \{x \in K \mid |x| \le 1\}$ be its valuation ring. Note that the valuation does not need to be discrete. For vectors $\mathbf{x} \in K^m$ let $|\mathbf{x}|$ be defined as before, and for $m \times m$ matrices J let

$$|J| = \min_{\mathbf{x} \neq 0} \frac{|J\mathbf{x}|}{|\mathbf{x}|}.$$

Note that if the entries of J lie in A, then $|J| \geq |\det J|$, so that the Hensel's lemma given here strengthens the one given in ([G], Prop. 5.20). Also, if $|J| \neq 0$ then J is nonsingular and

$$|J^{-1}\mathbf{x}| \leq \frac{|\mathbf{x}|}{|J|} \quad \text{for all } \mathbf{x} \in K^m. \tag{15.2}$$

Then the following is a multivariable Hensel's lemma.

Lemma 15.3. *Let $\mathbf{f} = (f_1, \ldots, f_m)$ be a vector of polynomials in $A[X_1, \ldots, X_m]^m$, and let*

$$J_{\mathbf{f}} := \left(\frac{\partial f_i}{\partial X_j} \right)_{i,j}$$

denote its jacobian matrix. Suppose $\alpha_0 \in A^m$ is such that

$$|\mathbf{f}(\alpha_0)| < |J_{\mathbf{f}}(\alpha_0)|^2. \tag{15.4}$$

Then there is a unique root α of $\mathbf{f}(X)$ in A^m with

$$|\alpha - \alpha_0| \leq \frac{|\mathbf{f}(\alpha_0)|}{|J_{\mathbf{f}}(\alpha_0)|} < 1, \tag{15.5}$$

and the sequence

$$\alpha_{i+1} = \alpha_i - J_{\mathbf{f}}(\alpha_i)^{-1} \cdot \mathbf{f}(\alpha_i) \tag{15.6}$$

converges to it.

Proof. First, consider α satisfying (15.5). Clearly $\alpha \in A^m$. Also, we show that $|J_{\mathbf{f}}(\alpha)| = |J_{\mathbf{f}}(\alpha_0)|$. Indeed, consider the Taylor expansion of $J_{\mathbf{f}}$: for $\mathbf{x} \in K^m$ with $|\mathbf{x}| = 1$,

$$J_{\mathbf{f}}(\alpha) \cdot \mathbf{x} = J_{\mathbf{f}}(\alpha_0) \cdot \mathbf{x} + \beta \cdot (\alpha - \alpha_0)$$

for some matrix β (depending on $\mathbf{x}$) with entries in A. Then by (15.5) and (15.4)

$$|\beta \cdot (\alpha - \alpha_0)| < |J_{\mathbf{f}}(\alpha_0)|.$$

In any case this implies that $|J_{\mathbf{f}}(\alpha)| \leq |J_{\mathbf{f}}(\alpha_0)|$; taking $\mathbf{x}$ so that $|J_{\mathbf{f}}(\alpha_0) \cdot \mathbf{x}|$ is minimal, we find the opposite inequality.

To prove uniqueness, assume α and α' are two roots satisfying (15.5). Then the Taylor expansion for $\mathbf{f}$ about α gives

$$\mathbf{f}(\alpha') = \mathbf{f}(\alpha) - J_{\mathbf{f}}(\alpha) \cdot (\alpha' - \alpha) + {}^t(\alpha' - \alpha) \cdot \beta \cdot (\alpha' - \alpha)$$

for some m-tuple β of matrices with entries in A. But $\mathbf{f}(\alpha') = \mathbf{f}(\alpha) = 0$, the second term on the right has absolute value at least $|J_{\mathbf{f}}(\alpha_0)||\alpha - \alpha'|$ by the result just proved, and the last term has absolute value at most $|\alpha - \alpha'|^2$. This contradicts (15.5) and (15.4) unless $\alpha = \alpha'$.

Finally, let $c = |\mathbf{f}(\alpha_0)|/|J_\mathbf{f}(\alpha_0)|^2 < 1$. We show inductively that

(i). α_i satisfies (15.5) and

(ii). $\dfrac{|\mathbf{f}(\alpha_i)|}{|J_\mathbf{f}(\alpha_i)|^2} \le c^{2^i}$.

These two conditions obviously imply the lemma: by (ii) the α_i converge to a root, and by (i) this root satisfies (15.5).

The conditions with $i = 0$ are obvious; then assume them for i. To show (i),

$$|\alpha_{i+1} - \alpha_i| \le c^{2^i}|J_\mathbf{f}(\alpha_i)| \le c|J_\mathbf{f}(\alpha_0)| = \frac{|\mathbf{f}(\alpha_0)|}{|J_\mathbf{f}(\alpha_0)|}.$$

Thus (15.5) holds.

The proof of (ii) again uses the Taylor expansion for $\mathbf{f}$:

$$\mathbf{f}(\alpha_{i+1}) = \mathbf{f}(\alpha_i) - J_\mathbf{f}(\alpha_i)\cdot J_\mathbf{f}(\alpha_i)^{-1}\cdot\mathbf{f}(\alpha_i) + {}^t(J_\mathbf{f}(\alpha_i)^{-1}\cdot\mathbf{f}(\alpha_i))\cdot\beta\cdot(J_\mathbf{f}(\alpha_i)^{-1}\cdot\mathbf{f}(\alpha_i))$$

for some m-tuple β of matrices with entries in A. The first two terms on the right cancel; by (15.2) this gives

(15.7) $$|\mathbf{f}(\alpha_{i+1})| \le \left(\frac{|\mathbf{f}(\alpha_i)|}{|J_\mathbf{f}(\alpha_i)|}\right)^2.$$

This gives (ii). □

The archimedean analogue of this lemma is slightly more complicated.

Lemma 15.8. *Let $\mathbf{f} = (f_1, \dots, f_m)$ be a vector of C^2 functions from $\mathbb{C}^m$ to $\mathbb{C}$ and let $J_\mathbf{f}$ be the Jacobian determinant, as in Lemma 15.3. Suppose $\alpha_0 \in \mathbb{C}^m$ and $B \in \mathbb{R}_{>0}$ satisfy*

(15.9) $$|\mathbf{f}(\alpha_0)| < \frac{|J_\mathbf{f}(\alpha_0)|^2}{2B};$$

and

(15.10) $$B > \sup\left|\sum_{i=1}^m\sum_{j=1}^m \frac{\partial^2\mathbf{f}(\alpha)}{\partial z_i\partial z_j}v_iw_j\right|$$

as α ranges over the ball $B_\rho(\alpha_0)$ of radius ρ centered at α_0, and $\mathbf{v}$ and $\mathbf{w}$ vary over the unit ball in $\mathbb{C}^m$; here

$$\rho = \frac{|\mathbf{f}(\alpha_0)|}{|J_\mathbf{f}(\alpha_0)|}\sum_{i=0}^\infty (1+c)^i\left(\frac{c}{2(1-c)^2}\right)^{2^i-1},$$

and

$$c = \frac{B|\mathbf{f}(\alpha_0)|}{|J_\mathbf{f}(\alpha_0)|^2}.$$

Then the sequence (15.6) converges to a root α of $\mathbf{f}(X)$ with

(15.11) $$|\alpha - \alpha_0| \le \frac{|\mathbf{f}(\alpha_0)|}{|J_\mathbf{f}(\alpha_0)|}\sum_{i=0}^\infty (1+c)^i\left(\frac{c}{2(1-c)^2}\right)^{2^i-1}$$

Proof. First we note that if α and α' both lie in $B_\rho(\alpha_0)$, then

$$|J_{\mathbf{f}}(\alpha)| - B|\alpha' - \alpha| \le |J_{\mathbf{f}}(\alpha')| \le |J_{\mathbf{f}}(\alpha)| + B|\alpha' - \alpha|. \tag{15.12}$$

To prove this, find some $\mathbf{x}$ with $|\mathbf{x}| = 1$ and $|J_{\mathbf{f}}(\alpha)\mathbf{x}| = |J_{\mathbf{f}}(\alpha)|$; then by Taylor's formula applied to $J_{\mathbf{f}} \cdot \mathbf{x}$ and by (15.10),

$$\begin{aligned} |J_{\mathbf{f}}(\alpha')| &\le |J_{\mathbf{f}}(\alpha')\mathbf{x}| \\ &\le |J_{\mathbf{f}}(\alpha)\mathbf{x}| + B|\alpha' - \alpha| \\ &\le |J_{\mathbf{f}}(\alpha)| + B|\alpha' - \alpha|. \end{aligned}$$

This proves the second half of the inequality; the first half follows by symmetry.

Next, as in the non-archimedean case, we prove inductively that

(i). $|\alpha_i - \alpha_0| \le \dfrac{|\mathbf{f}(\alpha_0)|}{|J_{\mathbf{f}}(\alpha_0)|} \displaystyle\sum_{j=1}^{i-1} (1+c)^j \left(\frac{c}{2(1-c)^2}\right)^{2^j - 1}$;

(ii). $|J_{\mathbf{f}}(\alpha_i)| \le (1+c)^i |J_{\mathbf{f}}(\alpha_0)|$; and

(iii). $\dfrac{B|\mathbf{f}(\alpha_i)|}{|J_{\mathbf{f}}(\alpha_i)|^2} \le c\left(\dfrac{c}{2(1-c)^2}\right)^{2^i - 1}$.

By (15.9), $c < 1/2$; thus $\frac{c}{2(1-c)^2} < 1$. Then (ii) and (iii) imply that the α_i converge to a root, and (i) implies (15.11).

To carry out the induction, we first observe that (ii) and (iii) for α_i imply (i) for α_{i+1}:

$$\begin{aligned} |\alpha_{i+1} - \alpha_i| &\le \frac{|\mathbf{f}(\alpha_i)|}{|J_{\mathbf{f}}(\alpha_i)|} \\ &\le \frac{c|J_{\mathbf{f}}(\alpha_0)|}{B}(1+c)^i \left(\frac{c}{2(1-c)^2}\right)^{2^i - 1} \\ &\le \frac{|\mathbf{f}(\alpha_0)|}{|J_{\mathbf{f}}(\alpha_0)|}(1+c)^i \left(\frac{c}{2(1-c)^2}\right)^{2^i - 1}. \end{aligned}$$

Next, by the second half of (15.12) and by (iii) for α_i,

$$\begin{aligned} |J_{\mathbf{f}}(\alpha_{i+1})| &\le |J_{\mathbf{f}}(\alpha_i)| + \frac{B|\mathbf{f}(\alpha_i)|}{|J_{\mathbf{f}}(\alpha_i)|} \\ &\le |J_{\mathbf{f}}(\alpha_i)|(1+c). \end{aligned}$$

This gives (ii).

To prove (iii), we first note that, as before,

$$|J_{\mathbf{f}}(\alpha_{i+1})| \ge |J_{\mathbf{f}}(\alpha_i)|(1-c).$$

As in the non-archimedean case, we obtain in place of (15.7) the formula

$$|\mathbf{f}(\alpha_{i+1})| \le \frac{B}{2}\left(\frac{|\mathbf{f}(\alpha_i)|}{|J_{\mathbf{f}}(\alpha_i)|}\right)^2.$$

Combining these two formulas gives

$$\begin{aligned}\frac{B|\mathbf{f}(\alpha_{i+1})|}{|J_{\mathbf{f}}(\alpha_{i+1})|^2} &\le \frac{B^2|\mathbf{f}(\alpha_i)|^2}{2(1-c)^2|J_{\mathbf{f}}(\alpha_i)|^4}\\ &\le \frac{c^2}{2(1-c)^2}\left(\frac{c}{2(1-c)^2}\right)^{2^{i+1}-2}.\end{aligned}$$

This implies (iii). □

Remark. In the archimedean case the uniqueness statement is not as clean as before: let $\rho' = |J_{\mathbf{f}}(\alpha_0)|/2B$; then if (15.10) holds for $\alpha \in B_{\rho'}(\alpha_0)$, then this ball contains at most one zero of $\mathbf{f}$. Note also that if c is sufficiently small, then $\rho' \ge \rho$. However, neither uniqueness statement will be needed here.

For this paper, Hensel's lemma will be applied to a finite morphism $q\colon X \to \mathbb{C}_v^m$, where X is an analytic subvariety of $\mathbb{C}_v^M$. Let P be a point where q is étale; then there exists a regular sequence

$$f_1,\dots,f_{M-m} \in A[X_1,\dots,X_M].$$

for some neighborhood of P in X, as a subset of $\mathbb{C}_v^M$. Then let $g_i = f_i$ for $i = 1,\dots,M-m$, and let $g_{M-m+1},\dots,g_M$ be the coordinates of q. Hensel's lemma gives a (pointwise) inverse function of q; moreover, the inverse function theorem implies that this function is locally given by power series in the variables. But what we need is a lower bound on the radius of a polydisc on which this power series both converges and gives a local inverse for q. This can be easily shown by applying Hensel's lemma to a different field, as follows.

Corollary 15.13. *Let $J_{\mathbf{g}}$ be the matrix $(\partial g_i/\partial x_j)_{i,j}$, and assume that $J_{\mathbf{g}}$ is nonsingular at a point $P \in \mathbb{C}_v^m$.*

(a). *Assume $P \in A^m$. Then the inverse of q is given by a single power series in the polydisc of radius $|J_{\mathbf{g}}(P)|^2$ about $q(P)$.*

(b). *Let ρ and B be as in Lemma 15.8; then the inverse of q is given by a single power series in the ball of radius ρ about $q(P)$.*

Proof. The power series is obtained by applying Hensel's lemma to the field

$$\mathbb{C}_v((X_1,\dots,X_M)),$$

using the index as a valuation (the weights of the variables may be chosen arbitrarily). It still must be shown that the power series converges in the indicated polydisc, and that evaluating it gives the function obtained earlier. This follows by comparing the α_i in this application of Hensel's lemma with the α_i obtained when applying Hensel's lemma pointwise. The two sets of convergents coincide, in the sense that evaluating the power series α_i at a given point gives the same value as the corresponding convergent in the method given above. Moreover, on any polydisc of strictly smaller radius, the bounds on $|\alpha_{i+1}-\alpha_i|$ are uniform; thus the power series converges on the given polydisc. □

§16. More derivatives

The higher dimensional generalization due to Faltings also uses partial derivatives on $X_1 \times \cdots \times X_n$, but now each X_i may be of higher dimension and may have singularities. The p-adic analysis of the preceding section is necessary in order to handle the singularities. But also, the varieties X_i may vary, so the theory of Chow coordinates is needed. Therefore let C be the Chow variety of subvarieties of $\mathbb{P}^m$ of a given degree, and let Γ be the universal variety sitting over C. For this purpose we use the chosen embedding of X (over k) into $\mathbb{P}^m$ given by the very ample divisor L. At finitely many places this fails to extend to an embedding over $\operatorname{Spec} R$, but this will be addressed later.

We assume that the varieties C are projective. Points in C not corresponding to an integral subvariety of $\mathbb{P}^m$ form an algebraic subset; we may assume that none of the X_i corresponds to such a point. If not, C can be replaced by a Chow variety of strictly smaller degree. Also, note that since the degrees of the X_i are bounded, finitely many such C will suffice. On each such C, the height of the point corresponding to X_i is related to the height $h(X_i)$ defined in Section 10; for details, see ([So 1], Théorème 3).

Proposition 16.1. *Let $\mathscr{L} = \mathscr{O}(dL_{-\epsilon,s})$, and let γ be a global section of $\mathscr{L}$ on the model $\mathscr{W}$ for $X_1 \times \cdots \times X_n$ obtained by taking the closure in the model $\mathscr{W}_0$ for A^n. Let $P_0 = (P_1, \ldots, P_n)$, $E = E_1 \times_B \cdots \times_B E_n$, and $d_1, \ldots, d_n$ be as in Corollary 6.3. Then there exist constants c_1, c_2, c_3, c_4, c_5, and c_6, depending only on $\mathscr{W}_0$, X, and $\deg X_1, \ldots, \deg X_n$, and subsets $Z_i \subsetneq X_i$ satisfying $\deg Z_i \le c_1$ and $h(Z_i) \le c_2 h(X_i)$ for all i, and such that either*

(a). *$P_i \in Z_i$ for some i, or*

(b). *the index $t = t(\gamma, P_0, d_1, \ldots, d_n)$ satisfies the inequality*

$$t \ge \frac{-\deg \mathscr{L}\big|_E - \sum_{v|\infty} \log \|\gamma\|_{\sup,v} - c_3 \sum d_i - c_4 \sum d_i h(X_i)}{c_5 \max_{1 \le i \le n} d_i h_L(P_i) + c_6 \sum d_i}.$$

Proof. For each $i = 1, \ldots, n$, recall that X_i is given with an embedding into $\mathbb{P}^m$. Let $m_i = \dim X_i$ and choose one of the standard projections from $\mathbb{P}^m$ to $\mathbb{P}^{m_i}$ such that the projection is well defined at P_i. Blow up $\mathbb{P}^m$ so that the projection becomes a morphism everywhere on the blowing-up. Let Z_i be the ramification locus of this projection. Clearly the degree and height of Z_i satisfy the required conditions. From now on, therefore, we assume that $P_i \notin Z_i$, and concentrate on part (b) of the lemma. Also, by permuting coordinates, we may assume that P_i does not lie over the hyperplane at infinity on $\mathbb{P}^{m_i}$. Thus we have projections $q_i \colon X_i \to \mathbb{A}^{m_i}$.

We now extend this picture, both over the Chow variety and over $\operatorname{Spec} R$. Let C_i denote the Chow variety appropriate for X_i, and restrict to the subset corresponding to subvarieties of X. Let $\mathscr{C}_i$ be defined similarly, as a Chow scheme over $\operatorname{Spec} R$. Let $\Gamma_i \subseteq \mathscr{C}_i \times_{\operatorname{Spec} R} \mathscr{X}$ be the universal family; we blow up as before to make Γ_i project to $\mathbb{P}^{m_i}$, and extend the model so that $\mathscr{L}$ extends to a line sheaf on Γ_i; this can be done independently of s, etc. Now extend C and $\mathscr{C}_i$ to make them proper over $\operatorname{Spec} k$ and $\operatorname{Spec} R$, respectively. Also, we assume Γ_i is projective over $\mathscr{C}_i$: $\Gamma_i \subseteq \mathbb{P}^M_{\mathscr{C}_i}$ for some M. And finally let the **bad subset** of C_i be the set whose corresponding subvariety is either not integral or does not map surjectively to $\mathbb{P}^{m_i}$.

Note that, so far, the choices do not depend on the place v. For each place v, q_i extends to a v-adic analytic morphism from a subset of $\mathscr{X}_i(\mathbb{C}_v)$ to $\mathbb{C}_v^{m_i}$. This is locally (by assumption on P_i) biholomorphic on some open polydisc of a radius $\rho_{i,v} > 0$. The next step consists of controlling this radius.

By a slight change of model, we regard $\mathscr{X}_i$ as a subset of Γ_i, and let E_i be the arithmetic curve in $\mathscr{X}_i$ corresponding to P_i. This does not affect the index, which is defined on the generic fiber.

Lemma 16.2. *For each i there exists an arithmetic divisor D_i on Γ_i, depending only on $\mathscr{C}_i$, the choice of projections to $\mathbb{P}^{m_i}$, and the permutation of coordinates on $\mathbb{P}^{m_i}$, such that for all places v of k, the map q_i gives a biholomorphic map from a neighborhood $U_{i,v}$ of $P_i(\mathbb{C}_v)$ (in $\mathscr{X}_i(\mathbb{C}_v)$) to the polydisc of radius $\exp(-(D_i \,.\, E_i)_v)$ and center $q_i(P_i)$ in $\mathbb{C}_v^{m_i}$. Moreover, E_i is not contained in the support of D_i. And finally, at archimedean places we can place an upper bound on the radius in $\mathbb{A}^M$ of the sets $U_{i,v}$.*

Proof. First consider non-archimedean places v. Since $P_i \notin Z_i$, q_i is étale at P_i, and therefore Γ_i is smooth over $\mathscr{C}_i$ at P_i (on the generic fiber over $\operatorname{Spec} k$). Recall that $E_{i,v}$ denotes the closed point in E_i lying over $v \in \operatorname{Spec} R$. By smoothness ([F-L], Ch. IV, Prop. 3.11), there exists a regular sequence $f_1, \dots, f_{M-m_i}$ for the ideal of some neighborhood of $E_{i,v}$ in Γ_i, as a subset of some affine space $\mathbb{A}^M_{\mathscr{C}_i}$. These are functions in local coordinates $x_1, \dots, x_M$ in $\mathbb{P}^M$, and in local coordinates on $\mathscr{C}_i$. Let $y_1, \dots, y_{m_i}$ denote the coordinates of q_i; then also $y_1 - y_1(P_i), \dots, y_{m_i} - y_{m_i}(P_i)$ form a regular sequence for P_i on X_i. Now specialize to $\mathscr{X}_i(\mathbb{C}_v)$, and note that these functions now describe $\mathscr{X}_i(\mathbb{C}_v)$ in the open unit disc about $P_i(\mathbb{C}_v)$. Since $f_1, \dots, f_{M-m_i}, y_1 - y_1(P_i), \dots, y_{m_i} - y_{m_i}(P_i)$ form a regular sequence, their Jacobian determinant is nonzero. Let the radius equal the square of the absolute value of this determinant. By Hensel's lemma (Corollary 15.13), the desired biholomorphic map exists.

As the points $E_{i,v}$ vary over all of a certain Zariski-open subset of Γ_i, only finitely many systems $f_1, \dots, f_{m_i}$ are needed (quasi-compactness of the Zariski topology), so there exists a divisor D_i on Γ_i (on the algebraic part; *i.e.*, not an arithmetic divisor yet) which dominates the squares of these determinants; moreover, it can be chosen so that its vertical (over C_i) components on the generic fiber correspond only to the bad set discussed earlier.

At archimedean places one can similarly construct a Green's function g_{D_i} for D_i such that the desired biholomorphic map exists, with a polydisc of radius $\exp(-g_{D_i})$. This is done using compactness of $\Gamma_i(\mathbb{C}_v)$. □

A second estimate is needed, but it is much easier. First, choose a hermitian metric on $\Omega_{\Gamma_i/\mathscr{C}_i}$.

Lemma 16.3. *For each i there exists an arithmetic divisor F_i on Γ_i, depending only on $\mathscr{C}_i$, the choice of projections to $\mathbb{P}^{m_i}$, and the permutation of coordinates on $\mathbb{P}^{m_i}$, such that for all places v of k, the metric of the element $q_i^* y_j - y_j(q_i(P_i))$ of the ideal sheaf of P_i in X_i satisfies*

$$-\log \|q_i^* y_j - y_j(q_i(P_i))\|_v \geq -(F_i \,.\, E_i)_v.$$

(At non-archimedean places, the above metric is given by the scheme structure. Also, we inject the ideal sheaf of P_i into $\Omega_{\Gamma_i/\mathscr{C}_i}$ via ([H 2], II 8.12).)

Proof. It will suffice to take a divisor F_i which dominates the torsion sheaf given by the relative differentials of q_i. Adding $q_i^* H_\infty$ (where H_∞ denotes the hyperplane at infinity in $\mathbb{P}^{m_i}$) corrects for the growth at infinity. □

For each place v, let U_v be the product of the neighborhoods $U_{i,v}$ from Lemma 16.2. Also let $\gamma_{0,v}$ be a local generator for $\mathscr{L}$ on U_v. This should be chosen uniformly in terms of generators on $\operatorname{pr}_i^* \mathscr{O}(L)$ and the Poincaré divisors $\mathscr{P}_{ij}$; then it will be possible to ensure that

$$\log \inf_{P \in U_v} \frac{\|\gamma_{0,v}(P)\|}{\|\gamma_{0,v}(P_0)\|_v} \geq -c_{3,v} \sum_{i=1}^n d_i. \tag{16.4}$$

At archimedean places this is possible since $\mathscr{L}$ is defined on X, and we can limit the radius of U_v, as noted following the proof of Lemma 16.2. At places where $\mathscr{A}$ has good reduction, this infimum is just 1. Other places should be treated as in the archimedean case; this is due to the change in model needed to define the Poincaré divisors. Also, let $\|\gamma\|_{\sup,v} = 1$ for non-archimedean v.

Now let $(\ell) = (\ell_{11}, \dots, \ell_{1m_1}, \dots, \ell_{nm_n})$ be a tuple such that

$$\sum_{i=1}^n \sum_{j=1}^{m_i} \frac{\ell_{ij}}{d_i} = t \tag{16.5}$$

and such that (letting y_{ij}, $j = 1, \dots, m_i$, denote the coordinates of $\mathbb{C}_v^{m_i}$)

$$D_{(\ell)}\gamma := \frac{1}{\ell_{11}! \cdots \ell_{nm_n}!} \left(\frac{\partial}{\partial y_{11}}\right)^{\ell_{11}} \cdots \left(\frac{\partial}{\partial y_{nm_n}}\right)^{\ell_{nm_n}} \gamma \neq 0,$$

where we define this partial derivative via the maps q_i. Then, as in the proof of Lemma 6.2, we find that the norm of the partial derivative satisfies

(16.6)

$$\begin{aligned} - \log \|D_{(\ell)}\gamma(P_0)\|_v & \\ \geq -\log \|\gamma\|_{\sup,v} &+ \sum_{i=1}^n \sum_{j=1}^{m_i} \ell_{ij} (\log \rho_{i,v} - (F_i \,.\, E_i)_v) + \log \inf_{P \in U_v} \frac{\|\gamma_{0,v}(P)\|}{\|\gamma_{0,v}(P_0)\|_v} \\ \geq -\log \|\gamma\|_{\sup,v} &+ \sum_{i=1}^n \sum_{j=1}^{m_i} \ell_{ij} (\log \rho_{i,v} - (F_i \,.\, E_i)_v) - c_{3,v} \sum_{i=1}^n d_i, \end{aligned}$$

by (16.4). Here $D_{(\ell)}\gamma(P_0)$ is regarded as a section of the vector sheaf

$$\left(\mathscr{L} \otimes S^{\ell_{11}+\dots+\ell_{1m_1}} \Omega_{\mathscr{X}_1/\operatorname{Spec} R} \otimes \dots \otimes S^{\ell_{n1}+\dots+\ell_{nm_n}} \Omega_{\mathscr{X}_n/\operatorname{Spec} R}\right)\Big|_E,$$

where E as usual is the arithmetic curve on $\Gamma_1 \times_{\operatorname{Spec} R} \cdots \times_{\operatorname{Spec} R} \Gamma_n$ corresponding to $P_0 = (P_1, \ldots, P_n)$. See also ([Laf], §4). Thus

$$\sum_v -\log \|D_{(\ell)}\gamma(P_0)\|_v \leq \deg \mathscr{L}|_E + \sum_{i=1}^n \sum_{j=1}^{m_i} \ell_{ij} \deg \mathscr{M}_i|_{E_i},$$

for some sufficiently large line sheaves $\mathscr{M}_i$ on Γ_i. Combining this with (16.6) then gives

$$(16.7)\quad \sum_{i=1}^n \sum_{j=1}^{m_i} \ell_{ij} \left(\deg \mathscr{M}_i|_{E_i} - \sum_v \log \rho_{i,v} + (F_i \,.\, E_i) \right) \geq -\deg \mathscr{L}|_E - \sum_{v|\infty} \log \|\gamma\|_{\sup,v} - c_3 \sum_{i=1}^n d_i.$$

But the quantity inside the parentheses on the left is an intersection number on Γ_i. From the structure of Γ_i as a subset of a product, there exist divisors G_i on $\mathbb{P}^M$ and H_i on $\mathscr{C}_i$ such that

$$\mathscr{M}_i \otimes \mathscr{O}(F_i + D_i) \leq \mathscr{O}(G_i + H_i)$$

relative to the cone of very ample line sheaves on the generic fiber of Γ_i over $\operatorname{Spec} k$. Thus

$$\deg \mathscr{M}_i|_{E_i} + (F_i \,.\, E_i) + (D_i \,.\, E_i) \leq (G_i \,.\, E_i) + (H_i \,.\, E_i) + O(1).$$

Also we have

$$(G_i \,.\, E_i) \leq c_5 h_L(P_i) + O(1)$$

and

$$(H_i \,.\, E_i) \leq c_4 h(X_i) + O(1).$$

Then by Lemma 16.2, (16.7) becomes

$$\sum_{i=1}^n \sum_{j=1}^{m_i} \ell_{ij} (c_5 h_L(P_i) + c_4 h(X_i) + c_6) \geq -\deg \mathscr{L}|_E - \sum_{v|\infty} \log \|\gamma\|_{\sup,v} - c_3 \sum_{i=1}^n d_i.$$

But now, for all i, $\sum_{j=1}^{m_i} \ell_{ij} \ll d_i$; therefore modifying c_4 gives (using also (16.5))

$$c_4 \sum_{i=1}^n d_i h(X_i) + c_5 t \max_{1 \leq i \leq n} d_i h_L(P_i) + c_6 t \sum_{i=1}^n d_i \geq -\deg \mathscr{L}|_E - \sum_{v|\infty} \log \|\gamma\|_{\sup,v} - c_3 \sum_{i=1}^n d_i.$$

Solving for t then gives the proposition. □

§17. Lower bound for the index

As was the case in Section 7, the Mordell-Weil theorem implies that $A(k)$ is a finitely generated abelian group; hence $A(k) \otimes_{\mathbb{Z}} \mathbb{R}$ is a finite dimensional vector space. Letting L replace the theta divisor, it follows that

$$(P_1, P_2)_L := \hat{h}_L(P_1 + P_2) - \hat{h}_L(P_1) - \hat{h}_L(P_2)$$

defines a nondegenerate dot product structure on $A(k) \otimes_{\mathbb{Z}} \mathbb{R}$. Also let $|P|^2 = (P,P)_L$. We now assume that $P_1, \dots, P_n$ have been chosen such that

$$(17.1) \qquad \begin{aligned}(P_i,P_j)_L &\ge (1-\epsilon_1)\sqrt{|P_i|^2|P_j|^2} \\ &\ge 2(1-\epsilon_1)\sqrt{\hat h_L(P_i)\hat h_L(P_j)}\end{aligned}$$

for some given $\epsilon_1 > 0$ and for all $i < j$. From fundamental properties of the canonical height,

$$\begin{aligned}\deg\left((s_i^2\cdot \mathrm{pr}_i - s_j^2 \cdot \mathrm{pr}_j)^*L\right)\big|_E &= s_i^2\hat h_L(P_i) + s_j^2 \hat h_L(P_j) - s_is_j(P_i,P_j)_L + O(s_i^2+s_j^2) \\ &\le \left(s_i\sqrt{\hat h_L(P_i)} - s_j\sqrt{\hat h_L(P_j)}\right)^2 \\ &\qquad + 2\epsilon_1 s_is_j\sqrt{\hat h_L(P_i)\hat h_L(P_j)} + O(s_i^2+s_j^2).\end{aligned}$$

Letting s_i be rational and close to $1/\sqrt{\hat h_L(P_i)}$, the square in the above expression approaches zero and we obtain

$$(17.2) \qquad \deg L_{-\epsilon,s}\big|_E \le n(n-1)\epsilon_1 - n\epsilon + O\left(\sum s_i^2\right).$$

We now apply Proposition 16.1. By (14.2) and (10.5.4), the second and fourth terms in the numerator of the fraction in (16.1) are bounded by $c\sum d_i$; hence by (17.2), the index $t = t(\gamma, (P_1,\dots,P_n), d_1, \dots, d_n)$ satisfies

$$t \ge \frac{n(\epsilon - (n-1)\epsilon_1) - c\sum s_i^2}{c_5 + c_6 \sum s_i^2}.$$

If the heights $\hat h_L(P_i)$ are sufficiently large and ϵ_1 sufficiently small, which we now assume, then the s_i^2 will be small, and the above inequality becomes

$$(17.3) \qquad t \ge \epsilon_2$$

for some $\epsilon_2 > 0$ depending only on the usual list X, $\mathscr{A}$, ..., $(\dim X_1, \dots, \dim X_n)$.

§18. The product theorem

The last step of the proof consists of applying the product theorem, as was done in ([F 1], §6) or [F 2].

Theorem 18.1 ([F 1], §3). *Let $\Pi = \mathbb{P}^{m_1} \times \cdots \times \mathbb{P}^{m_n}$ be a product of projective spaces over a field of characteristic zero, and let $\epsilon_3 > 0$ be given. Then there exist numbers r', c_1, c_2, and c_3 with the following property. Suppose γ' is a nonzero global section of the sheaf $\mathscr{O}(e_1,\dots,e_n)$ on Π which has index $\ge \epsilon_3$ relative to $(e_1,\dots,e_n)$ at some point $(x_1,\dots,x_n)$. If $e_i/e_{i+1} \ge r'$ for all $i = 1,\dots,n-1$, then there exist subvarieties $Y_i \subseteq \mathbb{P}^{m_i}$, not all of which are equal to $\mathbb{P}^{m_i}$, such that*

(i). *each Y_i contains x_i;*
(ii). *the degrees of Y_i are bounded by c_1; and*

(iii). *the heights* $h(Y_i)$ *satisfy the inequality*

$$\sum e_i h(Y_i) \le c_2 \sum_{v|\infty} \log \|\gamma'\|_{\sup,v} + c_3 \sum_i e_i. \tag{18.2}$$

Proof. See ([F 1], Thm. 3.1 and Thm. 3.3). □

We note that this theorem generalizes Roth's lemma (4.5), although without the explicit constants. Indeed, let $m_1 = \cdots = m_n = 1$. Then at least one of the resulting Y_i must equal x_i; thus (18.2) contradicts (4.6).

This theorem implies a result which is more suitable for the problem at hand:

Corollary 18.3. *Let* X *be a projective variety defined over a number field* k. *Fix a projective embedding of* X, *and let* L *be an ample divisor on* X. *Let* $X_1, \ldots, X_n$ *be geometrically irreducible subvarieties of* X *defined over* k *whose degrees (relative to the chosen projective embedding of* X *) are bounded. Let* $\epsilon > 0$ *and* $\epsilon_2 > 0$ *be given. Then there exist numbers* r, c_1, c_2, c_3, *and* c_4 *with the following property. Let* γ *be a nonzero global section of* $\mathcal{O}(dL_{-\epsilon,s})$ *which has index* $\ge \epsilon_2$ *relative to* $(d_1, \ldots, d_n)$ $(d_i = ds_i^2)$ *at some point* $(P_1, \ldots, P_n)$ *with* $P_i \in X_i(k)$ *for all* i. *If* $d_i/d_{i+1} \ge r$ *for all* $i = 1, \ldots, n-1$, *then there exist subvarieties* $X_i' \subseteq X_i$, *not all of which are equal to* X_i, *such that*

(i). *each* X_i' *contains* P_i;
(ii). *each* X_i' *is geometrically irreducible and defined over* k;
(iii). *the degrees of* X_i' *are bounded by* c_1; *and*
(iv). *the heights* $h(X_i')$ *satisfy the inequality*

$$\sum d_i h(X_i') \le c_2 \sum_{v|\infty} \log \|\gamma\|_{\sup,v} + c_3 \sum_i d_i h(X_i) + c_4 \sum_i d_i. \tag{18.4}$$

Proof. Let $m_i = \dim X_i$. Let $\mathbb{P}^m$ be the projective space in which X is embedded, and for each i fix a standard projection from $\mathbb{P}^m$ to $\mathbb{P}^{m_i}$ whose restriction to X_i is a generically finite rational map. Let N_i be its degree. In fact, $N_i = \deg X_i$, so that N_i is bounded. We would like to take the norm from $\prod X_i$ to Π. Therefore, for each i let K_i^* be a finite extension of $K(X_i)$ which is normal over $K(\mathbb{P}^{m_i})$. Let X_i^* be a model for K_i^* such that the rational maps $X_i^* \to X_i$ corresponding to all injections $K(X_i) \hookrightarrow K_i^*$ over $K(\mathbb{P}^{m_i})$ are morphisms. The product of the pull-backs of $\mathcal{O}(L)$ via these morphisms is a line sheaf on X_i^* which is isomorphic to the pull-back of some multiple of $\mathcal{O}(1)$ from $\mathbb{P}^{m_i}$. Expanding this to the Chow family Γ_i over $\mathscr{C}_i/\operatorname{Spec} R$ as in Section 16, the isomorphism holds up to a divisor on $\mathscr{C}_i$, fibral components over $\operatorname{Spec} R$, and, correspondingly, a change of metric at archimedean places. Here we metrize $\mathcal{O}(1)$ via the Fubini-Study metric. Thus, the isomorphism holds up to denominators bounded by $\exp(ch(X_i) + c')$, and the metrics correspond up to a factor bounded by a similar bound.

For the Poincaré divisors $\mathscr{P}_{ij}$, similar bounds are not good enough. This is due to the fact that they occur with coefficients of size $s_i s_j$. However, a more refined argument gives the required bounds. Let C_i denote the generic fiber of $\mathscr{C}_i$. Then the norm of $\mathscr{P}_{ij}$ is a divisor class on $\mathbb{P}^{m_i}_{\mathscr{C}_i} \times_{\operatorname{Spec} R} \mathbb{P}^{m_j}_{\mathscr{C}_j}$, and over each point of $C_i \times C_j$

it is trivial. But for each point $Q \in \mathbb{P}^{m_i}_{C_i}$, the restriction of this norm to $\{Q\} \times \mathbb{P}^{m_j}_{C_j}$ is algebraically equivalent to zero, because it is obtained from the restriction of $\mathscr{P}_{ij}$ to sets of the form $\{Q'\} \times X$, for $Q' \in \Gamma_i$ lying over $Q \in X$. The same argument holds after interchanging the factors; thus the difference is a divisor of Poincaré type on $C_i \times C_j$. Therefore the bounds on the ratio of norms and on the size of denominators are of the form $\exp\left(c\sqrt{h(X_i)h(X_j)} + c'\right)$. This is indeed fortunate, because the divisor classes $\mathscr{P}_{ij}$ occur with multiplicities $s_i s_j$ in $L_{-\epsilon,s}$, and

$$s_i s_j \sqrt{h(X_i)h(X_j)} \le \frac{1}{2}\left(s_i^2 h(X_i) + s_j^2 h(X_j)\right),$$

which is exactly the sort of bound that is needed!

After cancelling denominators, then, the norm γ' of γ has metrics satisfying

$$(18.5) \qquad \prod_{v|\infty} \|\gamma'\|_{\sup,v} \le \prod_{v|\infty} \|\gamma\|_{\sup,v} \cdot \exp\left(c\sum d_i h(X_i) + c'\sum d_i\right).$$

We now calculate e_i. First note that the map $\prod X_i \to \Pi$ has degree $N := \prod N_i$; then $e_i = d_i \cdot N/N_i$. Since the N_i are bounded, the index of γ' at the point $(x_1, \ldots, x_n)$ lying below $(P_1, \ldots, P_n)$ is bounded from below, even after switching from weights $(d_1, \ldots, d_n)$ to $(e_1, \ldots, e_n)$. The boundedness of the N_i also implies that there exists some r such that $d_i/d_{i+1} \ge r$ implies $e_i/e_{i+1} \ge r'$.

Then applying Theorem 18.1 to γ' gives subvarieties $Y_i \subseteq \mathbb{P}^{m_i}$. By (18.2) and (18.5),

$$\begin{aligned}\sum d_i h(Y_i) &\le c\sum e_i h(Y_i)\\ &\le c\sum_{v|\infty} \log\|\gamma\|_{\sup,v} + c'\sum d_i h(X_i) + c''\sum d_i + c'''\sum e_i\\ &\le c\sum_{v|\infty} \log\|\gamma\|_{\sup,v} + c'\sum d_i h(X_i) + c''\sum d_i.\end{aligned}$$

Now pull back these Y_i to subsets X_i' of X_i. Then $h(X_i')$ is bounded in terms of $h(Y_i)$ (using the definition from Section 10); hence Condition (iv) holds. Condition (i) holds by construction, and (iii) is easy to check. Since not all of the Y_i are equal to $\mathbb{P}^{m_i}$, not all of the X_i' equal X_i.

It remains only to ensure that (ii) holds. But we may intersect X_i' with finitely many (at most $\dim X_i$) of its conjugates over k until the geometrically irreducible component containing P_i is defined over k. Replacing X_i' with this irreducible component gives (ii), and since the number of intersections is bounded, (iii) and (iv) still hold (after adjusting the constants). □

This corollary can be applied directly to the situation of Theorem 0.3. Indeed, either (16.1a) holds, which gives the inductive step rather directly, or (16.1b) holds, so that by (17.3) and Corollary 18.3, we obtain subvarieties X_i' of X_i satisfying (10.5.1)–(10.5.3), possibly with a different set of constants. Also, (10.5.4) holds for X_i', by (14.2), (10.5.4) (for X_i), and (18.4). Moreover, at least one X_i' has dimension strictly smaller than $\dim X_i$. This concludes the main part of the proof of Theorem 0.3.

Returning to the overall plan of Section 10, then, we now choose

$$P_1, \ldots, P_n \in X(k) \setminus Z(X)(k)$$

such that

(a). The height $h_L(P_1)$ is sufficiently large to contradict (10.7) and to ensure that (17.3) holds.

(b). For $i = 1, \ldots, n-1$, $h(P_{i+1})/h(P_i) > r''$, where r'' is the largest of the r occurring in all applications of the product theorem; we also assume that $r'' \geq 1$.

(c). Condition (17.1) holds for all possible $(\dim X_1, \ldots, \dim X_n)$.

In particular, r and ϵ_1, as well as the various constants c, depend on the tuple $(\dim X_1, \ldots, \dim X_n)$, but only finitely many such tuples occur. Then the induction may proceed as outlined in Section 10, leading to a contradiction.

This concludes the proof of Theorem 0.3.

We conclude with a few remarks on how this proof differs from the proof of Theorem 0.2. In that case it is possible to show that $L_{-\epsilon,s}$ is ample. Shrinking ϵ a little, it is possible to obtain an upper bound on the dimension of the space of sections of $\mathcal{O}(dL_{-\epsilon,s})$ which have index $\geq \sigma$ at the point $(P_1, \ldots, P_n)$, for some suitable $\sigma > 0$. This bound is bounded away from $h^0(X^n, dY_{-\epsilon,s})$, so the more precise form (2.4) of Siegel's lemma allows us to construct a global section γ with index $\leq \sigma$ at $(P_1, \ldots, P_n)$. Thus Step 5 is incorporated into Step 2. One then obtains a contradiction in Step 4, without needing the induction on the subvarieties X_i.

REFERENCES

[Ab] S. S. Abhyankar, *Resolution of singularities of arithmetical surfaces*, Arithmetical algebraic geometry (O. F. G. Schilling, ed.), Harper & Row, New York, 1965, pp. 111–152.

[Ar] M. Artin, *Lipman's proof of resolution of singularities for surfaces*, Arithmetic geometry (G. Cornell and J. H. Silverman, eds.), Springer-Verlag, New York, 1986, pp. 267–287.

[Be] V. Berkovich, *Spectral theory and analytic geometry over non-Archimedean fields*, AMS Surveys and Monographs 33, Amer. Math. Soc., Providence, R. I., 1990.

[B-G-S] J.-B. Bost, H. Gillet, and C. Soulé, *Un analogue arithmétique du théorème de Bezout*, C. R. Acad. Sci, Paris, Sér. I **312** (1991), 845–848.

[Bo] E. Bombieri, *The Mordell conjecture revisited*, Ann. Sc. Norm. Super. Pisa, Cl. Sci., IV **17** (1990), 615–640.

[B-V] E. Bombieri and J. Vaaler, *On Siegel's lemma*, Invent. Math. **73** (1983), 11–32; *addendum*, Invent. Math. **75** (1984), 377.

[D] F. J. Dyson, *The approximation to algebraic numbers by rationals*, Acta Math. **79** (1947), 225–240.

[F 1] G. Faltings, *Diophantine approximation on abelian varieties*, Ann. Math. **133** (1991), 549–576.

[F 2] ———, *The general case of S. Lang's conjecture* (to appear).

[F-L] W. Fulton and S. Lang, *Riemann-Roch algebra*, Grundlehren der mathematischen Wissenschaften 277, Springer-Verlag, New York, 1985.

[G] M. J. Greenberg, *Lectures on forms in many variables*, Mathematics lecture note series, W. A. Benjamin, Inc., New York, 1969.

[G-S 1] H. Gillet and C. Soulé, *Arithmetic intersection theory*, Publ. Math. IHES **72** (1990), 93–174.

[G-S 2] ———, *Characteristic classes for algebraic vector bundles with hermitian metric*. I, Ann. Math. **131** (1990), 163–203; II, Ann. Math. **131** (1990), 205–238.

[G-S 3] ———, *Analytic torsion and the arithmetic Todd genus*, Topology **30** (1991), 21–54.

[G-S 4] ———, *Un théorème de Riemann-Roch-Grothendieck arithmétique*, C. R. Acad. Sci. Paris, Sér. I **309** (1989), 929–932.

[H 1] R. Hartshorne, *Ample subvarieties of algebraic varieties*, Lecture Notes in Mathematics 156, Springer-Verlag, New York, 1970.

[H 2] ———, *Algebraic geometry*, Graduate Texts in Mathematics 52, Springer-Verlag, New York, 1977.

[Ka] Y. Kawamata, *On Bloch's conjecture*, Invent. Math. **57** (1980), 97–100.

[Ko] N. Koblitz, *p-adic numbers, p-adic analysis, and zeta-functions*, Graduate Texts in Mathematics 58, Springer-Verlag, New York, 1977.

[Laf] L. Lafforgue, *Une version en géométrie diophantienne du "lemma de l'indice"* (to appear).

[L 1] S. Lang, *Some theorems and conjectures in diophantine equations*, Bull. AMS **66** (1960), 240–249.

[L 2] ———, *Integral points on curves*, Publ. Math. IHES **6** (1960), 27–43.

[L 3] ———, *Introduction to diophantine approximations*, Addison-Wesley, Reading, Mass., 1966.

[L 4] ———, *Algebraic number theory*, Addison-Wesley, Reading, Mass., 1970; reprinted, Springer-Verlag, Berlin-Heidelberg-New York, 1986.

[L 5] ———, *Fundamentals of diophantine geometry*, Springer-Verlag, New York, 1983.

[L 6] ———, *Hyperbolic and diophantine analysis*, Bull. AMS **14** (1986), 159–205.

[L 7] ———, *Introduction to Arakelov theory*, Springer-Verlag, New York, 1988.

[LeV] W. J. LeVeque, *Topics in Number Theory, Vol. II*, Addison-Wesley, Reading, Mass., 1956.

[M] D. Mumford, *A remark on Mordell's conjecture*, Amer. J. Math. **87** (1965), 1007–1016.

[N] J. Noguchi, *A higher dimensional analogue of Mordell's conjecture over function fields*, Math. Ann. **258** (1981), 207–212.

[Ra] M. Raynaud, *Courbes sur une variété abélienne et points de torsion*, Invent. Math. **71** (1983), 207–223.

[Ro] K. F. Roth, *Rational approximations to algebraic numbers*, Mathematika **2** (1955), 1–20; *corrigendum*, Mathematika **2** (1955), 168.

[Sch] W. M. Schmidt, *Diophantine approximation*, Lecture Notes in Mathematics 785, Springer-Verlag, Berlin Heidelberg, 1980.

[Sil] J. H. Silverman, *The theory of height functions*, Arithmetic geometry (G. Cornell and J. H. Silverman, eds.), Springer-Verlag, New York, 1986, pp. 151–166.

[So] C. Soulé, *Géométrie d'Arakelov et théorie des nombres transcendants*, Journées arithmétiques, Luminy, Astérisque (to appear).

[So 2] C. Soulé, D. Abramovich, J.-F. Burnol, and J. Kramer, *Lectures on Arakelov Geometry*, Cambridge studies in applied mathematics 33, Cambridge University Press, Cambridge, 1992.

[T] J. Tate, *Rigid analytic spaces*, Invent. Math. **12** (1971), 257–289.

[U] K. Ueno, *Classification theory of algebraic varieties and compact complex spaces*, Lecture Notes in Mathematics 439, Springer-Verlag, Berlin Heidelberg, 1975.

[V 1] P. Vojta, *Diophantine approximations and value distribution theory*, Lecture Notes in Mathematics 1239, Springer-Verlag, Berlin Heidelberg, 1987.

[V 2] ———, *Dyson's lemma for a product of two curves of arbitrary genus*, Invent. Math. **98** (1989), 107–113.

[V 3] ———, *Mordell's conjecture over function fields*, Invent. Math. **98** (1989), 115–138.

[V 4] ———, *Siegel's theorem in the compact case*, Ann. Math. **133** (1991), 509–548.

[V 5] ———, *A generalization of theorems of Faltings and Thue-Siegel-Roth-Wirsing*, J. Amer. Math. Soc. (to appear).

C.I.M.E. Session on "Arithmetic Algebraic Geometry"

List of Participants

D. ABRAMOVICH, Dept. of Math., Harvard Univ., One Oxford Str., Cambridge, MA 02138

M. ANDREATTA, Via Anzoletti 14, 38100 Trento

L. BARBIERI VIALE, Dip. di Mat., Via L.B. Alberti 4, 16132 Genova

A.-S. BASARAB, Inst. of Math. of the Romanian Academy, Str. Academiei 14, 70700 Bucharest

M. BERNI, Via del Cuore 6, 56100 Pisa

G. CANUTO, Dip. di Mat., Strada Nuova 65, 27100 Pavia

D. CHIRICA, Inst. of Math. of the Romanian Academy, Str. Academiei 14, 70700 Bucharest

J. COANDA, Dip. di Mat., Univ. di Trento, 38050 Povo, Trento

C.I. COBELI, Inst. of Math. of the Romanian Academy, Str. Academiei 14, 70700 Bucharest

C.-S. DALAWAT, Mathématique, Univ. de Paris-Sud, F-91405 Orsay Cedex

M.A. de CATALDO, Dept. of Math., Univ. of Notre Dame, Notre Dame, USA

T. EKEDAHL, Dept. of Math., Stockholm Univ., S-11385 Stockholm

G. ELENCWAJG, Lab. de Math., Univ. de Nice, Parc Valrose, F-06034 Nice Cedex

M. FLEXOR, Univ. de Paris IX, Orsay 91, Paris

N. GAVIOLI, Dip. di Mat., Univ. di Trento, 38050 Povo, Trento

W. GUBLER, ETH Zürich, Mathematik, 8092 Zürich

K. HA HUY, ICTP, Math. Sect., Box 586, 34100 Trieste

L. LAFFORGUE, 10 rue Louis Gaudry, 92 160 Antony, France

F. LECOMTE, Dept. de Math., Univ. Luis Pasteur, 7 rue René Descartes, 67084 Strasbourg Cedex

V. MONTI, Dip. di Mat., Univ. di Trento, 38050 Povo, Trento

M. NAKAMAYE, Dept. of Math., Yale Univ., Box 2155, Yale Station, New Haven, CT 06520

L. NARVAEZ-MACARRO, Fac. de Matematicas, Tarfia s/n, 4102 Sevilla

F. OORT, Math. Inst. Budapestlaan 6, Utrecht, NL

C. PEDRINI, Dip. di Mat., Via L.B. Alberti 4, 16132 Genova

D. PORTELLI, Dip. di Scienze Mat., Piazzale Europa 1, 34127 Trieste

A. PREVITALI, Via Sant'Eurosia 15, 22064 Casatenovo, Como

Q.-V. PHAM, Fehlinghohe 21, 2000 Hamburg 60

B. RUSSO, Via V. Veneto 821E, Bolzano

R. SALVATI MANNI, Dip. di Mat., Univ. "La Sapienza", P.le A. Moro 2, 00185 Roma

E. SCHIAVI, Dip. di Mat., Univ. di Trento, 38050 Povo, Trento

R. SCHOOF, Dip. di Mat., Univ. di Trento, 38050 Povo, Trento

C. SOULE', IHES, 35 Route de Chartres, F-91440 Bures-sur-Yvette

A. THORUP, Mat. Inst., Universitetsparken 5, DK-2100 Kobenhavn O

A. VISTOLI, Via Irma Bandiera 121, 40024 Crevalcore, Bologna

C. WIRSCHING, Math. Inst. d. Univ. München, Theresienstr. 39, D-8000 München 2

FONDAZIONE C.I.M.E.
CENTRO INTERNAZIONALE MATEMATICO ESTIVO
INTERNATIONAL MATHEMATICAL SUMMER CENTER

"Transition to Chaos in Classical and Quantum Mechanics"

is the subject of the Third 1991 C.I.M.E. Session.

The Session, sponsored by the Consiglio Nazionale delle Ricerche and by the Ministero dell'Università e della Ricerca Scientifica e Tecnologica, will take place under the scientific direction of Prof. Sandro GRAFFI (Università di Bologna, Dipartimento di Matematica, Piazza di Porta S. Donato, 40126 Bologna, E-mail: MK7BOG73 at ICINECA: graffi at dm.unibo.it.) at Villa "La Querceta", Montecatini (Pistoia), **from July 6 to July 13, 1991.**

Courses

a) **Non Commutative Method in Semi Classical Analysis.** (8 lectures in English).
Prof. Jean BELLISSARD (Wissenschaftskolleg zu Berlin).

Outline

1) Quantum phase space ad groupoid
2) Perturbation theory: Birkhoff versus Rayleigh Schrödinger in the semi classical limit
3) Semi classical expansion and tunneling effects
4) Analysis of resonances: KAM theory and quantum localization
5) Spectrum in the classically chaotic region

References

1. J. Bellissard. C* algebras in solid state physics. in "Operator algebras and applications", D.E. Evans and M. Takesuki Eds.. Cambridge Univ. Press 1988.
2. J. Bellissard. M. Vittot. Heisenberg's picture and non commutative geometry of the semi classical limit in quantum mechanics. Ann. Inst. H. Poincaré 52 (1990). 175-235.
3. J. Bellissard. R. Rammal. An algebraic semi-classical approach to Bloch electrons in a magnetic field. J. de Physique Paris. (March 1990).
4. J. Bellissard. Stability and instability in quantum mechanics. in "Trends and developments in the eighties", Albeverio. Blanchard. Eds. World Sc. Publ., 1985. 1-106.
5. O. Bohigas. M.J. Giannoni. C. Schmit. Spectral properties of the Laplacian and random matrix theory. Lecture notes in physics. vol. 209 (1984). p. 1.

b) **Stochastic Properties of Classical Dynamical Systems.** (8 lectures in English).
Prof. Anatole KATOK (Pennsylvania State University).

Outline

1. Review of stochastic properties of dynamical systems. Ergodicity, mixing, eigenfunctions. Entropy and asymptotic independence. K-property, Bernoulli property.

2. Hyperbolic systems with examples exhibiting stochastic properties and complicate dynamical behaviour.
3. Smooth hyperbolic systems. The Pesin theory. Lyapunov exponents, stable and unstable manifolds, stochastic behaviour. The Pesin entropy formula, local ergodicity and local Bernoulli property.
4. Application of Pesin's theory to specific systems. Invariant cone families and Lyapynov characteristic exponents. Wojtkoski's theorem. Symplectic cones.
5. From local to global ergodicity. Criteria of openness of ergodic components and ergodicity based on existence of virtually strictly invariant family of symplectic cones.
6. Dynamical systems with singularities. Billiards, collisions, elastic balls in a volume and multidimensional billiard. Generalization of the Pesin theory to include billiards.
7. Cone families for various special classes of classical dynamical systems, both smooth and with singularities.
8. Criteria of ergodicity for dynamical systems with singularities.

References

1. Walters, P.: An introduction to ergodic theory, Springer Verlag, 1981.
2. Katok, A.: Dynamical systems with hyperbolic structure, in: Three papers in dynamical systems. AMS Translations (2) Vol. 116, 1981.
3. Pesin, Ya. B.: Characteristic Lyapunov exponents and smooth ergodic theory, Russian Math. Surveys, Vol. 32 (1977), 54-114.
4. Katok, A.: Lyapunov exponents, entropy and periodic points of diffeomorphisms, Publ. Math. IHES, Vol. 51 (1980), 137-173.
5. Wojtkowski, M.: Invariant families of cones and Lyapunov exponents. Ergodic Theory and Dynamical Systems, Vol. 5 (1985), 145-161.

c) **Dynamics of Area Preserving Maps.** (8 lectures in English).
Prof. John N. MATHER (Princeton University).

Outline

1. Hyperbolic and elliptic fixed points. Birkhoff normal form
2. A brief discussion of KAM theory (no proofs)
3. A brief discussion on the "last invariant circle" with reference to the numerical results of Greene, Percival, McKay (no proofs)
4. An outline of Aubry's theory and of its generalization by Barget

5.6. Herman's method and the author's method of destroying invariant circles (with an outline of the proofs)

7.8. Orbits which pass close to a succession of Aubry-Mather sets (with an outline of the proofs)

References

Herman, M.R.: Sur les courbes invariantes par les difféomorphismes de l'anneau, Vol. 1, Astérisque, 103-104.
Bangert, V.: Mather sets for twist maps and geodesics on tori, Dynamics reported, Vol. 1, pp. 1-56.
Salamon, S. And Zehnder, E.: The Kolmogorov-Arnold-Moser theory in configuration space, Comm. Math. Helv., Vol. 64, pp. 84-132.
Mather, J.N.: Destruction of invariant circles, Ergodic Theory and Dynamical Systems, 8, 199-214.
Mather, J.N.: Variational construction of orbits of twist diffeomorphisms, Journal of the American Mathematical Society, to appear.

FONDAZIONE C.I.M.E.
CENTRO INTERNAZIONALE MATEMATICO ESTIVO
INTERNATIONAL MATHEMATICAL SUMMER CENTER

"Dirichlet Forms"

is the subject of the First 1992 C.I.M.E. Session.

The Session, sponsored by the Consiglio Nazionale delle Ricerche and the Ministero dell'Università e della Ricerca Scientifica e Tecnologica, will take place under the scientific direction of Prof. GIANFAUSTO DELL'ANTONIO (Università di Roma, La Sapienza), and Prof. UMBERTO MOSCO (Università di Roma, La Sapienza), at Villa Monastero, Varenna (Lake of Como), **from June 8 to June 19, 1992.**

Courses

a) **Parabolic Harnack inequalities and the behavior of fundamental solutions.** (5 lectures in English)
Prof. Eugen FABES (University of Minnesota)

Outline

A. Gaussian estimates for the fundamental solution of nondegenerate parabolic operators - applications to Harnack inequalities. (Ref. 4,6,7).
B. The Harnack inequality implies estimates for the fundamental solution - the nondegenerate and degenerate cases. (Ref. 1,2,5).
C. Gaussian estimates for heat flows on Riemannian manifolds. (Ref. 3).

References

1. Aronson, D.G., Bounds for the fundamental solution of a parabolic equation, Bulletin of the AMS 73 (1967), 890-896.
2. Chiarenza, F.M., Serapioni, R.P., A Harnack inequality for degenerate parabolic equations, Comm. in PDE 9 (1984), 719-749.
3. Davies, E.B., Heat Kernels and Spectral Theory, Cambridge Tracts in Mathematics 92, Cambridge University Press, 1990.
4. Fabes, E.B., Stroock, D.W., A new proof of Moser's parabolic Harnack inequality via the old ideas of Nash, Arch. Rat. Mech. Anal. 96 (1986), 327-338.
5. Gutierrez, C.E., Wheeden, R.L., Bounds for the fundamental solution of degenerate parabolic equations, to appear.
6. Moser, J., A Harnack inequality for parabolic differential equations, Comm. Pure and Applied Math. 17 (1964), 101-134, and also Correction to "A Harnack inequality for parabolic differential equations", Comm. Pure and Applied Math. 20 (1967), 232-236.
7. Nash, J., Continuity of solutions of parabolic and elliptic equations, Amer. J. Math. 80 (1958), 931-954.

b) **General Theory of Dirichlet forms: Part II.** (6 lectures in English).
Prof. Masatoshi FUKUSHIMA (Osaka University)

A short outline of the course

Chapter 1. Dirichlet forms in finite dimensional analysis.
In this chapter, are given some specific but basic examples of Dirichlet forms related to the finite dimensional analysis, e.g., spatially homogeneous Dirichlet forms, Dirichlet forms on fractal sets and Dirichlet forms generated by plurisubharmonic functions. As applications, quasi-everywhere convergences of Fourier series, point recurrence properties and spectral dimensions for fractal sets and characterizations of pluripolar sets are discussed.

Chapter 2. Stochastic analysis by additive functionals.
In this chapter, the theory of additive functionals associated with the regular Dirichlet form is presented; positive

continuous additive functionals and smooth measures, decompositions of additive functionals and their relations to Dirichlet forms. Various applications with specific emphasis on the roles of martingale additive functionals are given.

References

1. M. Fukushima, Dirichlet forms and Markov processes, North-Holland and Kodan-sha, 1980.
2. M. Fukushima and M. Takeda, A transformation of a symmetric Markov process and the Donsker-Varadhan theory, Osaka J. Math. 21 (1984), 311-326.
3. M. Fukushima and M. Okada, On Dirichlet forms for plurisubharmonic functions, Acta Math. 54 (1987), 171-213.
4. M. Fukushima, Dirichlet forms, diffusion processes and spectral dimensions for nested fractals, in "Ideas and Methods in Mathematical Analysis, Stochastics, and Applications, In Memory of R. Hoegh-Krohn, Vol. 1", Albeverio, Penstad, Holden, Lindstrom (eds.), Cambridge Univ. Press, to appear.
5. M. Takeda, On a martingale method for symmetric diffusion processes and its applications, Osaka J. Math. 26 (1989), 606-623.

c) **Logarithmic Sobolev inequalities over finite and infinite dimensional spaces.** (6 lectures in English).
Prof. Leonard GROSS (Cornell University).

Outline

I. Logarithmic Sobolev inequalities in L^2.
   A. The standard Gaussian L.S. inequality on R^n
   B. L.S. generators in L^2
   C. General properties
      1. Semiboundedness of perturbations
      2. Additivity
      3. Mass gap

II. Hypercontractive, supercontractive and ultracontractive semigroups.
   A. Definitions
   B. L.S. generators of index p in $(0,\infty)$
   C. Equivalence of L.S. generators with strong contraction properties
   D. Examples

III. Logarithmic Sobolev inequalities for Dirichlet forms.
   A. Index 2 implies index p
   B. Example (Nelson's best estimates for the number operator)

IV. Survey of applications.
   A. Heat kernel bounds - Davies et al.
   B. Semiboundedness of Hamiltonians

V. Some recent applications to Schrödinger operators loop groups.

Prerequisites

1. Spectral theorem for unbounded self-adjoint operator
2. The Beurling-Deny theorem - as in M. Fukushima, "Dirichlet Forms and Markov Processes", North-Holland, New York, 1980.

d) **Potential theory of non-divergence form elliptic operators.** (5 lectures in English)
Prof. Carlos KENIG (University of Chicago)

Outline

The purpose of these lectures will be to describe the potential theory of non divergence form elliptic equations and their adjoints, highlighting the connection with a natural degenerate Dirichlet form. We will study both local behavior and boundary behavior, including a Wiener test.

References

[B1] Bauman, P., Positive solutions of elliptic equations in nondivergence form and their adjoints, Arkiv für Mat. 22 (1984), 153-173.
[B2] Bauman, P., A Wiener test for nondivergence structure, second order elliptic equations, Indiana U. Math. J. 4 (1985), 825-844.
[F,G,M,S] Fabes, E., Garofalo, N., Marin Malave, S. and Salsa, S., Fatou theorems for some nonlinear elliptic equations, Revista Mat. Iberoamericana, Vol. 4 (1988), 227-251.
[F,S] Fabes, E. and Stroock, D., The L^p-integrability of Green's functions and fundamental solutions for elliptic and parabolic equations, Duke Math. J. 51 (1984), 977-1016.
[P] Pucci, C., Limitazioni per soluzioni di equazioni ellittiche, Ann. Mat. Pura ed Appl. 74 (1966), 15-30.
[S] Safonov, M.V., Harnack's inequality for elliptic equations and the Hölder property of their solutions, J. Soviet Math. 21 (1983), 851-863.

e) **General theory of Dirichlet forms: Part I.** (6 lectures in English).
Prof. Michael RÖCKNER (Universität Bonn)

Outline

The purpose of this part of the course is both to give an introduction to the theory of (not necessarily symmetric) Dirichlet forms and to describe their significance in infinite dimensional analysis. The first half will consist of a presentation of the basic theory on general state spaces and includes of the following topics:

1. Underlying L^2-theory and contraction properties
2. Potential theory of Dirichlet forms
3. Necessary and sufficient conditions for the existence of an associated Markov process
4. Compactification

The second half of the lectures is devoted to applications with a special emphasis on examples with infinite dimensional state spaces. It will cover the following topics:

5. Closability of classical (pre-) Dirichlet forms on topological vector spaces
6. Tightness of capacities
7. Existence and uniqueness of solutions for stochastic differential equations in infinite dimensional space
8. Girsanov transform in infinite dimensions

References

1. Albeverio, S., Röckner, M.: Stochastic differential equations in infinite dimensions: solutions via Dirichlet forms. Probab. Th. Rel. Fields 89 (1991), 347-386.
2. Fukushima, M.: Dirichlet forms and Markov processes. Amsterdam-Oxford-New York, North Holand (1980).
3. Ma, Z., Röckner, M.: An introduction to the theory of (non-symmetric) Dirichlet forms. Monograph, to appear.
4. Silverstein, M.L.: Symmetric Markov Processes. Lecture Notes in Math. 426. Berlin-Heidelberg-New York, Springer (1974).
5. Röckner, M., Zhang, T.S.: On uniqueness of generalized Schrödinger operators and applications. To appear in J. Funct. Anal.

f) **Dirichlet forms and ergodic properties.** (6 lectures in English).
Prof. Daniel W. STROOCK (MIT)

Outline

The course will emphasize the application of Dirichlet forms to the long time analysis of various stochastic processes. The topics which will be covered are as follows.

1. General comments about spectral gaps, Sobolev, and logarithmic Sobolev inequalities and their relationship to ergodic phenomena.
2. Analysis of the "simulated annealing" procedure.
3. Criteria for the existence of a logarithmic Sobolev inequality.
4. Preliminary discussion of Gibbs states and Glauber dynamics.
5. Conditions under which a Gibbs state admits a logarithmic Sobolev inequality.

References

Chapter VI of Large deviations by J.-D. Deuschel and D. Stroock, Academic Press, (1989).
R. Holley, D. Stroock, J. of Stat. Physics, 46 (1987), 1159-1194.
E. A. Carlen, S. Kusuoka, D.W. Stroock, Ann. Inst. H. Poincaré, Probab. et Stat., Supl. au n° 2, 1987, 245-287.
R. Holley, D. Stroock, Commun. Math. Phys., 115 (1988), 553-569.
R. Holley, S. Kusuoka, D. Stroock, J. Funct. Anal. 83 (1989), 333-334.
J. D. Deuschel, D.W. Stroock, J. Funct. Anal. 92 (1990), 30-48.
P. Diaconis, D. Stroock, Ann. of Appl. Prob. 1 (1991), 36-61.
D.W. Stroock, B. Zegarlinski, The logarithmic Sobolev inequality for continuous spin systems on a lattice, to appear J. Funct. Anal.
D.W. Stroock, B. Zegarlinski, The equivalence of the logarithmic Sobolev inequality and the Dobrushin-Shlosman mixing condition, to appear Comm. Math. Phys.

FONDAZIONE C.I.M.E
CENTRO INTERNAZIONALE MATEMATICO ESTIVO
INTERNATIONAL MATHEMATICAL SUMMER CENTER

"D-Modules and Representation Theory"

is the subject of the Second 1992 C.I.M.E. Session.

The Session, sponsored by the Consiglio Nazionale delle Ricerche and by the Ministero dell'Università e della Ricerca Scientifica e Tecnologica, will take place under the scientific direction of Prof. GIUSEPPE ZAMPIERI (Università di Padova) and Prof. ANDREA D'AGNOLO (Università di Padova) at Università di Venezia, Ca' Foscari, Venezia from **June 12 to June 20, 1992.**

Courses

a) **Formule de l'indice relative.** (5 lectures in English)
Prof. Louis BOUTET DE MONVEL (Université de Paris VI)

Résumé

Soient X et Y deux variétés analytiques, $f:Y \to X$ une application analytique, M un D_Y-module cohérent ("bien filtré"). Le théoréme de l'indice décrit l'élément de K-théorie $[f_+M]$ associé à l'image directe f_+M en fonction de [M] lorsque M est relativement elliptique (condition qui assure que $[f_+M]$ est cohérent et bien filtré, d'aprés Houzel et Schapira).

Les conférences comprendront deux parties:

1) une de rappels des éléments de K-théorie nécéssaires à la description de la formule, en particulier de la K-théorie à supports et de la description pour celle ci du théoréme de périodicité au moyen d'opérateurs de Toeplitz.
2) une description de la partie pertinente de la théorie des D-modules et des images directes de D-modules permettant d'énoncer le théoréme d'indice et d'en décrire la démonstration.

Reférences bibliographiques

[A] M.F. Atiyah, K-theory, Benjamin, Amsterdam.
[Bo] A. Borel et al., Algebraic D-modules, Perspect. in Math. n. 2, Academic Press.
[B-M] L. Boutet de Monvel, B. Malgrange, Le théorème de l'indice relatif. Ann. Sc. E.N.S., à paraître.
[BM1] L. Boutet de Monvel, On the index of Toeplitz operators of several complex variables, Inventiones Math. 50 (1979), 249-272.
[BM2] L. Boutet de Monvel, The index of almost elliptic systems. E. De Giorgi Colloquium, Research notes in Math. 125, Pitman 1985, 17-29.
[GRo] A. Grothendieck, SGA V, théorie des intersections et théorème de Riemann-Roch, Lecture Notes in Math. 225, Springer Verlag (1971)
[Hi] F. Hirzebruch, Neue topologische Methoden in der algebraische Geometrie. Springer Verlag, Berlin.
[Hö] L. Hörmander, The Analysis of Linear Partial Differential Operators, Vol. III et IV, Grundlehren der Math. Wiss. 124.
[H-Sch] Ch. Houzel, P. Schapira, Images directes de modules différentiels, C.R.A.S. 298 (1984), 461-464.
[M] B. Malgrange, Sur les images directes de D-modules, Manuscripta Math. 50 (1985), 49-71.
[K] M. Kashiwara, Cours Université Paris Nord, Birkhauser 1983.
[K-K-S] M. Kashiwara, T. Kawai, M. Sato, Microfunctions and pseudo-differential equations, Lecture Notes 287 (1973), Springer Verlag.

b) **Quantized Enveloping Algebras and Their Representations.** (10 lectures in English).
Prof. Corrado DE CONCINI (SNS, Pisa) and Prof. Claudio PROCESI (Univ. Roma La Sapienza).

Outline

1. Poisson Lie Groups.
2. Quantized enveloping algebras. The universal R-matrix. The center of quantized enveloping algebras, the Harish-Chandra isomorphism.
3. Representation for generic q. Their complete reducibility and their classification and characters.
4. Specializations at roots of unity. The center at roots of 1. The subalgebra Z_0 of the center as the coordinate ring of a Poisson algebraic group.
5. Representations at roots of 1. Their relations with coadjoint orbits.
6. The divided power algebra of Lusztig at a root of 1. Its representations.

c) **Index theorems for constructible sheaves and D-modules.** (5 lectures in English).
Prof. Pierre SCHAPIRA (Université Paris Nord)

Outline

1) Subanalytic stratifications and constructible sheaves
2) Lagrangian cycles
3) Characteristic cycle of constructible sheaves and index theorem
4) Applications to holonomic D-modules
5) Elliptic pairs and open problems

The main reference

M. Kashiwara and P. Schapira: Sheaves on manifolds. Grundlehren der Math. Wiss., Springer Verlag, 292 (1990).

One may also consult:

M. Kashiwara: Systems of microdifferential equations. Progress in Math. 34, Birkhäuser (1983).
P. Schapira: Microdifferential systems in the complex domain. Grundlehren der Math. Wiss., Springer Verlag, 269 (1985).

d) **Cohomologie équivariante et théorèmes d'indice.** (5 lectures in English).
Prof. Michèle VERGNE (ENS, Paris)

Outline

- Cohomologie de De Rham d'une variété
- Fibres, superconnections, classes caractéristiques
- Cohomologie équivariante d'une variété. Fibrés équivariants, classes caractéristiques en cohomologie équivariante
- Cohomologie équivariante et indice de l'opérateur de Dirac
- Cohomologie équivariante et théorème de l'indice pour les opérateurs transversalement elliptiques
- Représentations des groupes compacts. Formule d'Hermann Weyl et formule de Kirillov pour les caractères.

Reférences bibliographiques:

Berline-Getsler-Vergne, Heat kemels and Dirac operators, à paraitre Springer.

LIST OF C.I.M.E. SEMINARS | Publisher

1954 -	1. Analisi funzionale	C.I.M.E.
	2. Quadratura delle superficie e questioni connesse	"
	3. Equazioni differenziali non lineari	"
1955 -	4. Teorema di Riemann-Roch e questioni connesse	"
	5. Teoria dei numeri	"
	6. Topologia	"
	7. Teorie non linearizzate in elasticità, idrodinamica,aerodinamica	"
	8. Geometria proiettivo-differenziale	"
1956 -	9. Equazioni alle derivate parziali a caratteristiche reali	"
	10. Propagazione delle onde elettromagnetiche	"
	11. Teoria della funzioni di più variabili complesse e delle funzioni automorfe	"
1957 -	12. Geometria aritmetica e algebrica (2 vol.)	"
	13. Integrali singolari e questioni connesse	"
	14. Teoria della turbolenza (2 vol.)	"
1958 -	15. Vedute e problemi attuali in relatività generale	"
	16. Problemi di geometria differenziale in grande	"
	17. Il principio di minimo e le sue applicazioni alle equazioni funzionali	"
1959 -	18. Induzione e statistica	"
	19. Teoria algebrica dei meccanismi automatici (2 vol.)	"
	20. Gruppi, anelli di Lie e teoria della coomologia	"
1960 -	21. Sistemi dinamici e teoremi ergodici	"
	22. Forme differenziali e loro integrali	"
1961 -	23. Geometria del calcolo delle variazioni (2 vol.)	"
	24. Teoria delle distribuzioni	"
	25. Onde superficiali	"
1962 -	26. Topologia differenziale	"
	27. Autovalori e autosoluzioni	"
	28. Magnetofluidodinamica	"

1963 -	29. Equazioni differenziali astratte	"
	30. Funzioni e varietà complesse	"
	31. Proprietà di media e teoremi di confronto in Fisica Matematica	"
1964 -	32. Relatività generale	"
	33. Dinamica dei gas rarefatti	"
	34. Alcune questioni di analisi numerica	"
	35. Equazioni differenziali non lineari	"
1965 -	36. Non-linear continuum theories	"
	37. Some aspects of ring theory	"
	38. Mathematical optimization in economics	"
1966 -	39. Calculus of variations	Ed. Cremonese, Firenze
	40. Economia matematica	"
	41. Classi caratteristiche e questioni connesse	"
	42. Some aspects of diffusion theory	"
1967 -	43. Modern questions of celestial mechanics	"
	44. Numerical analysis of partial differential equations	"
	45. Geometry of homogeneous bounded domains	"
1968 -	46. Controllability and observability	"
	47. Pseudo-differential operators	"
	48. Aspects of mathematical logic	"
1969 -	49. Potential theory	"
	50. Non-linear continuum theories in mechanics and physics and their applications	"
	51. Questions of algebraic varieties	"
1970 -	52. Relativistic fluid dynamics	"
	53. Theory of group representations and Fourier analysis	"
	54. Functional equations and inequalities	"
	55. Problems in non-linear analysis	"
1971 -	56. Stereodynamics	"
	57. Constructive aspects of functional analysis (2 vol.)	"
	58. Categories and commutative algebra	"

1972 -	59. Non-linear mechanics		"
	60. Finite geometric structures and their applications		"
	61. Geometric measure theory and minimal surfaces		"
1973 -	62. Complex analysis		"
	63. New variational techniques in mathematical physics		"
	64. Spectral analysis		"
1974 -	65. Stability problems		"
	66. Singularities of analytic spaces		"
	67. Eigenvalues of non linear problems		"
1975 -	68. Theoretical computer sciences		"
	69. Model theory and applications		"
	70. Differential operators and manifolds		"
1976 -	71. Statistical Mechanics		Ed Liguori, Napoli
	72. Hyperbolicity		"
	73. Differential topology		"
1977 -	74. Materials with memory		"
	75. Pseudodifferential operators with applications		"
	76. Algebraic surfaces		"
1978 -	77. Stochastic differential equations		"
	78. Dynamical systems		Ed Liguori, Napoli and Birhäuser Verlag
1979 -	79. Recursion theory and computational complexity		"
	80. Mathematics of biology		"
1980 -	81. Wave propagation		"
	82. Harmonic analysis and group representations		"
	83. Matroid theory and its applications		"
1981 -	84. Kinetic Theories and the Boltzmann Equation	(LNM 1048)	Springer-Verlag
	85. Algebraic Threefolds	(LNM 947)	"
	86. Nonlinear Filtering and Stochastic Control	(LNM 972)	"
1982 -	87. Invariant Theory	(LNM 996)	"
	88. Thermodynamics and Constitutive Equations	(LN Physics 228)	"
	89. Fluid Dynamics	(LNM 1047)	"

1983 -	90. Complete Intersections	(LNM 1092)	"
	91. Bifurcation Theory and Applications	(LNM 1057)	"
	92. Numerical Methods in Fluid Dynamics	(LNM 1127)	"
1984 -	93. Harmonic Mappings and Minimal Immersions	(LNM 1161)	"
	94. Schrödinger Operators	(LNM 1159)	"
	95. Buildings and the Geometry of Diagrams	(LNM 1181)	"
1985 -	96. Probability and Analysis	(LNM 1206)	"
	97. Some Problems in Nonlinear Diffusion	(LNM 1224)	"
	98. Theory of Moduli	(LNM 1337)	"
1986 -	99. Inverse Problems	(LNM 1225)	"
	100. Mathematical Economics	(LNM 1330)	"
	101. Combinatorial Optimization	(LNM 1403)	"
1987 -	102. Relativistic Fluid Dynamics	(LNM 1385)	"
	103. Topics in Calculus of Variations	(LNM 1365)	"
1988 -	104. Logic and Computer Science	(LNM 1429)	"
	105. Global Geometry and Mathematical Physics	(LNM 1451)	"
1989 -	106. Methods of nonconvex analysis	(LNM 1446)	"
	107. Microlocal Analysis and Applications	(LNM 1495)	"
1990 -	108. Geoemtric Topology: Recent Developments	(LNM 1504)	"
	109. H_∞ Control Theory	(LNM 1496)	"
	110. Mathematical Modelling of Industrical Processes	(LNM 1521)	"
1991 -	111. Topological Methods for Ordinary Differential Equations	(LNM 1537)	"
	112. Arithmetic Algebraic Geometry	(LNM 1553)	"
	113. Transition to Chaos in Classical and Quantum Mechanics	to appear	"
1992 -	114. Dirichlet Forms	to appear	"
	115. D-Modules and Representation Theory	to appear	"
	116. Nonequilibrium Problems in Many-Particle Systems	(LNM 1551)	"

1993 -	117. Integrable Systems and Quantum Groups	to appear	"
	118. Algebraic Cycles and Hodge Theories	to appear	"
	119. Modelling and Analysis of Phase Transition and Hysteresis Phenomena	to appear	"

Printing: Weihert-Druck GmbH, Darmstadt
Binding: Buchbinderei Schäffer, Grünstadt

Lecture Notes in Mathematics Vol. 1553
ISBN 978-3-540-57110-0

J.-L. Colliot-Thélène, K. Kato, P. Vojta

Arithmetic algebraic geometry, Trento, Italy 1991

Article *Cycles algébriques de torsion et K-théorie algébrique* (pages 1–49) par Jean-Louis Colliot-Thélène

Errata et commentaires

p. 14, bas de la page : Chad Schoen (J. Algebraic Geom. **11** (2002), 41–100) a montré que cette application n'est en général pas injective.

p. 18, ligne −5 : réduction

p. 24, ligne 19 : D'après

p. 32, ligne 2 : apparue

p. 38, ligne 24 : [CR3]

p. 40, Théorème 8.4 et Remarque 8.4.1 : On ne peut pas supprimer l'hypothèse de bonne réduction de la variété d'Albanese dans le théorème 8.4, comme le montre un exemple de R. Parimala et V. Suresh (Inventiones math. **122** (1995), 83–117).

p. 41, ligne 6 : lire $H^2(K, H^1_{et}(\overline{X}, \mathbf{Q}/\mathbf{Z}(2))^0)$

p. 42, lignes −5 à −1 : Dans le théorème 8.6, ajouter l'hypothèse que $\mathbf{X}/R$ est de dimension relative 2.

p. 45, ligne 10 : enlever un "le"

p. 45, ligne −2. Ajouter : En utilisant la trace de L à k on voit que la flèche

$$H^1(X, \mathcal{K}_2) \otimes \mathbb{Q} \to \oplus_{p \in \mathrm{Spec}\ A^{(1)}} \mathrm{Pic}(\mathbb{X}_p) \otimes \mathbb{Q}$$

est surjective.

p. 46, ligne 2, remplacer $CH_2(\mathbb{X}_B)_{tors} \to CH^2(X_L)_{tors}$ par $CH_2(\mathbb{X})_{tors} \to CH^2(X)_{tors}$

p. 46, ligne −14 : exacte

p. 46, ligne −12 : remplacer $H^i(\mathbf{X}, \mathbf{Z}_l(j))$ par le quotient de ce groupe par son sous-groupe de torsion.

p. 46, ligne −6 : enlever "les"

p. 48 L'article [Sb2] de Per Salberger est paru : Torsion cycles of codimension 2 and l-adic realizations of motivic cohomology, Séminaire de théorie des nombres, Paris 1991-1992 (Sinnou David, éd.), Progress in Mathematics **116**, Birkhäuser (1993), p. 247–277.

Voici des références complémentaires :

J.-L. Colliot-Thélène, L'arithmétique des zéro-cycles (exposé aux Journées arithmétiques de Bordeaux, septembre 93), Journal de théorie des nombres de Bordeaux **7** (1995) 51–73.

J.-L. Colliot-Thélène, Zéro-cycles sur les surfaces sur un corps p-adique : quelques résultats obtenus au moyen de la K-théorie algébrique, notes préparées pour les conférences de Morelia (juin-juillet 2003) et de Sestri Levante (juin-juillet 2004), disponible sur la page

`http://www.math.u-psud.fr/ colliot/liste-cours-exposes.html`

Masanori Asakura et Shuji Saito, Surfaces over a p-adic field with infinite torsion in the Chow group of 0-cycles, Algebra and Number Theory, vol. **1** (2007) 163–181.

Shuji Saito et Kanetomo Sato (with an Appendix by Uwe Jannsen), A finiteness theorem on zero-cycles over p-adic fields, à paraître dans Annals of Math.

Shuji Saito et Kanetomo Sato, Torsion cycle class maps in codimension two of arithmetic schemes, prépublication arXiv:math/0612081